The Foundations of Statistics

LEONARD J. SAVAGE

Late Eugene Higgins Professor of Statistics
Yale University

SECOND REVISED EDITION

DOVER PUBLICATIONS, INC.
NEW YORK

International Standard Book Number: 0-486-62349-1
Library of Congress Catalog Card Number: 79-188245

Manufactured in the United States of America
Dover Publications, Inc.
180 Varick Street
New York, N.Y. 10014

Theory of Decision

P2   For every **f**, **g**, and B, **f** $\leq$ **g** given B or **g** $\leq$ **f** given B (page 23).

D2 $g \leq g'$; if and only if **f** $\leq$ **f**′, when $f(s) = g$, $f'(s) = g'$ for every $s \, \varepsilon \, S$ (page 25).

D3 B is null, if and only if **f** $\leq$ **g** given B for every **f**, **g** (page 24).

P3   If $f(s) = g$, $f'(s) = g'$ for every $s \, \varepsilon \, B$, and B is not null; then **f** $\leq$ **f**′ given B, if and only if $g \leq g'$ (page 26).

D4 $A \leq B$; if and only if $f_A \leq f_B$ or $g \leq g'$ for every f_A, f_B, g, g' such that: $f_A(s) = g$ for $s \, \varepsilon \, A$, $f_A(s) = g'$ for $s \, \varepsilon \sim A$, $f_B(s) = g$, for $s \, \varepsilon \, B$, $f_B(s) = g'$ for $s \, \varepsilon \sim B$ (page 31).

P4   For every A, B, $A \leq B$ or $B \leq A$ (page 31).

P5   It is false that, for every f, f', $f \leq f'$ (page 31).

P6   Suppose it false that **g** $\leq$ **h**; then, for every f, there is a (finite) partition of S such that, if **g**′ agrees with **g** and **h**′ agrees with **h** except on an arbitrary element of the partition, **g**′ and **h**′ being equal to f there, then it will be false that **g**′ $\leq$ **h** or **g** $\leq$ **h**′ (page 39).

D5   **f** $\leq$ g given B (g $\leq$ **f** given B); if and only if **f** $\leq$ **h** given B (**h** $\leq$ **f** given B), when $h(s) = g$ for every s (page 72).

P7   If **f** $\leq$ $g(s)$ given B ($g(s) \leq$ **f** given B) for every $s \, \varepsilon \, B$, then **f** $\leq$ **g** given B (**g** $\leq$ **f** given B) (page 77).

The Foundations
of Statistics

TO MY FATHER

Preface to the Dover Edition

CONTINUING INTEREST HAS ENCOURAGED PUBLICATION OF A SECOND edition of this book. Because revising it to fit my present thinking and the new climate of opinion about the foundations of statistics would obliterate rather than restore, I have limited myself in the preparation of this edition much as though dealing with the work of another.

The objective errors that have come to my attention, mainly through the generosity of readers, of whom Peter Fishburn has my special thanks, have been corrected, of course. Minor and mechanical ones, such as a name misspelled or an inequality that had persisted in pointing in the wrong direction, have been silently eliminated. Other changes are conspicuous as additions. They consist mainly of this Preface, Appendix 4: Bibliographic Supplement, and several footnotes identified as new by the sign+. To enable you to pursue the many new developments since 1954 according to the intensity and direction of your own interests, a number of new references leading to many more are listed in the Bibliographic Supplement, and the principle advances known to me are pointed out in new footnotes or in comments on the new references.

Citations to the bibliography in the original Appendix 3 are made by a compact, but otherwise ill-advised, letter and number code; those to the new Appendix 4 are made by a now popular system, which is effective, informative, and flexible. Example: The historic papers (Borel 1924) and [D2] have been translated by Kyburg and Smokler (1964).

The following paragraphs are intended to help you approach this book with a more current perspective. To some extent, they will be intelligible and useful even to a novice in the foundations of statistics, but they are necessarily somewhat technical and will therefore take on new meaning if you return to them as your reading in this book and elsewhere progresses.

The book falls into two parts. The first, ending with Chapter 7, is a general introduction to the personalistic tradition in probability and utility. Were this part to be done over, radical revision would not be required, though I would now supplement the line of argument centering around a system of postulates by other less formal approaches, each convincing in its own way, that converge to the general conclusion that personal (or subjective) probability is a good key, and the best yet

iii

known, to all our valid ideas about the applications of probability. There would also be many new works to report on and analyze more thoroughly than can be done in footnotes.

The original aim of the second part of the book, beginning with Chapter 8, is all too plainly stated in the second complete paragraph on page 4. There, a personalistic justification is promised for the popular body of devices developed by the enthusiastically frequentistic schools that then occupied almost the whole statistical scene and still dominate it, though less completely. The second part of the book is indeed devoted to personalistic discussion of frequentistic devices, but for one after another it reluctantly admits that justification has not been found. Freud alone could explain how the rash and unfulfilled promise on page 4 went unamended through so many revisions of the manuscript.

Today, as I see it, the theory of personal probability applied to statistics shows that many of the prominent frequentistic devices can at best lead to accidental and approximate, not systematic and cogent, success, as is expanded upon, perhaps more optimistically, by Pratt (1965). Among the ill-founded frequentistic devices are minimax rules, almost all tail-area tests, tolerance intervals, and, in a sort of class by itself, fiducial probability.

If I have lost faith in the devices of the frequentistic schools, I have learned new respect for some of their general theoretical ideas. Let me amplify first in connection with the Neyman-Pearson school. While insisting on long-run frequency as the basis of probability, that school wisely emphasizes the ultimate subjectivity of statistical inference or behavior within the objective constraint of "admissibility," as in (Lehmann 1958; Wolfowitz 1962). But careful study of admissibility leads almost inexorably to the recognition of personal probabilities and their central role in statistics (Savage 1961, Section 4; 1962, pp. 170-175), so personalistic statistics appears as a natural late development of the Neyman-Pearson ideas.

One consequence of this sort of analysis of admissibility is the extremely important likelihood principle, a corollary of Bayes' theorem, of which I was not even aware when writing the first edition of this book. This principle, inferable from, though nominally at variance with, Neyman-Pearson ideas (Birnbaum 1962), was first put forward by Barnard (1947) and by Fisher (1955), members of what might be called the Fisher school of frequentists. See also (Barnard 1965; Barnard et al. 1962; Cornfield 1966).

The views just expressed are evidently controversial, and if I have permitted myself such expressions as "show" and "inexorably," they are not meant with mathematical finality. Yet, controversial though they may be, they are today shared by a number of statisticians, who may be called personalistic Bayesians, or simply personalists. This book has played—and continues to play—a role in the personalistic movement, but the movement itself has other sources apart from those from which this book itself was drawn. One with great impact on practical statistics and scientific management is a book by Robert Schlaifer (1959). This is a welcome opportunity to say that his ideas were developed wholly independently of the present book, and indeed of other personalistic literature. They are in full harmony with the ideas in this book but are more down to earth and less spellbound by tradition.

L. J. SAVAGE

Yale University
June, 1971

Preface to the First Edition

A BOOK ABOUT SO CONTROVERSIAL A SUBJECT AS THE FOUNDATIONS of statistics may have some value in the classroom, as I hope this one will; but it cannot be a textbook, or manual of instruction, stating the accepted facts about its subject, for there scarcely are any. Openly, or coyly screened behind the polite conventions of what we call a disinterested approach, it must, even more than other books, be an airing of its author's current opinions.

One who so airs his opinions has serious misgivings that (as may be judged from other prefaces) he often tries to communicate along with his book. First, he longs to know, for reasons that are not altogether noble, whether he is really making a valuable contribution. His own conceit, the encouragement of friends, and the confidence of his publisher have given him hope, but he knows that the hopes of others in his position have seldom been fully realized.

Again, what he has written is far from perfect, even to his biased eye. He has stopped revising and called the book finished, because one must sooner or later.

Finally, he fears that he himself, and still more such public as he has, will forget that the book is tentative, that an author's most recent word need not be his last word.

The application of statistics interests some workers in almost every field of empirical investigation—not only in science, but also in commerce and industry. Moreover, the foundations of statistics are connected conceptually with many disciplines outside of statistics itself, particularly mathematics, philosophy, economics, and psychology—a situation that, incidentally, must augment the natural misgivings of an author in this field about his own competence. Those who read in this book may, therefore, be diverse in background and interests. With this consideration in mind, I have endeavored to keep the book as free from technical prerequisites as its subject matter and its restriction to a reasonable size permit.

Technical knowledge of statistics is nowhere assumed, but the reader who has some general knowledge of statistics will be much better prepared to understand and appraise this book. The books *Statistics*, by L. H. C. Tippett, and *On the Principles of Statistical Inference* by

A. Wald, listed in the Bibliography at the end of Appendix 3, are short authoritative introductions to statistics, either of which would provide some statistical background for this book. The books of Tippett and Wald are so different in tone and emphasis that it would by no means be wasteful to read them both, in that order.

Any but the most casual reader should have some formal preparation in the theory of mathematical probability. Those acquainted with moderately advanced theoretical statistics will automatically have this preparation; others may acquire it, for example, by reading *Theory of Probability*, by M. E. Munroe, or selected parts of *An Introduction to Probability Theory and Its Applications*, by W. Feller, according to their taste. In Feller's book, a thorough reading of the Introduction and Chapter 1, and a casual reading of Chapters 5, 7, and 8 would be sufficient.

The explicit mathematical prerequisites are not great; a year of calculus would in principle be more than enough. But, in practice, readers without some training in formal logic or one of the abstract branches of mathematics usually taught only after calculus will, I fear, find some of the long though elementary mathematical deductions quite forbidding. For the sake of such readers, I therefore take the liberty of giving some pedagogical advice here and elsewhere that mathematically more mature readers will find superfluous and possibly irritating. In the first place, it cannot be too strongly emphasized that a long mathematical argument can be fully understood on first reading only when it is very elementary indeed, relative to the reader's mathematical knowledge. If one wants only the gist of it, he may read such material once only; but otherwise he must expect to read it at least once again. Serious reading of mathematics is best done sitting bolt upright on a hard chair at a desk. Pencil and paper are nearly indispensable; for there are always figures to be sketched and steps in the argument to be verified by calculation. In this book, as in many mathematical books, when exercises are indicated, it is absolutely essential that they be read and nearly essential that they be worked, because they constitute part of the exposition, the exercise form being adopted where it seems to the author best for conveying the particular information at hand.

To some mathematicians, and even more to logicians, I must say a word of apology for what they may consider lapses of rigor, such as using the same symbol with more than one meaning and failing to distinguish uniformly between the use and the mention of a symbol; but they will understand that these lapses are sacrifices to what I take to be general intelligibility and will have, I hope, no real difficulty in repairing them.

Few will wish to read the whole book; therefore introductions to the chapters and sections have been so written as not only to provide orientation but also to facilitate skipping. In particular, safe detours are indicated around mathematically advanced topics and other digressions.

A few words in explanation of the conventions, such as those by which internal and external references are made in this book, may be useful. The abbreviation § 3.4 means Section 4 of Chapter 3; within Chapter 3 itself, this would be abbreviated still further to § 4. The abbreviation (3.4.1) means the first numbered and displayed equation or other expression in § 3.4; within Chapter 3, this would be abbreviated still further to (4.1) and within § 3.4 simply to (1). Theorems, lemmas, exercises, corollaries, figures, and tables are named by a similar system, e.g., Theorem 3.4.1, Theorem 4.1, Theorem 1. Incidentally, the proofs of theorems are terminated with the special punctuation mark ◆, a device borrowed from Halmos's *Measure Theory*.

Seven postulates, P1, P2, etc., are introduced over the course of several chapters. For ready reference these are, with some explanatory material, reproduced on the end papers.

Entries in the Bibliography at the end of Appendix 3 are designated by a self-explanatory notation in square brackets. For example, the works of Tippett, Wald, Munroe, Feller, and Halmos, already referred to, are [T2], [W1], [M6], [F1], and [H2], respectively.

I often allude to a set of *key references* to a given topic. This means a set of external references intended to lead the reader that wishes to pursue that particular topic to the fullest and most recent bibliographies; it has nothing to do with the merit or importance of the works referred to.

Technical terms (except for non-verbal symbols) that are defined in this book are printed in bold face or italics (depending on the importance of the term for this book or for established usage) in the context where the term is defined. These special fonts are occasionally used for other purposes as well. Terms are sometimes used informally— even in unofficial definitions—before being officially defined. Even the official definitions are sometimes of necessity very loose, corresponding to the well-known principle that, in a formal theory, some terms must in strict logic be left undefined.

L. J. SAVAGE

University of Chicago
April, 1954

Acknowledgement

I HAVE MANY FRIENDS, FEW OF WHOM SHARE MY PRESENT OPINions, to thank for criticism and encouragement. Though the list seems long, I cannot refrain from explicitly mentioning: I. Bross, A. Burks, R. Carnap, B. de Finetti, M. Flood, I. J. Good, P. R. Halmos, O. Helmer, C. Hildreth, T. Koopmans, W. Kruskal, C. F. Mosteller, I. R. Savage, W. A. Wallis, and M. A. Woodbury. Wallis as chairman of my department and close friend has particularly encouraged me to write the book and facilitated my doing so in many ways. Mrs. Janet Lowrey and Miss Louise Forsyth typed and retyped and did so many other painstaking tasks so well that it would be inadequate to call their help secretarial.

My work on the book was made possible by four organizations to which I herewith express thanks. During the years 1950 through 1954 I worked on it at the University of Chicago, where the work was supported by the Office of Naval Research and the University itself, which also supported it during the summer of 1952. During the academic year 1951–52 I worked on it as a research scholar in France under the Fulbright Act (Public Law 584, 79th Congress), and during the whole of that year as a fellow of the John Simon Guggenheim Memorial Foundation.

L. J. S.

Contents

CHAPTER 1

Introduction

1 The role of foundations

It is often argued academically that no science can be more secure than its foundations, and that, if there is controversy about the foundations, there must be even greater controversy about the higher parts of the science. As a matter of fact, the foundations are the most controversial parts of many, if not all, sciences. Physics and pure mathematics are excellent examples of this phenomenon. As for statistics, the foundations include, on any interpretation of which I have ever heard, the foundations of probability, as controversial a subject as one could name. As in other sciences, controversies over the foundations of statistics reflect themselves to some extent in everyday practice, but not nearly so catastrophically as one might imagine. I believe that here, as elsewhere, catastrophe is avoided, primarily because in practical situations common sense generally saves all but the most pedantic of us from flagrant error. It is hard to judge, however, to what extent the relative calm of modern statistics is due to its domination by a vigorous school relatively well agreed within itself about the foundations.

Although study of the foundations of a science does not have the role that would be assigned to it by naive first-things-firstism, it has a certain continuing importance as the science develops, influencing, and being influenced by, the more immediately practical parts of the science.

2 Historical background

The concept and problem of inductive inference have been prominent in philosophy at least since Aristotle. Mathematical work on some aspects of the problem of inference dates back at least to the early eighteenth century. Leibniz is said to be the first to publish a suggestion in that direction, but Jacob Bernoulli's posthumous *Ars Conjectandi* (1713) [B12] seems to be the first concerted effort.† This mathe-

† Valuable information on this and other topics of the early philosophic history of probability is attractively presented in Keynes' treatise [K4], especially in Chapters VII, XXIII, and the bibliography.

1

matical work has always revolved around the concept of probability; but, though there was active interest in probability for nearly a century before the publication of *Ars Conjectandi*, earlier activity seems not to have been concerned with inductive inference.

In the present century there has been and continues to be extraordinary interest in mathematical treatment of problems of inductive inference. For reasons I cannot and need not analyze here, this activity has been strikingly concentrated in the English-speaking world. It is known under several names, most of which stress some aspect of the subject that seemed of overwhelming importance at the moment when the name was coined. "Mathematical statistics," one of its earliest names, is still the most popular. In this name, "mathematical" seems to be intended to connote rational, theoretical, or perhaps mathematically advanced, to distinguish the subject from those problems of gathering and condensing numerical data that can be considered apart from the problem of inductive inference, the mathematical treatment of which is generally relatively trivial. The name "statistical inference" recognizes that the subject is concerned with inductive inference. The name "statistical decision" reflects the idea that inductive inference is not always, if ever, concerned with what to believe in the face of inconclusive evidence, but that at least sometimes it is concerned with what action to decide upon under such circumstances. Within this book, there will be no harm in adopting the shortest possible name, "statistics."

It is unanimously agreed that statistics depends somehow on probability. But, as to what probability is and how it is connected with statistics, there has seldom been such complete disagreement and breakdown of communication since the Tower of Babel. There must be dozens of different interpretations of probability defended by living authorities, and some authorities hold that several different interpretations may be useful, that is, that the concept of probability may have different meaningful senses in different contexts. Doubtless, much of the disagreement is merely terminological and would disappear under sufficiently sharp analysis. Some believe that it would all disappear, or even that they have themselves already made the necessary analysis.

Considering the confusion about the foundations of statistics, it is surprising, and certainly gratifying, to find that almost everyone is agreed on what the purely mathematical properties of probability are. Virtually all controversy therefore centers on questions of interpreting the generally accepted axiomatic concept of probability, that is, of determining the extramathematical properties of probability.

The widely accepted axiomatic concept referred to is commonly ascribed to Kolmogoroff [K7] and goes by his name. It should be mentioned that there is some dissension from it on the part of a small group led by von Mises [V2]. There are also a few minor technical variations on the Kolmogoroff system that are sometimes of interest; they will be discussed in § 3.4.

I would distinguish three main classes of views on the interpretation of probability, for the purposes of this book, calling them objectivistic, personalistic, and necessary. Condensed descriptions of these three classes of views seem called for here. If some readers find these descriptions condensed to the point of unintelligibility, let them be assured that fuller ones will gradually be developed as the book proceeds.

Objectivistic views hold that some repetitive events, such as tosses of a penny, prove to be in reasonably close agreement with the mathematical concept of independently repeated random events, all with the same probability. According to such views, evidence for the quality of agreement between the behavior of the repetitive event and the mathematical concept, and for the magnitude of the probability that applies (in case any does), is to be obtained by observation of some repetitions of the event, and from no other source whatsoever.

Personalistic views hold that probability measures the confidence that a particular individual has in the truth of a particular proposition, for example, the proposition that it will rain tomorrow. These views postulate that the individual concerned is in some ways "reasonable," but they do not deny the possibility that two reasonable individuals faced with the same evidence may have different degrees of confidence in the truth of the same proposition.

Necessary views hold that probability measures the extent to which one set of propositions, out of logical necessity and apart from human opinion, confirms the truth of another. They are generally regarded by their holders as extensions of logic, which tells when one set of propositions necessitates the truth of another.

After what has been said about the intensity and complexity of the controversy over the probability concept, you must realize that the short taxonomy above is bound to infuriate any expert on the foundations of probability, but I trust it may do the less learned more good than harm.

The great burst of statistical research in the English-speaking world in the present century has revolved around objectivistic views on the interpretation of probability. As will shortly be explained, any purely objectivistic view entails a severe difficulty for statistics. This difficulty is recognized by members of the **British-American School,** if I may use that name without its being taken too literally or at all nationalistically, and is regarded by them as a great, though not insurmountable, obstacle; indeed, some of them see it as the central problem of statistics.

The difficulty in the objectivistic position is this. In any objectivistic view, probabilities can apply fruitfully only to repetitive events, that is, to certain processes; and (depending on the view in question) it is either meaningless to talk about the probability that a given proposition is true, or this probability can be only 1 or 0, according as the proposition is in fact true or false. Under neither interpretation can probability serve as a measure of the trust to be put in the proposition. Thus the existence of evidence for a proposition can never, on an objectivistic view, be expressed by saying that the proposition is true with a certain probability. Again, if one must choose among several courses of action in the light of experimental evidence, it is not meaningful, in terms of objective probability, to compute which of these actions is most promising, that is, which has the highest expected income. Holders of objectivistic views have, therefore, no recourse but to argue that it is not reasonable to assign probabilities to the truth of propositions or to calculate which of several actions is the most promising, and that the need expressed by the attempt to set up such concepts must be met in other ways, if at all.

The British-American School has had great success in several respects. The number of its adherents has rapidly increased. It has contributed many procedures of strong intuitive appeal and (one feels) of lasting worth. These have found widespread application in many sciences, in industry, and in commerce. The success of the school may pragmatically be taken as evidence for the correctness of the general view on which it is based. Indeed, anyone who overthrows that view must either discredit the procedures to which it has led, or show, as I hope to show in this book, that they are on the whole consistent with the alternative proposed.

Some, I among them, hold that the grounds for adopting an objectivistic view are not overwhelmingly strong; that there are serious logical objections to any such view; and, most important of all, that the difficulty a strictly objectivistic view meets in statistics reflects real inadequacy.

3   General outline of this book

This book presents a theory of the foundations of statistics which is based on a personalistic view of probability derived mainly from the work of Bruno de Finetti, as expressed for example in [D2]. The theory is presented in a tentative spirit, for I realize that the serious blemishes in it apparent to me are not the only ones that will be discovered by critical readers. A theory of the foundations of statistics that appears contrary to the teaching of the most productive statisticians will properly be regarded with extraordinary caution. Other views on probability will, of course, be discussed in this book, partly for their own interest and partly to explain the relationship between the personalistic view on which this book is based and other views.

The book is organized into seventeen chapters, of which the present introduction is the first. Chapters 2–7 are, so to speak, concerned with the foundations at a relatively deep level. They develop, explain, and defend a certain abstract theory of the behavior of a highly idealized person faced with uncertainty. That theory is shown to have as implications a theory of personal probability, corresponding to the personalistic view of probability basic to this book, and also a theory of utility due, in its modern form, to von Neumann and Morgenstern [V4].

There is a transition, occurring in Chapter 8 and maintained throughout the rest of the book, to a shallower level of the foundations of statistics; I might say from pre-statistics to statistics proper. In those later chapters, it is recognized that the theory developed in the earlier ones is too highly idealized for immediate application. Some compromises have to be made, and the appropriate ones are sought in an analysis of some of the inventions and ideas of the British-American School. It will, I hope, be demonstrated thereby that the superficially incompatible systems of ideas associated on the one hand with a personalistic view of probability and on the other with the objectivistically inspired developments of the British-American School do in fact lend each other mutual support and clarification.

CHAPTER 2

Preliminary Considerations
on Decision in
the Face of Uncertainty

1 Introduction

Decisions made in the face of uncertainty pervade the life of every individual and organization. Even animals might be said continually to make such decisions, and the psychological mechanisms by which men decide may have much in common with those by which animals do so. But formal reasoning presumably plays no role in the decisions of animals, little in those of children, and less than might be wished in those of men. It may be said to be the purpose of this book, and indeed of statistics generally, to discuss the implications of reasoning for the making of decisions.

Reasoning is commonly associated with logic, but it is obvious, as many have pointed out, that the implications of what is ordinarily called logic are meager indeed when uncertainty is to be faced. It has therefore often been asked whether logic cannot be extended, by principles as acceptable as those of logic itself, to bear more fully on uncertainty. An attempt to extend logic in this way will be begun in this chapter, differing in two important respects from most, but not all, other attempts.

First, since logic is concerned with implications among propositions, many have thought it natural to extend logic by setting up criteria for the extent to which one proposition tends to imply, or provide evidence for, another. It seems to me obvious, however, that what is ultimately wanted is criteria for deciding among possible courses of action; and, therefore, generalization of the relation of implication seems at best a roundabout method of attack. It must be admitted that logic itself does lead to some criteria for decision, because what is implied by a proposition known to be true is in turn true and sometimes relevant to making a decision. Should some notion of partial implication be demonstrably even better articulated with decision than is implication it-

6

self, that would be excellent; but how is such a notion to be sought except by explicitly studying decision? Ramsey's discussion in [R1] of the point at issue here is especially forceful.

Second, it is appealing to suppose that, if two individuals in the same situation, having the same tastes and supplied with the same information, act reasonably, they will act in the same way. Such agreement, belief in which amounts to a necessary (as opposed to a personalistic) view of probability, is certainly worth looking for. Personally, I believe that it does not correspond even roughly with reality, but, having at the moment no strong argument behind my pessimism on this point, I do not insist on it. But I do insist that, until the contrary be demonstrated, we must be prepared to find reasoning inadequate to bring about complete agreement. In particular, the extensions of logic to be adduced in this book will not bring about complete agreement; and whether enough additional principles to do so, or indeed any additional principles of much consequence, can be adduced, I do not know. It may be, and indeed I believe, that there is an element in decision apart from taste, about which, like taste itself, there is no disputing.

The next four sections of this chapter build up a formal model, or scheme, of the situation in which a person is faced with uncertainty; the final two, in terms of this model, motivate and state some of the few principles that seem to me entitled to be taken as postulates for rational decision.

2　The person

I am about to build up a highly idealized theory of the behavior of a "rational" person with respect to decisions. In doing so I will, of course, have to ask you to agree with me that such and such maxims of behavior are "rational." In so far as "rational" means logical, there is no live question; and, if I ask your leave there at all, it is only as a matter of form.† But our person is going to have to make up his mind in situations in which criteria beyond the ordinary ones of logic will be necessary. So, when certain maxims are presented for your consideration, you must ask yourself whether you try to behave in accordance with them, or, to put it differently, how you would react if you noticed yourself violating them.

---

† The assumption that a person's behavior is logical is, of course, far from vacuous. In particular, such a person cannot be uncertain about decidable mathematical propositions. This suggests, at least to me, that the tempting program sketched by Polya [P6] of establishing a theory of the probability of mathematical conjectures cannot be fully successful in that it cannot lead to a truly formal theory, but de Finetti [D5] seems more optimistic about the program.+

+ Polya has greatly elaborated his program, but not in the direction of seeking a formal theory. A curious early work by Cérésole (1915) is somewhat pertinent, and Hacking (1967) argues for the possibility of including mathematical uncertainty in a formal theory.

It is brought out in economic theory that organizations sometimes behave like individual people, so that a theory originally intended to apply to people may also apply to (or may even apply better to) such units as families, corporations, or nations. In view of this possibility, economic theorists are sometimes reluctant to use the word "person," or even "individual," for the behaving units to which they refer; but for our purpose "person" threatens no confusion, though the possibility of using it in an extended sense may well be borne in mind.

3 The world, and states of the world

A formal description, or model, of what the person is uncertain about will be needed. To motivate this formal description, let me begin informally by considering a list of examples. The person might be uncertain about:

1. Whether a particular egg is rotten.
2. Which, if any, in a particular dozen eggs are rotten.
3. The temperature at noon in Chicago yesterday.
4. What the temperature was and will be in the place now covered by Chicago each noon from January 1, 1 A.D., to January 1, 4000 A.D.
5. The infinite sequence of heads and tails that will result from repeated tosses of a particular (everlasting) coin.
6. The complete decimal expansion of π.
7. The exact and entire past, present, and future history of the universe, understood in any sense, however wide.

These examples have a few features in common, though, if there are more than a few, it is a discredit to my imagination. Thus, in each there is some object about which the person is uncertain, an egg, a dozen eggs, a temperature, a sequence of temperatures, etc. Each object admits a certain class of descriptions that might thinkably apply to it. To illustrate, the egg of Example 1 might be rotten or not; and the terms of the example are meant to exclude any other description from consideration, though, of course, a real egg has many other features. Again, since any subset of the dozen eggs (including the extreme cases of all and none at all) might be rotten, there are 2^{12} descriptions associated with Example 2. For Example 3 and each subsequent one, there are an infinite number of descriptions, though the array of descriptions is more complicated in some than in others, reaching the ultimate of complexity in Example 7. Example 6 is a little anomalous in that anything the person does not know about the description of π he could know in principle by thinking sufficiently hard about it, that is, by logic alone. This point, banal to some readers, needs explanation for others. If, for example, π is understood to be the area of a circle of unit radius, it follows by logic alone that π is not greater than the area of a square circumscribing the unit circle, that is, $\pi \leq 4$. By an elaboration of this method π can be computed to any degree of accuracy, and by other purely logical methods many other facts about π can be established, such as the fact that π is not a rational number.

In connection with the concepts suggested by the preceding paragraph, the following nomenclature is proposed as brief, suggestive, and in reasonable harmony with the usages of statistics and ordinary discourse.

Term	Definition
the **world**	the object about which the person is concerned
a **state** (of the world)	a description of the world, leaving no relevant aspect undescribed
the **true** state (of the world)	the state that does in fact obtain, i.e., the true description of the world

In application of the theory, the question will arise as to which world to use in a given context. Thus, if the person is interested in the only brown egg in a dozen, should that egg or the whole dozen be taken as the world? It will be seen as the theory is developed that in principle no harm is done by taking the larger of two worlds as a model of the situation. One is therefore tempted to adopt, once and for all, one world sufficiently large, say Example 7. The most serious objection to this is that Example 7 is vague, and some mathematical and philosophical experience suggests that the vagueness cannot be removed without ruining the universality of the example. It may also be added that the use of modest little worlds, tailored to particular contexts, is often a simplification, the advantage of which is justified by a considerable body of mathematical experience with related ideas.

The sense in which the world of a dozen eggs is larger than the world of the one brown egg in the dozen is in some respects obvious. It may be well, however, to emphasize that a state of the smaller world corresponds not to one state of the larger, but to a set of states. Thus, "The brown egg is rotten" describes the smaller world completely, and therefore is a state of it; but the same statement leaves much about the larger world unsaid and corresponds to a set of 2^{11} states of it. In the sense under discussion a smaller world is derived from a larger by neglecting some distinctions between states, not by ignoring some states outright. The latter sort of contraction may be useful in case certain states are regarded by the person as virtually impossible so that they can be ignored.

4 Events

An **event** is a set of states. For example, in connection with the world of Example 2, the person might well be concerned with the event that exactly one egg in the dozen is rotten (an event having 12 states as elements), or, a little less academically, that at least one of the eggs is rotten (an event having $2^{12} - 1$ states as elements, i.e., all the states in the world but one). In connection with the world of Example 3, the person might be concerned with the event, having an infinite number of states, that the temperature at noon in Chicago yesterday was below freezing. To give a final illustration, of a more mathematical flavor, consider in connection with Example 5 the event that the ratio of the number of heads to tails approaches 3 as the sequence progresses to infinity.

In connection with any given world, there are two events that are of the utmost logical importance, though in ordinary discourse it may seem banal even to mention their existence. These are the universal and the vacuous events. The **universal** event, here to be symbolized by S, is the event having every state of the world as element. In so far as "world" has a real technical meaning, S is the world. The **vacuous** event, which can here be safely enough symbolized by the 0 of arithmetic, is the event having no states as elements. To illustrate, in Example 1 the event that the egg is rotten or good is the universal event, and that it is both rotten and good is the vacuous event.

It is important to be able to express the idea that a given event contains the true state among its elements. English usage seems to offer no alternative to the rather stuffy expression, "the event **obtains**."

The theory under development makes no formal reference to time. In particular, the concept of event as here formulated is timeless, though temporal ideas may be employed in the description of particular events. Thus, it would not be said that Lincoln's assassination is an event that occurred in 1865 and that the next return of Halley's comet is one that will occur in 1985, but that Lincoln's assassination in 1865 and the return of Halley's comet in, but not before, 1985 are events that obtain.

Modern mathematical usage, especially that of a branch of mathematics called Boolean algebra, suggests the following table of definitions in connection with the concepts of state and event. Some of these are synonyms, others abbreviations, and still others new terms compounded out of old.

Though the notations introduced in Table 1 are very elementary and of great utility, they are not ordinarily taught except in connection with logic or relatively advanced mathematics. A set of exercises illustrating their use is therefore given below in the form of a numbered list of statements. These statements are true whatever the sets A, B,

TABLE 1. MATHEMATICAL NOMENCLATURE PERTAINING TO STATE AND EVENTS

Term	Definition
(Basic terms)	
set	event
$A, B, C, \cdots$	generic symbols for events
s, s', s''	generic symbols for states
S	the universal event
0	the vacuous event
(Relations)	
$s \; \varepsilon \; A.$	s is an element of A, i.e., a state in A.†
$A \subset B$ (or $B \supset A$).	A is contained in B, i.e., every element of A is an element of B.
$A = B.$	A equals B, i.e., A is the same set as B, i.e., A and B have exactly the same elements.
(Constructs)	
the **complement** of A with respect to S	those elements of S that are not in A
$\sim A$	the complement of A with respect to S
the **union** of the A_i's	those elements of S that are elements of at least one of the sets A_1, A_2, etc.
$\bigcup_i A_i$	the union of the A_i's
$A \cup B$	the union of A and B, i.e., those elements of S that are elements of A or B (possibly of both)
the **intersection** of the A_i's	those elements of S that are elements of each of the sets A_1, A_2, etc.
$\bigcap_i A_i$	the intersection of the A_i's
$A \cap B$	the intersection of A and B, i.e., those elements of S that are elements of both A and B

† Typographical note: The Porson font of the Greek alphabet (α, β, γ, δ, ϵ, ζ, $\cdots$) is the one almost always printed, at least in America, when mathematical constants and variables are denoted by Greek letters. The symbol ε used in this and some other publications to denote "element of" is, however, the epsilon of the Vertical font (α, β, γ, δ, ε, ζ, $\cdots$). Some publications use the special symbol $\in$; and some use ϵ, the Porson epsilon, presumably because of its resemblance to $\in$. The latter usage entails either using ϵ for two different purposes or else changing fonts in mid alphabet (α, β, γ, δ, ε, ζ, $\cdots$) when constants and variables are denoted by Greek letters.

C may be. Mathematicians would for the most part verify them by translating them into English and appealing to common sense, though in complicated cases explicit use might be made of Exercise 9. Diagrams, called **Venn diagrams,** in which sets are symbolized by areas, as illustrated by Figure 1, are often suggestive.

 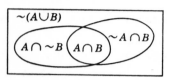

Figure 1

It is a remarkable and useful fact that any universally valid statement about sets remains so if, throughout, $\cup$ is interchanged with $\cap$, 0 with S, and $\subset$ with $\supset$. The dual in this sense of each exercise should be studied along with the exercise itself. For example, the dual of Exercise 7 is: $A \supset B$, if and only if $A = A \cup B$. Note that the first parts of Exercises 1 through 6 are dual to the second parts.

It may be remarked that, if Exercises 1–6 are taken as axioms and 7 as a definition, Exercises 8–21 and also the duality principle follow formally from them. For example, 10 can be proved thus: By 7, if $A \cap B$ is A, then $A \subset B$; but, by 1, $A \cap A$ is A; therefore $A \subset A$. Again, 8 can be proved, using 6, 3, 2, 1, 3, and 6 in that order, thus:

$$(1) \quad 0 \cap A = (A \cap \sim A) \cap A = (\sim A \cap A) \cap A$$
$$= \sim A \cap (A \cap A) = \sim A \cap A = A \cap \sim A = 0.$$

Such formal demonstration is fun and helps develop mathematical skill. In the present exercises the novice, however, should consider it as a possible supplement to, but not as a substitute for, demonstration by interpretation.

If the exercises fail to render the notations familiar, it would be best to talk with someone to whom they are already familiar or failing that, to read in any elementary book where the subject is treated, for example, Chapter II, "The Boole-Schroeder Algebra," in the text of Lewis and Langford [L7].

Exercises illustrating Boolean algebra

   1. $A \cap A = A = A \cup A$.
   2. $(A \cap B) \cap C = A \cap (B \cap C)$; $(A \cup B) \cup C = A \cup (B \cup C)$. (These facts often render parentheses superfluous.)
   3. $A \cap B = B \cap A; A \cup B = B \cup A$.

4. $A \cap (B \cup C) = (A \cap B) \cup (A \cap C); A \cup (B \cap C) =$
$(A \cup B) \cap (A \cup C)$.

5. $S \cap A = A; 0 \cup A = A$.

6. $A \cap (\sim A) = 0; A \cup (\sim A) = S$.

7. $A \subset B$, if and only if $A = A \cap B$.

8. $0 \cap A = 0$.

9. $A = B$, if and only if $A \subset B$ and $B \subset A$.

10. $A \subset A$.

11. $(A \cap B) \subset A$.

12. If $A \subset B$, then $(A \cap C) \subset (B \cap C)$, and $(A \cup C) \subset (B \cup C)$.

13. $(A \cup B) \subset C$, if and only if $A \subset C$ and $B \subset C$.

14. $0 \subset A \subset S$.

15. $A \cap (A \cup B) = A$.

16. $\sim(\sim A) = A$.

17. $\sim(A \cup B) = (\sim A) \cap (\sim B)$ (De Morgan's theorem).

18. $\sim 0 = S$.

19. $A \cap (\sim A \cup B) = A \cap B$.

20. $A \subset B$, if and only if $(\sim B) \subset (\sim A)$.

21. $A \subset B$, if and only if $A \cap (\sim B) = 0$.

22. $\sim(\bigcup_i A_i) = \bigcap_i (\sim A_i)$ (General De Morgan's theorem).

23. $A \cup (\bigcap_i B_i) = \bigcap_i (A \cup B_i)$.

24. $A \cap (\bigcap_i B_i) = \bigcap_i (A \cap B_i)$.

25. $(\bigcup_i A_i) \cup (\bigcup_j B_j) = \bigcup_{i,j} (A_i \cup B_j)$.

26. $(\bigcap_i A_i) \cup (\bigcap_j B_j) = \bigcap_{i,j} (A_i \cup B_j)$.

27. $A \subset (\bigcap_i B_i)$, if and only if $A \subset B_i$ for every i.

28. $(\bigcap_i B_i) \subset B_j \subset (\bigcup_i B_i)$ for every j.

5 Consequences, acts, and decisions

To say that a decision is to be made is to say that one of two or more acts is to be chosen, or decided on. In deciding on an act, account must be taken of the possible states of the world, and also of the consequences implicit in each act for each possible state of the world. A **consequence** is anything that may happen to the person.

Consider an example. Your wife has just broken five good eggs into a bowl when you come in and volunteer to finish making the omelet. A sixth egg, which for some reason must either be used for the omelet or wasted altogether, lies unbroken beside the bowl. You must decide what to do with this unbroken egg. Perhaps it is not too great an oversimplification to say that you must decide among three acts only, namely, to break it into the bowl containing the other five, to break it into a saucer for inspection, or to throw it away without inspection.

Depending on the state of the egg, each of these three acts will have some consequence of concern to you, say that indicated by Table 1.

TABLE 1. AN EXAMPLE ILLUSTRATING ACTS, STATES, AND CONSEQUENCES

Act	State	
	Good	Rotten
break into bowl	six-egg omelet	no omelet, and five good eggs destroyed
break into saucer	six-egg omelet, and a saucer to wash	five-egg omelet, and a saucer to wash
throw away	five-egg omelet, and one good egg destroyed	five-egg omelet

Even the little example concerning the omelet suggests how varied the things, or experiences, regarded as consequences, can be. They might in general involve money, life, state of health, approval of friends, well-being of others, the will of God, or anything at all about which the person could possibly be concerned. Consequences might appropriately be called states of the person, as opposed to states of the world. They might also be referred to, with some extension of the economic notion of income, as the possible incomes of the person. In any one problem, the set of consequences envisaged will be denoted by F, and the individual consequences will be denoted by f, g, h, etc. In the omelet example, F consists of the six consequences tabulated in Table 1: six-egg omelet; no omelet, and five good eggs destroyed; etc.

If two different acts had the same consequences in every state of the world, there would from the present point of view be no point in considering them two different acts at all. An act may therefore be identified with its possible consequences. Or, more formally, an **act** is a function attaching a consequence to each state of the world. The notation **f** will be used to denote an act, that is, a function, attaching the consequence $f(s)$ to the state s. The notation **f** is logically a better name for a function than the more customary $f(s)$ for exactly the same reason that the word "logarithm" is a better term for logarithm than "logarithm of x" would be. The notational distinction involved here is often justifiably neglected in mathematical work, but we will have special need to observe it, at least in connection with acts, as will soon be explained. When several acts are to be discussed at once, they may be denoted by different letters thus: **f**, **g**, **h**; by the use of primes thus: **f**, **f′**, **f″**; or by subscripts thus: f_1, f_i. The set of all acts available in a given situation will be denoted by **F** or a similar symbol. In the example of the omelet, **F** has three acts as elements. If, for example, **f** denotes the first of the three acts listed in Table 1, then **f** is defined thus:

$$f(\text{good}) = \text{six-egg omelet;}$$

(1)

$$f(\text{rotten}) = \text{no omelet, and five good eggs destroyed.}$$

The argument might be raised that the formal description of decision that has thus been erected seems inadequate because a person may not know the consequences of the acts open to him in each state of the world. He might be so ignorant, for example, as not to be sure whether one rotten egg will spoil a six-egg omelet. But in that case nothing could be simpler than to admit that there are four states in the world corresponding to the two states of the egg and the two conceivable answers to the culinary question whether one bad egg will spoil a six-egg omelet. It seems to me obvious that this solution works in the greatest generality, though a thoroughgoing analysis might not be trivial. A reader interested in the technicalities of this point or that of the succeeding paragraph will find an extensive discussion of a similar problem in Chapter II of [V4], where von Neumann and Morgenstern discuss the reduction of a general game to its reduced form.

Again, the formal description might seem inadequate in that it does not provide explicitly for the possibility that one decision may lead to another. Thus, if the omelet should be spoiled by breaking a rotten egg into it, new questions might arise about what to substitute for breakfast and how to appease your justifiably furious wife. But, just as in the preceding paragraph an apparent shortcoming of the proposed mode of description was attributed to an incomplete analysis of the possible states, here I would say that the list of available acts envisaged in Table 1 is inadequate for the interpretation that has just been put on the problem. Where the single act "break into bowl" now stands, there should be several, such as: "break into bowl, and in case of disaster have toast," "break into bowl, and in case of disaster take family to a neighboring restaurant for breakfast." Appropriate consequences of these new acts can easily be imagined.

As has just been suggested, what in the ordinary way of thinking might be regarded as a chain of decisions, one leading to the other in time, is in the formal description proposed here regarded as a single decision. To put it a little differently, it is proposed that the choice of a policy or plan be regarded as a single decision. This point of view, though not always in so explicit a form, has played a prominent role in the statistical advances of the present century. For example, the great majority of experimentalists, even today, suppose that the function of statistics and of statisticians is to decide what conclusions to draw from data gathered in an experiment or other observational program. But statisticians hold it to be lacking in foresight to gather data without a view to the method of analysis to be employed, that is, they hold that the design and analysis of an experiment should be decided upon as an articulated whole.

The point of view under discussion may be symbolized by the proverb, "Look before you leap," and the one to which it is opposed by the proverb, "You can cross that bridge when you come to it." When two proverbs conflict in this way, it is proverbially true that there is some truth in both of them, but rarely, if ever, can their common truth be captured by a single pat proverb. One must indeed look before he leaps, in so far as the looking is not unreasonably time-consuming and otherwise expensive; but there are innumerable bridges one cannot afford to cross, unless he happens to come to them.

Carried to its logical extreme, the "Look before you leap" principle demands that one envisage every conceivable policy for the government of his whole life (at least from now on) in its most minute details, in the light of the vast number of unknown states of the world, and decide here and now on one policy. This is utterly ridiculous, not—as some might think—because there might later be cause for regret, if things did not turn out as had been anticipated, but because the task implied in making such a decision is not even remotely resembled by human possibility. It is even utterly beyond our power to plan a picnic or to play a game of chess in accordance with the principle, even when the world of states and the set of available acts to be envisaged are artificially reduced to the narrowest reasonable limits.

Though the "Look before you leap" principle is preposterous if carried to extremes, I would none the less argue that it is the proper subject of our further discussion, because to cross one's bridges when one comes to them means to attack relatively simple problems of decision by artificially confining attention to so small a world that the "Look before you leap" principle can be applied there. I am unable to formulate criteria for selecting these small worlds and indeed believe that their selection may be a matter of judgment and experience about which it is impossible to enunciate complete and sharply defined general principles, though something more will be said in this connection in § 5.5. On the other hand, it is an operation in which we all necessarily have much experience, and one in which there is in practice considerable agreement.

In view of the "Look before you leap" principle, acts and decisions, like events, are timeless. The person decides "now" once for all; there is nothing for him to wait for, because his one decision provides for all contingencies. None the less, temporal modes of description, though translatable into atemporal ones, are often suggestive. Thus, there will be occasion to analyze and make frequent use of the idea of deferring a decision until an observation relevant to it has been made.

6   The simple ordering of acts with respect to preference

Of two acts **f** and **g**, it is possible that the person **prefers f** to **g**. Loosely speaking, this means that, if he were required to decide between **f** and **g**, no other acts being available, he would decide on **f**.

This procedure for testing preference is not entirely adequate, if only because it fails to take account of, or even define, the possibility that the person may not really have any preference between **f** and **g**, regarding them as equivalent; in which case his choice of **f** should not be regarded as significant. If the person really does regard **f** and **g** as equivalent, that is, if he is indifferent between them, then, if **f** or **g** were modified by attaching an arbitrarily small bonus to its consequences in every state, the person's decision would presumably be for whichever act was thus modified. This test for indifference does not provide an altogether satisfactory definition, since it begs the question to some extent by postulating in effect that the tester knows what constitutes a small bonus. Another attempted solution would be to say that the person knows by introspection whether he has decided haphazardly or in response to a definite feeling of preference. This sort of solution seems to me especially objectionable, because I think it of great importance that preference, and indifference, between **f** and **g** be determined, at least in principle, by decisions between acts and not by response to introspective questions. In spite of the difficulty of distinguishing between preference and indifference, I think enough has been said for us to proceed to a postulational treatment of them.

The very meaning of the relationship of preference that I have attempted to establish in the preceding paragraph implies that the person cannot simultaneously prefer **f** to **g** and **g** to **f**. In the postulational treatment of the relationships of preference and indifference, it will be technically convenient to work with the relation "is not preferred to" rather than directly with its complementary relation "is preferred to." Thus, rather than say that it is impossible that both **f** is preferred to **g** and **g** to **f**, I might say that, of any two acts **f** and **g**, **f** is not preferred to **g** or **g** is not preferred to **f**, possibly both. Again, the definition of preference suggests that, if **f** is not preferred to **g**, and **g** is not preferred to **h**, then it is impossible that **f** should be preferred to **h**.

The two assumptions just made about the relation "is not preferred to" is sometimes expressed in ordinary mathematical usage by saying that the relation is a simple ordering among acts. Formally, a relation $\leq\cdot$ among a set of elements x, y, z $\cdots$, is called a **simple ordering**, in this book, if and only if for every x, y, and z:

1. Either $x \leq\cdot y$, or $y \leq\cdot x$.
2. If $x \leq\cdot y$, and $y \leq\cdot z$, then $x \leq\cdot z$.

Borrowing from arithmetic the suggestive abbreviation $\leq$ for the relation "is not preferred to," the assumption that $\leq$ is a simple ordering can be expressed formally by a postulate, thus:

P1   The relation $\leq$ is a simple ordering among acts.

It is noteworthy that P1 makes no explicit reference to states of the world. Except possibly for mathematical refinements,† it seems to me that no additional postulates can be formulated without making such reference—at any rate none will be in this book.

P1 by itself is not very rich in consequences, but one easily proved theorem following from it may be mentioned.

THEOREM 1   If **F** is a finite set of acts, there exist **f** and **h** in **F** such that for all **g** in **F**
$$\mathbf{f} \leq \mathbf{g} \leq \mathbf{h}.$$

Theorem 1 is especially relevant to application of the theory of decision, because I interpret the theory to imply that, if **F** is finite, the person will decide on an act **h** in **F** to which no other act in **F** is preferred, the existence of at least one such **h** being guaranteed by the theorem.

It is often appropriate to consider infinite sets of available acts. In economic contexts, for example, it is generally an inappropriate complication to take explicit account of the possibility that all transactions must be in integral numbers of pennies. If infinite sets of available acts are set up and interpreted without some mathematical tact, unrealistic conclusions are likely to follow. Suppose, for example, that you were free to choose any income, provided it be definitely less than $100,000 per year. Precisely which income would you choose, abstracting from the indivisibility of pennies?

---

† For example, such topological assumptions about the space with neighborhoods defined in terms of $\leq$ as connectedness, local compactnesss, or density.

It is sometimes convenient to supplement the relation $\leq$ by other relations derived from it in accordance with the definitions in Table 1, analogous definitions being applicable to any simple ordering. The assumption of simple ordering, P1, has several implications for the derived relations $\geq$, $<$, $>$, and $\doteq$. These are generally strongly suggested by the properties of the corresponding relations in arithmetic.

TABLE 1.   TABLE OF RELATIONS DERIVED FROM $\leq$

New Relation	Definition
$f \geq g$.	$g \leq f$.
$f < g$, i.e., **g is preferred to f.**	It is false that $g \leq f$.
$f > g$.	$g < f$.
$f \doteq g$, i.e., **f is equivalent to (or indifferent with respect to) g.**	$f \leq g$, and $g \leq f$.
g is between f and h.	$f \leq g \leq h$, or $h \leq g \leq f$.

A few such implications of P1 are listed below, with no intention of completeness, as exercises for those who may not already be familiar with the elementary properties of simple ordering.

Exercises

1. The relation $\geq$ is also a simple ordering.
2. All the relations $\leq$, $\geq$, $<$, $>$, and $\doteq$ are **transitive,** that is, they can be validly substituted for $\leq$ in the second part of the definition of simple ordering.
3. Between any pair of acts **f, g**, one and only one of the three relations $<$, $\doteq$, and $>$ holds.
4. If $f < g$, and $g \doteq h$, then $f < h$.
5. If $f \doteq g$, then $g \doteq f$.
6. For any f, $f \doteq f$.
7. At least one of three acts **f, g, h** is between the other two. When can there be more than one such?

Two very different sorts of interpretations can be made of P1 and the other postulates to be adduced later. First, P1 can be regarded as a prediction about the behavior of people, or animals, in decision situations. Second, it can be regarded as a logic-like criterion of consistency in decision situations. For us, the second interpretation is the only one of direct relevance, but it may be fruitful to discuss both, calling the first **empirical** and the second **normative.**

Logic itself admits an empirical as well as a normative interpretation. Thus, if an experimental subject believes certain propositions, it is to be expected that he will also believe their logical consequences and disbelieve the negations of these consequences. This theory of human psychology has some validity and is of great practical utility in our everyday dealings with other people, though it is very crude and approximate. For one thing, people often do make elementary mistakes in logic; more refined theories would attribute these mistakes to such things as accident or subconscious motivation. For another, if anyone who believed the axioms of mathematics also believed all that they imply and nothing that they contradict, mathematical study would be superfluous for him; such a person would, as has been explained, be able to state the ten-thousandth or any other term in the decimal expansion of π on demand. To summarize, logic can be interpreted as a crude but sometimes handy empirical psychological theory.

The principal value of logic, however, is in connection with its normative interpretation, that is, as a set of criteria by which to detect, with sufficient trouble, any inconsistencies there may be among our beliefs, and to derive from the beliefs we already hold such new ones as consistency demands. It does not seem appropriate here to attempt an analysis of why and in what contexts we wish to be consistent; it is sufficient to allude to the fact that we often do wish to be so.

Analogously, P1 together with the postulates to be adduced later can be interpreted as a crude and shallow empirical theory predicting the behavior of people making decisions. This theory is practical in suitably limited domains, and everyone in fact makes use of at least some aspects of it in predicting the behavior of others. At the same time, the behavior of people is often at variance with the theory. The departure is sometimes flagrant, in which case our attitude toward it is much like that we hold toward a slip in logic, calling the departure a mistake and attributing it to such things as accident and subconscious motivation. Or, the departure may be detectable only by a long chain of argument or calculation, the possibilities becoming increasingly complicated as new postulates are brought to stand beside P1.

Pursuing the analogy with logic, the main use I would make of P1 and its successors is normative, to police my own decisions for consistency and, where possible, to make complicated decisions depend on simpler ones.

Here it is more pertinent than it was in connection with logic that something be said of why and when consistency is a desideratum, though I cannot say much. Suppose someone says to me, "I am a rational person, that is to say, I seldom, if ever, make mistakes in logic. But I behave in flagrant disagreement with your postulates, because they violate my personal taste, and it seems to me more sensible to cater to my taste than to a theory arbitrarily concocted by you." I don't see how I could really controvert him, but I would be inclined to match his introspection with some of my own. I would, in particular, tell him that, when it is explicitly brought to my attention that I have shown a preference for **f** as compared with **g**, for **g** as compared with **h**, and for **h** as compared with **f**, I feel uncomfortable in much the same way that I do when it is brought to my attention that some of my beliefs are logically contradictory. Whenever I examine such a triple of preferences on my own part, I find that it is not at all difficult to reverse one of them. In fact, I find on contemplating the three alleged preferences side by side that at least one among them is not a preference at all, at any rate not any more.

There is some temptation to explore the possibilities of analyzing preference among acts as a **partial ordering,** that is, in effect to replace part 1 of the definition of simple ordering by the very weak proposition **f** ≤ **f**, admitting that some pairs of acts are incomparable. This would seem to give expression to introspective sensations of indecision or vacillation, which we may be reluctant to identify with indifference. My own conjecture is that it would prove a blind alley losing much in power and advancing little, if at all, in realism; but only an enthusiastic exploration could shed real light on the question.

7 The sure-thing principle

A businessman contemplates buying a certain piece of property. He considers the outcome of the next presidential election relevant to the attractiveness of the purchase. So, to clarify the matter for himself, he asks whether he would buy if he knew that the Republican candidate were going to win, and decides that he would do so. Similarly, he considers whether he would buy if he knew that the Democratic candidate were going to win, and again finds that he would do so. Seeing that he would buy in either event, he decides that he should buy, even though he does not know which event obtains, or will obtain, as we would ordinarily say. It is all too seldom that a decision can be arrived at on the basis of the principle used by this businessman, but, except possibly for the assumption of simple ordering, I know of no other extralogical principle governing decisions that finds such ready acceptance.

Having suggested what I shall tentatively call the **sure-thing principle,** let me give it relatively formal statement thus: If the person would not prefer **f** to **g**, either knowing that the event B obtained, or knowing that the event $\sim B$ obtained, then he does not prefer **f** to **g**.

Moreover (provided he does not regard B as virtually impossible) if he would definitely prefer **g** to **f**, knowing that B obtained, and, if he would not prefer **f** to **g**, knowing that B did not obtain, then he definitely prefers **g** to **f**.

The sure-thing principle cannot appropriately be accepted as a postulate in the sense that P1 is, because it would introduce new undefined technical terms referring to knowledge and possibility that would render it mathematically useless without still more postulates governing these terms. It will be preferable to regard the principle as a loose one that suggests certain formal postulates well articulated with P1.

What technical interpretation can be attached to the idea that **f** would be preferred to **g**, if B were known to obtain? Under any reasonable interpretation, the matter would seem not to depend on the values **f** and **g** assume at states outside of B. There is, then, no loss of generality in supposing that **f** and **g** **agree** with each other except in B, that is, that $f(s) = g(s)$ for all $s \in \sim B$. Under this unrestrictive assumption, **f** and **g** are surely to be regarded as equivalent given $\sim B$; that is, they would be considered equivalent, if it were known that B did not obtain. The first part of the sure-thing principle can now be interpreted thus: If, after being modified so as to agree with one another outside of B, **f** is not preferred to **g**; then **f** would not be preferred to **g**, if B were known. The notion will be expressed formally by saying that $f \leq g$ **given** B.+

It is implicit in the argument that has just led to the definition of $f \leq g$ given B that, if two acts **f** and **g** are so modified in $\sim B$ as to agree with each other, then the order of preference obtaining between the modified acts will not depend on which of the permitted modifications was actually carried out. Equivalently, if **f** and **g** are two acts that do agree with each other in $\sim B$, and $f \leq g$; then, if **f** and **g** are modified in $\sim B$ in any way such that the modified acts **f′** and **g′** continue to agree with each other in $\sim B$, it will also be so that $f' \leq g'$. This assumption is made formally in the postulate P2 below and illustrated schematically in Figure 1, a kind of diagram I find suggestive in many such contexts.

In Figure 1, the set S of all states s and the set F of all consequences f are represented by horizontal and vertical intervals respectively. In any such diagram an act **f**, being a function attaching a value $f(s) \in F$ to each $s \in S$ is represented by a graph. This particular diagram graphs two acts **f** and **g** that agree with each other in $\sim B$, and two other acts **f′** and **g′** that also agree with each other in $\sim B$ and arise by modifying **f** and **g** respectively only in $\sim B$, that is, acts agreeing with **f** and **g** respectively in B.

+ In this edition, the corresponding definition D1 on the end papers has been slightly strengthened to compensate an inadvertent weakness in the end paper version of P2, pointed out to me by Peter Fishburn.

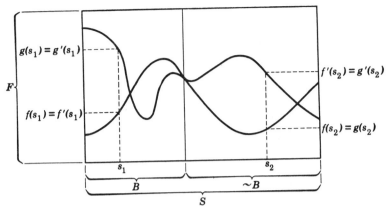

Figure 1

P2   If **f**, **g**, and **f′**, **g′** are such that:
  1. in $\sim B$, **f** agrees with **g**, and **f′** agrees with **g′**,
  2. in B, **f** agrees with **f′**, and **g** agrees with **g′**,
  3. **f** $\leq$ **g**;
then **f′** $\leq$ **g′**.

Each of the relations "$\leq$ given B" is now easily seen to be a simple ordering, and the relations "$\geq$, $<$, $>$, $\doteq$ given B" are to be defined mutatis mutandis. It is noteworthy though obvious that, if $f(s) = g(s)$ for all $s \, \varepsilon \, B$, then **f** $\doteq$ **g** given B.

It is now possible and instructive to give an atemporal analysis of the following temporally described decision situation: The person must decide between **f** and **g** after he finds out, that is, observes, whether B obtains; what will his decision be if he finds out that B does in fact obtain?

Atemporally, the person can submit himself to the consequences of **f** or else of **g** for all $s \, \varepsilon \, B$, and, independently, he can submit himself to the consequences of **f** or else of **g** for all $s \, \varepsilon \sim B$; which alternative will he decide upon for the s's in B?

Finally, describing the situation not only atemporally but also quite formally, the person must decide among four acts defined thus:

$$\mathbf{h}_{00} \text{ agrees with } \mathbf{f} \text{ on } B \text{ and with } \mathbf{f} \text{ on } \sim B,$$
$$\mathbf{h}_{01} \text{ agrees with } \mathbf{f} \text{ on } B \text{ and with } \mathbf{g} \text{ on } \sim B,$$
$$\mathbf{h}_{10} \text{ agrees with } \mathbf{g} \text{ on } B \text{ and with } \mathbf{f} \text{ on } \sim B,$$
$$\mathbf{h}_{11} \text{ agrees with } \mathbf{g} \text{ on } B \text{ and with } \mathbf{g} \text{ on } \sim B.$$

The question at issue now takes this form. Supposing that none of the four functions is preferred to the particular one $\mathbf{h}_{ij}$, is $i = 0$, or is $i = 1$; that is, does $\mathbf{h}_{ij}$ agree with $\mathbf{f}$ on B or with $\mathbf{g}$ on B?

It is not hard to see that i can be 1, if and only if $\mathbf{f} \leq \mathbf{g}$ given B. Indeed, if $i = 1$, $\mathbf{h}_{0j} \leq \mathbf{h}_{ij}$, which means that $\mathbf{f} \leq \mathbf{g}$ given B. Arguing in the opposite direction, if $\mathbf{f} \leq \mathbf{g}$ given B; then $\mathbf{h}_{00} \leq \mathbf{h}_{10}$, and $\mathbf{h}_{01} \leq \mathbf{h}_{11}$. Suppose now, for definiteness, $\mathbf{h}_{10} \leq \mathbf{h}_{11}$, then none of the four possibilities is preferred to $\mathbf{h}_{11}$; this proves the point in question.

It may fairly be said that the person considers B **virtually impossible,** or that B is **null;** if and only if, for all $\mathbf{f}$ and $\mathbf{g}$, $\mathbf{f} \leq \mathbf{g}$ given B. Indeed, if B is null in this sense, the values acts take on elements of B are irrelevant to all decisions.

Several trivial conclusions about null events are listed as a compound theorem, all components but the last of which have immediate intuitive interpretations.

THEOREM 1

1. The vacuous event, 0, is null.
2. B is null, if and only if, for every $\mathbf{f}$ and $\mathbf{g}$, $\mathbf{f} \doteq \mathbf{g}$ given B.
3. If B is null, and $B \supset C$; then C is null.
4. If $\sim B$ is null; $\mathbf{f} \leq \mathbf{g}$ given B, if and only if $\mathbf{f} \leq \mathbf{g}$.
5. $\mathbf{f} \leq \mathbf{g}$ given S, if and only if $\mathbf{f} \leq \mathbf{g}$.
6. If S is null, $\mathbf{f} \doteq \mathbf{g}$ for every $\mathbf{f}$ and $\mathbf{g}$.

Component 6 of Theorem 1 requires comment, because it corresponds to a pathological situation. In case S is null, it is not really intuitive to say that S (and therefore every event) is virtually impossible. The interpretation is rather that the person simply doesn't care what happens to him. This is imaginable, especially under a suitably restricted interpretation of F, but it is uninteresting and will accordingly be ruled out by a later postulate, P5.

A finite set of events B_i is a **partition** of B; if $B_i \cap B_j = 0$, for $i \neq j$, and $\bigcup_i B_i = B$. With this definition, it is easily proved by arithmetic induction that

THEOREM 2   If B_i is a partition of B, and $\mathbf{f} \leq \mathbf{g}$ given B_i for each i, then $\mathbf{f} \leq \mathbf{g}$ given B. If, in addition, $\mathbf{f} < \mathbf{g}$ given B_j for at least one j, then $\mathbf{f} < \mathbf{g}$ given B.

COROLLARY 1   The union of any finite number of null events is null.

There are still other interesting consequences of Theorem 2, which may be most conveniently mentioned informally. If, in Theorem 2, $B = S$ (or, more generally, if $\sim B$ is null), it is superfluous to say "given

B'' in the conclusions of the theorem. If $\mathbf{f} \doteq \mathbf{g}$ given B_i for each i, then $\mathbf{f} \doteq \mathbf{g}$ given B. So much for the consequences of P2.

Acts that are **constant**, that is, acts whose consequences are independent of the state of the world, are of special interest. In particular, they lead to a natural definition of preference among consequences in terms of preference among acts. Following ordinary mathematical usage, $\mathbf{f} \equiv g$ will mean that $\mathbf{f}$ is identically g, that is, for every s, $f(s) = g$. A formal definition of **preference among consequences** can now conveniently be expressed thus. For any consequences g and g', $g \leq g'$; if and only if, when $\mathbf{f} \equiv g$ and $\mathbf{f}' \equiv g'$, $\mathbf{f} \leq \mathbf{f}'$.

In the same spirit, meaning can be assigned to such expressions as $\mathbf{f} \leq g$, $g \leq \mathbf{f}$ given B, etc., and I will freely use such expressions without defining them explicitly. In particular, $f \leq g$ given B has a natural meaning, but one that is rendered superfluous by the next postulate, P3.

Incidentally, it is now evident how awkward for us it would be to use $f(s)$ for $\mathbf{f}$; because $f(s) \leq g(s)$ is a statement about the consequences $f(s)$ and $g(s)$, whereas $\mathbf{f} \leq \mathbf{g}$ is a statement about acts, and we will have frequent need for both sorts of statements.

Suppose that $\mathbf{f} \equiv g$, and $\mathbf{f}' \equiv g'$, and that $g \leq g'$, is it reasonable to admit that, for some B, $\mathbf{f} > \mathbf{f}'$ given B? That depends largely on the interpretation we choose to make of our technical terms, as an example helps to bring out.+

Before going on a picnic with friends, a person decides to buy a bathing suit or a tennis racket, not having at the moment enough money for both. If we call possession of the tennis racket and possession of the bathing suit consequences, then we must say that the consequences of his decision will be independent of where the picnic is actually held. If the person prefers the bathing suit, this decision would presumably be reversed, if he learned that the picnic were not going to be held near water. Thus the question whether it can happen that $\mathbf{f} > \mathbf{f}'$ given B would be answered in the affirmative. But, under the interpretation of "act" and "consequence" I am trying to formulate, this is not the correct analysis of the situation. The possession of the tennis racket and the possession of the bathing suit are to be regarded as acts, not consequences. (It would be equivalent and more in accordance with ordinary discourse to say that the coming into possession, or the buying, of them are acts.) The consequences relevant to the decision are such as these: a refreshing swim with friends, sitting on a shadeless beach twiddling a brand-new tennis racket while one's friends swim, etc. It seems clear that, if this analysis is carried to its limit, the question at issue must be answered in the negative; and I therefore propose

+ The role of such freedom throughout science is brilliantly discussed by Quine (1951).

to assume the negative answer as a postulate. The postulate is so couched as not only to assert that knowledge of an event cannot establish a new preference among consequences or reverse an old one, but also to assert that, if the event is not null, no preference among consequences can be reduced to indifference by knowledge of an event.

P3　　If $\mathbf{f} \equiv g$, $\mathbf{f}' \equiv g'$, and B is not null; then $\mathbf{f} \leq \mathbf{f}'$ given B, if and only if $g \leq g'$.

Applying Theorem 2, it is obvious that

THEOREM 3　　If B_i is a partition of B; and if (for all i and s) $f_i \leq g_i$, $f(s) = f_i$, and $g(s) = g_i$ when $s \; \varepsilon \; B_i$; then $\mathbf{f} \leq \mathbf{g}$ given B. If, in addition, $f_j < g_j$ for some j for which B_j is not null, then $\mathbf{f} < \mathbf{g}$ given B.

Theorem 3 is logically equivalent to P3 in the presence of P1 and P2, and Theorem 3 can as easily be given an intuitive basis as the postulate P3. Therefore the assumption of P3 as a postulate instead of Theorem 3 is only a matter of taste.

Theorem 3 has been widely accepted by the British-American School of statisticians, special emphasis having been given to it, in connection with his notion of admissibility, by the late Abraham Wald. I believe, as will be more fully explained later, that much of its particular significance for that school stems from the implication that, if several different people agree in their preferences among consequences, then they must also agree in their preferences among certain acts.

This brings the present chapter to a natural conclusion, since the further postulates to be proposed can be more conveniently introduced in connection with the uses to which they are put in later chapters.

CHAPTER 3

Personal Probability

1 Introduction

I personally consider it more probable that a Republican president will be elected in 1996 than that it will snow in Chicago sometime in the month of May, 1994. But even this late spring snow seems to me more probable than that Adolf Hitler is still alive. Many, after careful consideration, are convinced that such statements about probability to a person mean precisely nothing, or at any rate that they mean nothing precisely. At the opposite extreme, others hold the meaning to be so self-evident as to be unanalyzable. An intermediate position is taken in this chapter, where a particular interpretation of probability to a person is given in terms of the theory of consistent decision in the face of uncertainty, the exposition of which was begun in the last chapter. Much as I hope that the notion of probability defined here is consistent with ordinary usage, it should be judged by the contribution it makes to the theory of decision, not by the accuracy with which it analyzes ordinary usage.

Perhaps the first way that suggests itself to find out which of two events a person considers more probable is simply to ask him. It might even be argued, though I think fallaciously, that, since the question concerns what is inside the person's head, there can be no other method, just as we have little, if any, access to a person's dreams except through his verbal report. Attempts to define the relative probability of a pair of events in terms of the answers people give to direct interrogation has justifiably met with antipathy from most statistical theorists. In the first place, many doubt that the concept "more probable to me than" is an intuitive one, open to no ambiguity and yet admitting no further analysis. Even if the concept were so completely intuitive, which might justify direct interrogation as a subject worthy of some psychological study, what could such interrogation have to do with the behavior of a person in the face of uncertainty, except of course for his verbal behavior under interrogation? If the state of mind in question is not capable of manifesting itself in some sort of extraverbal behavior,

27

it is extraneous to our main interest. If, on the other hand, it does manifest itself through more material behavior, that should, at least in principle, imply the possibility of testing whether a person holds one event to be more probable than another, by some behavior expressing, and giving meaning to, his judgment. It would, in short, be preferable, at least in principle, to interrogate the person, not literally through his verbal answer to verbal questions, but rather in a figurative sense somewhat reminiscent of that in which a scientific experiment is sometimes spoken of as an interrogation of nature. Several schemes of behavioral, as opposed to direct, interrogation have been proposed. The one introduced below was suggested to me by a passage of de Finetti's (on pp. 5–6 of [D2]), though the passage itself does not emphasize behavioral interrogation.

To illustrate the scheme, our idealized person has just taken two eggs from his icebox and holds them unbroken in his hand. We wonder whether he thinks it more probable that the brown one is good than that the white one is. Our curiosity being real, we are prepared to pay, if necessary, to have it satisfied. We therefore address him thus: "We see that you are about to open those eggs. If you will be so cooperative as to guess that one or the other egg is good, we will pay you a dollar, should your guess prove correct. If incorrect, you and we are quits, except that we will in any event exchange your two eggs for two of guaranteed goodness." If under these circumstances the person stakes his chance for the dollar on the brown egg, it seems to me to correspond well with ordinary usage to say that it is more probable to him that the brown one is good than that the white one is. Though, of course, I hope for your agreement on this analysis of ordinary usage, I repeat that it is not really fundamental to the subsequent argument, as indeed no such lexicographical point could be; for the utility of a construct or definition depends only secondarily on the aptness of the expression in terms of which it is couched.

There is a mode of interrogation intermediate between what I have called the behavioral and the direct. One can, namely, ask the person, not how he feels, but what he would do in such and such a situation. In so far as the theory of decision under development is regarded as an empirical one, the intermediate mode is a compromise between economy and rigor. But, in the theory's more important normative interpretation as a set of criteria of consistency for us to apply to our own decisions, the intermediate mode seems to me to be just the right one.

Though it entails digression from the main theme, some readers may be interested in a few words about actual experimentation on strictly

empirical behavioral interrogation. Some key references bearing on the subject are [M4], [R3], and [W8].

In the first place, a little reflection shows that an experiment in which human subjects are required to decide among actual acts may be very expensive in time, money, and effort, especially if the consequences envisaged are expensive to provide, a point discussed in detail in [W8]. Questions of morality, and even of legality, toward the subject may further complicate the investigation. For example, Mosteller and Nogee, as described in Section 3B of [M4], made certain that every subject in one experiment of theirs would be financially benefited, though they kept this security secret from the subjects.

There is also a difficulty in principle. Suppose that I wish to discover a person's preferences among several acts—three acts **f**, **g**, and **h** are sufficient to bring out the difficulty. If I in good faith offer him the opportunity to decide among all three, and he decides on **f**; then there is no further possibility of discovering what his preference was between **g** and **h**. Suppose, for example, that a hot man actually prefers a swim, a shower, and a glass of beer, in that order. Once he decides on, and thereby becomes entitled to, the swim, he can no longer appropriately be asked to decide between shower and beer. A naive attempt to do so would result in his deciding between a swim and shower on the one hand, and a swim and beer on the other—an altogether different situation from the one intended.

The difficulty can sometimes be met by special devices. For example, the investigator might wait for a different but "similar" occasion. But W. Allen Wallis has mentioned to me an interesting and very general device, which will now be described, with his permission.†

Suppose that the hot man is instructed to rank the three acts in order, subject to the consideration that two of them will be drawn at random (e.g., by card drawing or dice rolling), and that he is then to have whichever of these two acts he has assigned the lower rank. He is thus called on to select one of six acts, that is, one of the six possible rankings. If he does, for example, select the ranking {swim, shower, beer}, it follows easily from the theory of decision thus far developed that for him swim ≥ shower ≥ beer, barring the farfetched possibility that he regards one or more of the three drawings as virtually impossible and provided that his preference among the three acts swim, shower, beer given any of the three drawings is the same as his original preference. The investigator could in practice design the drawing in such a

† I have since seen this same device used by M. Allais.

way as to be well satisfied that the required "irrelevance" obtained, except for very "superstitious" people. This ends the present digression on actual behavioral interrogation.

The purpose of this chapter is to explore the concept of personal probability † that was indicated in the example about the two eggs. The concept will be put on a formal basis in § 2 by introducing two new postulates, P4 and P5, to be used in conjunction with P1–3. This will lead to a formal analysis of the notion that one event is no more probable than another. Several deductions about this notion reminiscent of mathematical properties ordinarily attributed to probability will be made; but only in § 3, after adjunction of still another postulate, P6, can the notion be connected quantitatively with what mathematicians ordinarily call mathematical probability. Section 4 is devoted to some mathematically technical criticisms of the notion of personal probability, which can safely be skipped or skimmed by those not interested in such matters. Section 5 discusses conditional personal probability; 6, the approach to certainty through a long sequence of conditionally independent relevant observations; and 7, an extension of the concept of a sequence of independent events, particularly interesting from the viewpoint of personal probability.

2 Qualitative personal probability

When I spoke in the introductory section of offering the person a dollar if his guess about the egg proved correct, it was tacitly assumed that his guess would not be affected by the amount of the prize offered. That seems to me correct in principle. It would, for example, seem unreasonable for the person with the two eggs to reverse his decision if the prize were reduced from a dollar to a penny. He might reverse himself in going from a penny to a dollar, because he might not have found it worth his trouble to give careful consideration for too small a prize. I think the anomaly can best be met by deliberately pretending that consideration costs the person nothing, though that is far from the truth in actual complicated situations. It might, on the other hand, be stimulating, and it is certainly more realistic, to think of consideration or calculation as itself an act on which the person must decide. Though I have not explored the latter possibility carefully, I suspect that any attempt to do so formally leads to fruitless and endless regression.

† The term "personal probability" was suggested to me orally by Thornton C. Fry. Some other terms suggested for the same concept are "subjective probability," "psychological probability," and "degree of conviction."

To offer a *prize* in case A obtains means to make available to the person an act $\mathbf{f}_A$ such that

(1)
$$f_A(s) = f \qquad \text{for } s \,\varepsilon\, A,$$
$$f_A(s) = f' \qquad \text{for } s \,\varepsilon\, {\sim}A,$$

where $f' < f$. The assumption that on which of two events the person will choose to stake a given prize does not depend on the prize itself is expressed by the following postulate, which looks formidable only because it contains four definitions like (1). The reader may find it helpful to graph an instance of the postulate in the spirit of Figure 2.7.1.

P4   If f, f', g, g'; A, B; $\mathbf{f}_A, \mathbf{f}_B, \mathbf{g}_A, \mathbf{g}_B$ are such that:

1. $f' < f, \qquad g' < g;$

2a. $f_A(s) = f, \qquad g_A(s) = g \qquad \text{for } s \,\varepsilon\, A,$

            $f_A(s) = f', \qquad g_A(s) = g' \qquad \text{for } s \,\varepsilon\, {\sim}A;$

2b. $f_B(s) = f, \qquad g_B(s) = g \qquad \text{for } s \,\varepsilon\, B,$

            $f_B(s) = f', \qquad g_B(s) = g' \qquad \text{for } s \,\varepsilon\, {\sim}B;$

3. $\mathbf{f}_A \leq \mathbf{f}_B;$

then $\mathbf{g}_A \leq \mathbf{g}_B$.

In the light of P4, it will be said that A is **not more probable** than B, abbreviated $A \leq B$; if and only if when $f' < f$ and $\mathbf{f}_A, \mathbf{f}_B$ are such that

$$f_A(s) = f \quad \text{for } s \,\varepsilon\, A, \qquad f_A(s) = f' \quad \text{for } s \,\varepsilon\, {\sim}A,$$
$$f_B(s) = f \quad \text{for } s \,\varepsilon\, B, \qquad f_B(s) = f' \quad \text{for } s \,\varepsilon\, {\sim}B;$$

then $\mathbf{f}_A \leq \mathbf{f}_B$.

The assumption that there is at least one worth-while prize is innocuous; for, though a context failing to satisfy it might arise, such a context would be too trivial to merit study. I therefore propose the following postulate.

P5   There is at least one pair of consequences f, f' such that $f' < f$.

All the implications to be deduced from P1–5 for some time to come are themselves implications of the three easily established conclusions, which are introduced by the following definition and theorem.

A relation $\leq\cdot$ between events is a **qualitative probability**; if and only if, for all events B, C, D,

1. $\leq\cdot$ is a simple ordering,
2. $B \leq\cdot C$, if and only if $B \cup D \leq\cdot C \cup D$, provided $B \cap D = C \cap D = 0$,
3. $0 \leq\cdot B$, $0 <\cdot S$.

It may be helpful to remark that the second part of the above definition says, in effect, that it will not affect the person's guess to offer him a consolation prize in case neither B nor C obtains, but D happens to.

THEOREM 1   The relation $\leq$ as applied to events is a qualitative probability.

You will have no difficulty in proving that Theorem 1 follows from P1–5. Theorem 1 has many consequences of the sort one would expect if $\leq$ meant "not more probable than" in any sense having the mathematical properties ordinarily attributed to numerical probability. This is illustrated by the following list of exercises, which should not only be proved formally, but also interpreted intuitively. One easy exercise not included in the list below, because it is not strictly a consequence of Theorem 1 alone, is to show that $B \doteq 0$, if and only if B is a null event.

Exercises

1. If $B \subset C$, then $0 \leq B \leq C \leq S$.
2a. If $B \cap D = C \cap D = 0$; then $B < C$, if and only if $B \cup D < C \cup D$.
2b. If $0 < C$, and $B \cap C = 0$; then $B < B \cup C$.
3. If $B \leq C$, then $\sim C \leq \sim B$; and conversely. Hint: Draw a Venn diagram of the fourfold partition $B \cap C$, $\sim B \cap C$, $B \cap \sim C$, $\sim B \cap \sim C$.
4a. If $B \leq C$, and $C \cap D = 0$; then $B \cup D \leq C \cup D$.
4b. If $B \leq 0$; then $B \cup C \doteq C$, and $B \doteq 0$.
4c. If $S \leq B$; then $B \cap C \doteq C$, and $B \doteq S$.
4d. If $B \cup D \leq C \cup D$, and $B \cap D = 0$; then $B \leq C$.
5a. If $B_1 \leq C_1$, $B_2 \leq C_2$, and $C_1 \cap C_2 = 0$; then $B_1 \cup B_2 \leq C_1 \cup C_2$. Hint: Exhibit B_2 and C_1 in the form $B_2 = B_2' \cup Q$, $C_1 = C_1' \cup Q$ with B_2', C_1', Q disjoint. Justify the following calculation, step by step.

$$B_1 \cup B_2' \leq C_1 \cup B_2' = C_1' \cup B_2 \leq C_1' \cup C_2,$$

whence $B_1 \cup B_2 \leq C_1 \cup C_2$.

5b. If $B_1 \cup B_2 \leq C_1 \cup C_2$ and $B_1 \cap B_2 = 0$; then $B_1 \leq C_1$ or $B_2 \leq C_2$.

6. If $B \leq \sim B$ and $C \geq \sim C$, then $B \leq C$; equality holding in the conclusion, if and only if it holds in both parts of the hypothesis.

3 Quantitative personal probability

As I have said, the exercises terminating the preceding section suggest a close mathematical parallelism between personal probability and the mathematical properties ordinarily attributed to probability, though the postulates assumed thus far do not (as could easily be demonstrated) make it possible to deduce from this parallelism the unambiguous assignment of a numerical probability to each event. But, if, for example (following de Finetti [D2]), a new postulate asserting that S can be partitioned into an arbitrarily large number of equivalent subsets were assumed, it is pretty clear (and de Finetti explicitly shows in [D2]) that numerical probabilities could be so assigned. It might fairly be objected that such a postulate would be flagrantly ad hoc. On the other hand, such a postulate could be made relatively acceptable by observing that it will obtain if, for example, in all the world there is a coin that the person is firmly convinced is fair, that is, a coin such that any finite sequence of heads and tails is for him no more probable than any other sequence of the same length; though such a coin is, to be sure, a considerable idealization.

After some general and abstract discussion of the mathematical connection between qualitative and quantitative probability, a postulate, P6, will be proposed, which, though logically actually stronger than the assumption that there are partitions of S into equivalent events, seems to me even easier to accept. Once P6 is accepted, there will scarcely again be any need to refer directly to qualitative probability.

To begin with, let me say precisely what is meant, in the present context, by a probability measure, this being the standard term for what I would here otherwise prefer to call a quantitative probability, and what it means for a probability measure to be in agreement with a qualitative probability.

A **probability measure** on a set S is a function $P(B)$ attaching to each $B \subset S$ a real number such that:

1. $P(B) \geq 0$ for every B.
2. If $B \cap C = 0$, $P(B \cup C) = P(B) + P(C)$.
3. $P(S) = 1$.

This definition, or something very like it, is at the root of all ordinary mathematical work in probability.

If S carries a probability measure P and a qualitative probability
$\leq\cdot$ such that, for every B, C, $P(B) \leq P(C)$, if and only if $B \leq\cdot C$;
then P **(strictly) agrees** with $\leq\cdot$. If $B \leq\cdot C$ implies $P(B) \leq P(C)$,
then P **almost agrees with** $\leq\cdot$. This terminology is obviously consistent in that, if P agrees, that is, strictly agrees, with $\leq\cdot$, P also almost agrees with $\leq\cdot$. It is also easily seen that, if P agrees with $\leq\cdot$,
then knowledge of P implies knowledge of $\leq\cdot$. But, if P only almost
agrees with $\leq\cdot$, it may happen, as examples in § 4 show, that $P(B) =$
$P(C)$, though $B <\cdot C$, so that knowledge of P may imply only imperfect
knowledge of $\leq\cdot$.

The rest of this section is mainly a study of qualitative probabilities
generally, with a view to discovering interesting conditions under which
there is a probability measure that agrees, either strictly or almost,
with a given qualitative probability. These conditions suggest a new
postulate governing the special qualitative probability $\leq$. The work
is necessarily rather tedious and burdened with detail. It will, therefore, be wise for most readers to skim over the material, omitting the
proofs but noticing the more obvious logical connections among the
theorems and definitions. Some may then find themselves sufficiently
interested in the details to return and read or supply the proofs, as the
case may require. Others may safely go forward. Here, as elsewhere,
technical terms of interest for the moment only are introduced with
italics rather than boldface.

An *n-fold almost uniform partition* of B is an n-fold partition of B
such that the union of no r elements of the partition is more probable
than that of any $r + 1$ elements.

THEOREM 1   If there exist n-fold almost uniform partitions of B for
arbitrarily large values of n, then there exist m-fold almost uniform partitions for every positive integer m.

PROOF.   Let B_i, $i = 1, \cdots, n$, be an n-fold almost uniform partition
(of B) with $n \geq m^2$. Using the euclidean algorithm, let n be written
$n = am + b$, where a and b are integers such that $m \leq a$ and $0 \leq b <$
m. Now let C_j, $j = 1, \cdots, m$, be any m-fold partition such that each
C_j is the union of a or $a + 1$ of the B_i's. The union of any r of the C_j's,
$r < m$, is the union of from ar to $(a + 1)r$ of the B_i's and the union of
$r + 1$ of the C_j's is that of from $a(r + 1)$ to $(a + 1)(r + 1)$ of the B_i's.
Since $r < m \leq a$, $(a + 1)r = ar + r < ar + a = a(r + 1)$. ◆

THEOREM 2   If there exist n-fold almost uniform partitions of S for
arbitrarily large values of n, then there is one and only one probability
measure P that almost agrees with $\leq\cdot$. Furthermore, for any ρ, $0 \leq \rho$

≤ 1, any $B \subset S$, and the unique P just defined, there exists $C \subset B$ such that $P(C) = \rho P(B)$.†

PROOF. The proof is broken into a sequence of easy steps, left, for the most part, to the reader. These steps are grouped in blocks, only the last step in each being needed in the proof of later steps.

1. There exist n-fold almost uniform partitions of S for every positive n.

2a. If $p_1, \cdots, p_n$ are real numbers such that $0 \leq p_1 \leq p_2 \leq \cdots \leq p_n$, and $\Sigma p_i = 1$; then

(1) $$\sum_{1}^{r} p_i \leq r/n, \qquad r = 1, \cdots, n.$$

2b. If further

$$\sum_{1}^{r+1} p_i \geq \sum_{n-r+1}^{n} p_i \qquad \text{for } r = 1, \cdots, n-1;$$

then

(2) $$\sum_{1}^{r} p_i \geq (r-1)/n, \quad \text{and} \quad \sum_{n-r+1}^{n} p_i \leq (r+1)/n.$$

2c. The sum of any r of the p_i's lies between $(r-1)/n$ and $(r+1)/n$.

2d. If P almost agrees with $\leq\cdot$, and $C(r, n)$ denotes here and later in this proof any union of r elements of any n-fold almost uniform partition (not necessarily the same from one context to another), then

(3) $$(r-1)/n \leq P(C(r, n)) \leq (r+1)/n.$$

3. Let $k(B, n)$ denote the largest integer r (possibly zero) such that some $C(r, n)$ is not more probable than B. The function $k(B, n)$ is well-defined, and $0 \leq k(B, n) \leq n$.

4a. For any P that almost agrees with $\leq\cdot$,

(4) $$(k(B, n) - 1)/n \leq P(B) \leq (k(B, n) + 2)/n.$$

4b. At most one P can almost agree with $\leq\cdot$

5a. If B_i and C_i are n-fold partitions (not necessarily almost uniform) so indexed that $B_1 \leq\cdot B_2 \leq\cdots \leq\cdot B_n$, and $C_1 \geq\cdot C_2 \geq\cdots \geq\cdot C_n$; then

(5) $$\bigcup_{n-r}^{n} B_i \geq\cdot \bigcup_{n-r}^{n} C_i, \qquad r = 0, \cdots, n-1.$$

† Technical note: The mathematical essence of the terminal conclusion of this theorem, and other conclusions related to it, are given by Sobczyk and Hammer [S15]. It might be conjectured, in analogy with countably additive measures, that this conclusion means only that P is non-atomic, but that conjecture is false [N5].+

+ A key reference for further information on the structure of finitely additive measures is (Dubins 1969). Sustained use of finitely additive probability is illustrated in (Dubins and Savage 1965).

5b. If in addition the two partitions are almost uniform, then

$$(6) \qquad \bigcup_1^r C_i \leq \cdot \bigcup_1^{r+2} B_i, \qquad r = 1, \cdots, n-2.$$

(Proof. $\bigcup_1^{r+2} B_i \geq \cdot \bigcup_{n-r}^n B_i \geq \cdot \bigcup_{n-r}^n C_i \geq \cdot \bigcup_1^r C_i.$)

5c. The union of any r elements of one almost uniform n-fold partition is not more probable than the union of any $r+2$ elements of another.

5d. If $B \cap C = 0$, then

$$(7) \quad k(B, n) + k(C, n) - 2 \leq k(B \cup C, n) \leq k(B, n) + k(C, n) + 1.$$

6a. If a $C(r, m)$ is not more probable than a $C(s, n)$, then

$$(8) \qquad \left(\frac{r-2}{m} \right) \leq \left(\frac{s+2}{n} \right) + \frac{1}{mn}.$$

(Consider an mn-fold almost uniform partition, and use the easily established fact that the union of any $t+2$ elements of an almost uniform partition is actually more probable than that of any t elements.)

6b. $\qquad \left| \dfrac{k(B, m)}{m} - \dfrac{k(B, n)}{n} \right| \leq \dfrac{3}{m} + \dfrac{3}{n} + \dfrac{1}{mn}.$

6c. It is meaningful to define $P(B)$ by

$$(9) \qquad P(B) =_{\text{Df}} \lim_{n \to \infty} \frac{k(B, n)}{n},$$

that is, the limit exists.

7. $P(B)$, as just defined, is a probability measure, and the only one that almost agrees with $\leq \cdot$.

8a. There exist two infinite sequences of sets C_n and D_n contained in B such that:

1. $C_n \cap D_n = 0$,
2. $C_n \subset C_{n+1}$, and $D_n \subset D_{n+1}$,
3. $P(C_n) \geq \rho P(B) - n^{-1}$,
4. $P(D_n) \geq (1 - \rho)P(B) - n^{-1}$.

8b. $P(\bigcup_n C_n) \geq \rho P(B)$, $P(\bigcup_n D_n) \geq (1 - \rho)P(B)$, and $(\bigcup_n C_n) \cap (\bigcup_n D_n) = 0$.

8c. $P(\bigcup_n C_n) = \rho P(B)$. ◆

A few technical terms of localized interest only are now introduced. If and only if, for every $B > \cdot 0$, there is a partition of S, no element of which is as probable as B; $\leq\cdot$ is *fine*.[+] B and C are *almost equivalent*, written $B \approx\cdot C$; if and only if for all non-null G and H such that $B \cap G = C \cap H = 0$, $B \cup G \geq\cdot C$ and $C \cup H \geq\cdot B$. It is obvious that equivalent events are also almost equivalent. Finally, if and only if every pair of almost equivalent events are equivalent, $\leq\cdot$ is *tight*.

THEOREM 3

HYP. $\leq\cdot$ is fine.

CONCL.   1. If $B >\cdot 0$, and $C >\cdot 0$; there exists $D \subset C$ such that $0 <\cdot D <\cdot B$.
2. If $B \approx\cdot G$, $C \approx\cdot H$, and $B \cap C = G \cap H = 0$; then $B \cup C \approx\cdot G \cup H$.
3. If $B \approx\cdot C$, $G \approx\cdot H$, $B \cup C \approx\cdot G \cup H$, and $B \cap C = G \cap H = 0$; then $B \approx\cdot G$.
4. Any partition of S into almost equivalent events is an almost uniform partition.
5. Any event can be partitioned into two almost equivalent events.
6. Any event can be partitioned into 2^n almost equivalent events, for any non-negative integer n.
7. There exists one and only one P that almost agrees with $\leq\cdot$. For any B, ρ $(0 \leq \rho \leq 1)$, and the unique P just defined, there exists $C \subset B$ such that $P(C) = \rho P(B)$. If $B >\cdot 0$, $P(B) > 0$. Finally, $B \approx\cdot C$, if and only if $P(B) = P(C)$.

PROOF. The parts of the conclusion are so arranged that each is easy to prove in the light of its predecessors, but proofs for Parts 3 and 5 are given below. It may be remarked that all parts are trivial consequences of the last one and have therefore relatively little importance in themselves.
Part 3. Suppose, for example, $B \cup E <\cdot G$, $B \cap E = 0$, and $E >\cdot 0$; and consider two cases:
(a) If $B \cup C <\cdot S$, it may be assumed without loss of generality that $C \cap E = 0$, whence $(B \cup C) \cup E \geq\cdot G \cup H$. Therefore, $C >\cdot H$. Let E be partitioned into two non-null events E_1 and E_2; then (since it is absurd to suppose that the part of G outside of C is null, which would imply $C \geq\cdot G >\cdot B \cup E$) there is in G an E' such that $C \cap E' = 0 <\cdot E' \leq\cdot E_2$. Now $C \cup E' >\cdot H \cup E' \geq\cdot G >\cdot (B \cup E_1) \cup E_2$, whence $C >\cdot B \cup E_1$, which is absurd.
(b) If $B \cup C \doteq\cdot S$, it can (setting aside the easy special case $C \cap G \doteq\cdot 0$) be shown successively that: $H \cup G \doteq\cdot S$; $C <\cdot B \cup E <\cdot G$, where $E >\cdot 0$ and $E \subset C \cap G$; $(B \cap H) \cup E <\cdot (G \cap C)$; $(C \cap H) <\cdot (G \cap B)$; and $H \cup E <\cdot G$, which establishes a contradiction.

[+] In the first edition, this definition was a trifle too weak, as pointed out by Malcolm Pike.

Part 5. There exists a sequence of threefold partitions of B, say C_n, D_n, and G_n, such that:

1. $C_n \cup G_n \geq \cdot D_n$, and $D_n \cup G_n \geq \cdot C_n$,
2. $C_{n+1} \supset C_n$, $D_{n+1} \supset D_n$, and $G_{n+1} \subset G_n$,
3. $\sim G_{n+1} \cap G_n \geq \cdot G_{n+1}$; whence $G_n \cdot$ contains two disjoint events each at least as probable as G_{n+1}.

For any $H > \cdot 0$, $G_n \leq \cdot H$ for sufficiently large n, as may be seen by considering some m-fold partition no element of which is more probable than H, and letting n be such that $2^{n-1} > m$. If G_n were more probable than H and therefore more probable than each element of the partition, it would follow that the union of all elements of the partition, namely S, is less probable than G_1, which would be absurd.

The two events $B_1 = \bigcup_n C_n$, $B_2 = (\bigcup_n D_n) \cup (\bigcap_n G_n)$ partition B in the required fashion. ◆

COROLLARY 1   If $\leq \cdot$ is both fine and tight; the only probability measure that almost agrees with $\leq \cdot$ strictly agrees with it, and there exist partitions of S into arbitrarily many equivalent events.

THEOREM 4 $\leq \cdot$ is both fine and tight, if and only if, for every $B < \cdot C$, there exists a partition of S the union of each element of which with B is less probable than C.

The proof of this theorem is easy.

In the light of Theorems 3 and 4, I tentatively propose the following postulate, P6′, governing the relation $\leq$ among events, and thereby the relation $\leq$ among acts.

P6′   If $B < C$, there exists a partition of S the union of each element of which with B is less probable than C.

It seems to me rather easier to justify the assumption of P6′, which says in effect that $\leq$ is both fine and tight, than to justify the assumption, which was made by de Finetti [D2] and by Koopman [K9], [K10], [K11] in closely related contexts, that there exist partitions of S into arbitrarily many equivalent events, though logically P6′ implies that assumption and somewhat more. Suppose, for example, that you yourself consider $B < C$, that is, that you would definitely rather stake a gain in your fortune on C than on B. Consider the partition of your own world into 2^n events each of which corresponds to a particular sequence of n heads and tails, thrown by yourself, with a coin of your own choosing. It seems to me that you could easily choose such a coin and choose n sufficiently large so that you would continue to prefer to stake your gain on C, rather than on the union of B and any particular sequence of n heads and tails. For you to be able to do so, you need by no means consider every sequence of heads and tails equally probable.

It would, however, be disingenuous not to mention that some who have worked on a closely related concept of probability, notably Keynes [K4] and Koopman [K9], [K10], [K11], would object to P6′ precisely because it implies that the agreement between numerical probability and qualitative probability is strict. Koopman, for example, holds that, if $A \supset B$ and $A \neq B$, then A is necessarily more probable than B, though the numerical probability of A may well be the same as that of B. Thus, if a marksman shoots at a wall, it is logically contradictory that his bullet should fall nowhere at all, but it is logically consistent that a prescribed mathematically ideal point on the bullet should strike a prescribed mathematically ideal line on the wall. Since the event of the prescribed point hitting a prescribed line is logically possible, Koopman would insist that the event is more probable than the vacuous event, namely that the bullet goes nowhere, though the numerical probability of both events is zero. I do not take direct issue with Koopman, because he is presumably talking about a somewhat different concept of probability from the particular relation $\leq$; but I do not think it appropriate to suppose that the person would distinctly rather stake a gain on the line than on the null set. The issue is not really either an empirical or a normative one, because the point and line in question are mathematical idealizations. If the point and line are replaced by a dot and a band, respectively, then, of course, no matter how small the dot and band may be, the probability of the one hitting the other is greater than that of the vacuous event. But it seems to me entirely a matter of taste, conditioned by mathematical experience, to decide what idealization to make if the dot and band are replaced by their idealized limits. So much for hair splitting.

As far as the theory of probability per se is concerned, postulate P6′ is all that need be assumed, but in Chapter 5 a slightly stronger assumption will be needed that bears on acts generally, not only on those very special acts by which probability is defined. Therefore, I am about to propose a postulate, P6, that obviously implies P6′ and will therefore supersede it. This stronger postulate seems to me acceptable for the same reason that P6′ itself does.

P6   If $\mathbf{g} < \mathbf{h}$, and f is any consequence; then there exists a partition of S such that, if $\mathbf{g}$ or $\mathbf{h}$ is so modified on any one element of the partition as to take the value f at every s there, other values being undisturbed; then the modified **g** remains less than **h**, or **g** remains less than the modified **h**, as the case may require.

4 Some mathematical details

Are there qualitative probabilities that are both fine and tight, that are fine but not tight, that are tight but not fine, that are neither fine nor tight but do have one and only one almost agreeing probability measure? Examples answering all these questions in the affirmative will be exhibited in this section.

To indicate a different topic that will also be treated here, those of you who have had more than elementary experience with mathematical treatments of probability know that it is not usual to suppose, as has been done here, that *all* sets have a numerical probability, but rather that a sufficiently rich class of sets do so, the remainder being considered unmeasurable. Again, it is usual to suppose that, if each of an infinite sequence of disjoint sets is measurable, the probability of their union is the sum of their probabilities, that is, probability measures are generally assumed to be countably additive. But the theory being developed here does assume that probability is defined for all events, that is, for all sets of states, and it does not imply countable additivity, but only finite additivity. The present section not only answers the questions raised in the preceding paragraph, but also discusses the relation of the notions of limited domain of definition and of countable additivity to the theory of probability developed here. The general conclusions of this discussion are: First, there is no technical obstacle to working with a limited domain of definition, and, except for expository complications, it might have been mildly preferable to have done so throughout. Second, it is a little better not to assume countable additivity as a postulate, but rather as a special hypothesis in certain contexts. A different and much more extensive treatment of these questions has been given by de Finetti [D4].

Finally, before entering upon the main technical work of this section, one easy question about the relation between qualitative and quantitative probability will be answered and several as yet unanswered ones will be raised.

Are there qualitative probabilities without any strictly agreeing measure? Yes, because any qualitative probability that is fine but not tight is easily shown to provide an example. It is, however, an open question, stressed by de Finetti [D5], whether a qualitative probability on a finite S always has a strictly agreeing measure. It would also be technically interesting to know about the existence of almost agreeing measures in the same context.[+]

[+] Even this has since been answered in the negative by **Kraft, Pratt, and Seidenberg** (1959). See also (Fishburn 1970, pp. 210-211).

The matters to be treated in the rest of this section are rather technical mathematically, and, though I would not delete them altogether, it does not seem justifiable to lay the necessary groundwork for presenting them in an elementary fashion. Some may, therefore, find it necessary to skip the rest of this section altogether, or to skim it rather lightly.

It is well known that there does not exist a countably additive probability measure defined for every subset of the unit interval, agreeing with Lebesgue measure on those sets where Lebesgue measure is defined, and assigning the same measure to each pair of congruent sets[+] (Problem (b), p. 276 of [H2]). On the other hand, there do exist finitely additive probability measures agreeing with Lebesgue measure on those sets for which Lebesgue measure is defined, and assigning the same measure to each of any pairs of congruent sets; cf. p. 32 of [B4]. The existence of such measures shows, among other things, that a finitely additive measure need not be countably additive. Again, calling such a finitely additive extension of Lebesgue measure P and defining $B \leq \cdot C$ to mean $P(B) \leq P(C)$, we see an example of a qualitative probability that is both fine and tight.

An example of a qualitative probability that is tight but not fine may be constructed by taking for S two unit intervals, S_1 and S_2, in each of which finitely additive extensions of Lebesgue measure, P_1 and P_2, are defined. The generic set B in this example is therefore partitioned into $B_1 = B \cap S_1$ and $B_2 = B \cap S_2$, respectively. For this example, let $B \leq \cdot C$; if, and only if, $P_1(B_1) < P_1(C_1)$, or else $P_1(B_1) = P_1(C_1)$, and $P_2(B_2) \leq P_2(C_2)$. This $\leq \cdot$ is not fine, because, for example, S cannot be partitioned into events none of which is more probable than S_2. On the other hand, it is easily seen to be tight.

Next, take S to be the union of S_1 and S_2 with the measures of P_1 and P_2 as defined in the preceding example, but modify the definition of $\leq \cdot$, saying $B \leq \cdot C$; if and only if $P_1(B_1) + P_2(B_2) < P_1(C_1) + P_2(C_2)$, or else $P_1(B_1) + P_2(B_2) = P_1(C_1) + P_2(C_2)$, and $P_1(B_1) \leq P_1(C_1)$. This is an example of a qualitative probability that is fine but not tight.

Combining the ideas of the two preceding examples, it is easy to exhibit a qualitative probability that is neither fine nor tight but is such that S can be divided into arbitrarily many equally probable events. Thus all the questions raised in the opening paragraph of this section are answered in the affirmative.

+ S. Ulam (1930) proves that any nonatomic, countably additive probability measure defined on all the subsets of the unit interval is inconsistent with the continuum hypothesis.

To get a feeling for the question whether literally all sets should be regarded as measurable, suppose that S is a cube of unit volume and that the probability measure P that strictly agrees with $\leq$ is such that the probability of a parallelepiped is equal to its volume. It follows that the probability of any set having Jordan content is its Jordan content, but, if a set has not Jordan content, a continuum of possibilities is still open. Though other possibilities are conceivable, it is not unnatural to consider an idealized person for whom the numerical probability attached to each Borel set, or even each Lebesgue measurable set, is its Lebesgue measure. To go further and take seriously comparisons between sets that are not Lebesgue measurable, or even between those that are not Borel measurable, seems to me to be without any implication bearing on reality. I suppose it might be argued, on the contrary, that there is no feature of reality that can properly be interpreted by postulating that the person is able to compare only sets from a sufficiently narrow field, so that it is simpler and more elegant to admit all sets. The question seems to be one of taste, but the following remark illustrates what I consider an awkwardness in supposing probability to be attached to all sets. It would seem, at first glance, that the person should be able, if he is so constituted, to regard all pairs of geometrically congruent sets for which he makes any comparison at all as equivalent, but the famous Banach-Tarski paradox [B5] shows that this cannot be done if all sets are regarded as measurable. I think it a little more graceful to abstain from comparison between the more bizarre sets than to give up, or even much modify, my everyday notions about the symmetry of such probability problems associated with geometry.

If one is unwilling to insist on comparison between every pair of sets, or events; then, in the same spirit, it is inappropriate to insist on comparison between every pair of acts. All that has been, or is to be, formally deduced in this book concerning preferences among sets, could be modified, mutatis mutandis, so that the class of events would not be the class of all subsets of S, but rather a Borel field, that is, a σ-algebra, on S; the set of all consequences would be a measurable space, that is, a set with a particular σ-algebra singled out; and an act would be a measurable function from the measurable space of events to the measurable space of consequences. Indeed, the whole thing could be done for abstract σ-algebras without reference to sets at all, and this might have some actual advantage, since it would make possible the identification of events with propositions in almost any formal language, even one unable to formulate at all the complete descriptions I call states.

It may seem peculiar to insist on σ-algebras as opposed to finitely additive algebras even in a context where finitely additive measures are the central object, but countable unions do seem to be essential to some of the theorems of § 3—for example, the terminal conclusions of Theorem 3.2 and Part 5 of Theorem 3.3. So much of the modern mathematical theory of probability depends on the assumption that the probability measures at hand are countably additive that one is strongly tempted to assume countable additivity, or its logical equivalent, as a postulate to be adjoined to P1–6.[+] But I am inclined to agree with de Finetti [D2], [D4] and Koopman [K9], [K10], [K11] that, however convenient countable additivity may be, it, like any other assumption, ought not be listed among the postulates for a concept of personal probability unless we actually feel that its violation deserves to be called inconsistent or unreasonable. I know of no argument leading to the requirement of countable additivity, and many of us have a strong intuitive tendency to regard as natural probability problems about the necessarily only finitely additive uniform probability densities on the integers, on the line, and on the plane. It therefore seems better not to assume countable additivity outright as a postulate, but to recognize it as a special hypothesis yielding, where applicable, a large class of useful theorems.

5 Conditional probability, qualitative and quantitative

Conditional preferences among acts in the light of a given event were introduced in § 2.7. Since the relation ≤ among events has been defined in terms of the corresponding relation among acts, we may well expect to attach meaning to statements of the form $B \leq C$ given D, provided that D is not null. The natural way to do so is to take a pair of acts **f** and **g** that test whether $B \leq C$ (as prescribed by the definition of ≤ between acts in § 2) and say that $B \leq C$ given D, if and only if $\mathbf{f} \leq \mathbf{g}$ given D. Since there is more than one pair of acts **f**, **g** by which the proposition $B \leq C$ can be tested, it is at first sight conceivable that not all such pairs would be in the same order given D, which would frustrate the proposed definition of ≤ given D. However, it is easily seen that for any **f**, **g** testing $B \leq C$, $\mathbf{f} \leq \mathbf{g}$ given D (D not null) is equivalent to $B \cap D \leq C \cap D$. Thus it is seen not only that the proposed definition is unambiguous, but also that it is expressible in terms of probability comparisons among sets, without direct reference to acts at all, and, still further, that the postulates P1–6 apply to the conditional preference relation ≤ given D among acts. This preamble sufficiently motivates the following definition and easy theorem about qualitative probability relations generally.

─────────

[+] Carried out by Villegas (1964).

If $\leq\cdot$ is a qualitative probability, and $0 <\cdot D$; then $B \leq\cdot C$ **given** D, if and only if $B \cap D \leq\cdot C \cap D$.

THEOREM 1   If $\leq\cdot$ is a qualitative probability, then so is $\leq\cdot$ given D. If in addition $\leq\cdot$ is fine or tight, then $\leq\cdot$ given D is correspondingly fine or tight.

If $\leq\cdot$ is fine, then, for any D that is not null, there exists, in view of Theorem 3.3, one and only one probability measure $P(B \mid D)$, the **(conditional) probability of** B **given** D, that almost agrees with $\leq\cdot$. But, just as one would expect from the traditional study of numerical probability, and as may be easily verified, $P(B \cap D)/P(D)$ considered as a function of B for fixed D is a probability measure that almost agrees with $\leq\cdot$ given D. Therefore,

(1) $P(B \mid D) = P(B \cap D)/P(D).$

As was explained in § 2.7, preference among acts given B can suggestively be expressed in temporal terms. Analogously, the comparison among events given B and, therefore, conditional probability given B can be expressed temporally. Thus $P(C \mid B)$ can be regarded as the probability the person would assign to C after he had observed that B obtains. It is conditional probability that gives expression in the theory of personal probability to the phenomenon of learning by experience.

In accordance with established usage, a pair of events B, C are called independent if $P(B \cap C) = P(B)P(C)$. More generally, a set of events are called **independent,** if for every finite set of them, say $B_1, \cdots, B_n$,

(2) $P \left(\bigcap_i B_i \right) = \prod_i P(B_i).$

Obviously, if D is not null, B and D are independent; if and only if $P(B \mid D) = P(B)$, in which case D may fairly be called **irrelevant to** B.

The notions of independence and irrelevance have, so far as I can see, no analogues in qualitative probability; this is surprising and unfortunate, for these notions seem to evoke a strong intuitive response. The absence of these analogues is traceable to the absence of a qualitative analogue for propositions of the form $P(B \mid C) \leq P(G \mid H)$. Working under a rather different motivation from that which guides this book, B. O. Koopman [K9], [K10], and [K11] has developed a system of qualitative possibility in which it is meaningful to compare B given C with G given H. It is true also that for qualitative probability, even as it is defined here, some interconditional comparisons might be naturally defined. If, for example, $B \leq\cdot \sim B$ given C and $\sim G \leq\cdot G$ given H, it would not be unreasonable to establish the convention that B given $C \leq\cdot G$ given H. This sort of extension is not, however, highly pertinent to my purpose, for here I have little interest in qualitative probabilities, except as a foundation for quantitative probability. The following *partition formula* is well known and easy to prove:

$$(3) \qquad P(C) = \sum_j P(C \mid B_j)P(B_j)$$

where B_i is a partition of S into non-null sets. If, further, C is not null, it is also trivial to derive the celebrated **Bayes' rule** (or **theorem**),

$$(4) \qquad P(B_i \mid C) = \frac{P(C \mid B_i)P(B_i)}{P(C)}$$

$$= \frac{P(C \mid B_i)P(B_i)}{\sum_j P(C \mid B_j)P(B_j)}.$$

Illustrations of these formulas are found in all elementary texbooks on probability, as well as in later sections of this book.

Finally, if neither B nor C is null,

$$(5) \qquad \frac{P(B \mid C)}{P(B)} = \frac{P(C \mid B)}{P(C)} = \frac{P(B \cap C)}{P(B)P(C)},$$

which may be given the suggestive reading: Knowledge of C modifies the probability of B by the same factor by which knowledge of B modifies the probability of C.

The concept of random variable enters into almost any discussion of probability. Experts are fairly well agreed on the following definition. A **random variable** is a function $\mathbf{x}$ attaching a value $x(s)$ in some set X to every s in a set S on which a probability measure P is defined.† Such an S together with the measure P is called a **probability space.**

Real-valued random variables are the most familiar, though in general the values X can be things of any sort. If, for example, $\mathbf{x}$ and $\mathbf{y}$, with values in X and Y, respectively, are random variables on the same measure space, a new random variable $\mathbf{z} = \{\mathbf{x}, \mathbf{y}\}$ is defined by setting $z(s) = \{x(s), y(s)\}$. The values of $\mathbf{z}$ are thus elements of what is called $X \times Y$ (read the **cartesian product** of X and Y), the set of ordered pairs with first element in X and second in Y. The same sort of thing can be done, of course, for ordered n-tuples and also for infinite sequences of random variables.

† In many applications of the theory of probability, not all subsets of S or of X are considered measurable. It is then required as part of the definition of random variable that $\mathbf{x}$ be **measurable,** i.e., that for every measurable $Y \subset X$, the set of s's such that $x(s) \,\varepsilon\, Y$ be measurable.

Two random variables **x** and **y** defined on the same measure space S are called **(statistically) independent**; if and only if, for every $X_0 \subset X$ and $Y_0 \subset Y$, the two events (i.e., subsets of S) defined by the conditions $x(s) \, \varepsilon \, X_0$ and $y(s) \, \varepsilon \, Y_0$, respectively, are independent.† The extension of this definition from pairs to any number of random variables is obvious.

6 The approach to certainty through experience

In § 3, the theory of personal probability was, from the purely mathematical point of view, reduced to that of probability measures, a subject that has been elaborately studied, more or less explicitly, for centuries. Any mathematical problem concerning personal probability is necessarily a problem concerning probability measures—the study of which is currently called by mathematicians **mathematical probability** —and conversely. The particular outlook and interpretation implicit in a personalistic concept of probability leads, however, to problems that, though perfectly meaningful for mathematical probability, might not otherwise have been emphasized. This section and the succeeding one each briefly discuss one such problem. These two problems are selected from among many possibilities for the insight they provide into the concept of personal probability.

Before studying these problems, it is necessary to be conversant with the material in Appendixes 1 and 2, which is used in the immediate sequel and often throughout the rest of this book.

As was brought out in § 5, the person learns by experience. The purpose of the present section is to explore with a moderate degree of generality how he typically becomes almost certain of the truth, when the amount of his experience increases indefinitely. To be specific, suppose that the person is about to observe a large number of random variables, all of which are independent given B_i for each i, where the B_i are a partition of S. It is to be expected intuitively, and will soon be shown, that under general conditions the person is very sure that after making the observation he will attach a probability of nearly 1 to whichever element of the partition actually obtains.

To describe the situation formally, let B_i be a partition of S with $P(B_i) = \beta(i)$. Let x_r, $r = 1, 2, \cdots$, be a sequence of random variables, each taking on only a finite number of values (which can without loss of generality be thought of as integers). The restriction to a finite set of values could be removed, but to do so would raise problems of mathematical technique that, however interesting, are rather beside the point

† Where not all sets are measurable, X_0 and Y_0 must, of course, be required to be measurable.

of this book. Let $\mathbf{x}$ denote the first n of the random variables $\mathbf{x}_r$. It is to be borne in mind that $\mathbf{x}$ depends on n, so, strictly speaking, it should be written $\mathbf{x}(n)$. The assumption that, given B, the $\mathbf{x}_r$'s all have the same distribution is expressed by

(1) $$P(x_r(s) = x_r \mid B_i) = \xi(x_r \mid i),$$

where $\xi(x_r \mid i)$ is defined by the context. Combining (1) with the assumption that the $\mathbf{x}_r$'s are independent given B_i,

(2) $$P(x \mid B_i) =_{\mathrm{Df}} P(x(s) = \{x_1, \cdots, x_n\} \mid B_i) = \prod_{r=1}^{n} \xi(x_r \mid i),$$

where a conventional symbol has been used for **equal by definition**.

These hypotheses having been laid down, it follows from Bayes' rule and the partition formula (5.3) and (5.2), that

(3) $$P(B_i \mid x) = \frac{P(x \mid B_i)P(B_i)}{P(x)}$$

$$= \frac{\beta(i) \prod_r \xi(x_r \mid i)}{P(x)}$$

and

(4) $$P(x) = \sum_i \beta(i) \prod_r \xi(x_r \mid i).$$

In connection with (3), it may be observed in passing that, if the **a priori probability**, $\beta(i)$, of B_i is 0, then, no matter what value x is observed, the **a posteriori probability** of B_i, $P(B_i \mid x)$, is also 0. This is an example of the general principle that, if some event is regarded as virtually impossible, then no evidence whatsoever can lend it credibility. Similarly, (3) implies the equally common-sense principle that, if an observation x is virtually impossible on the hypothesis (i.e., given) B_i, and x is observed, then B_i becomes virtually impossible a posteriori.

It is particularly interesting to compare the probability of two elements of the partition, say B_1 and B_2 for definiteness, in the light of x.

(5) $$\frac{P(B_1 \mid x)}{P(B_2 \mid x)} = \frac{\beta(1)}{\beta(2)} \prod_r \frac{\xi(x_r \mid 1)}{\xi(x_r \mid 2)}$$

$$= \frac{\beta(1)}{\beta(2)} \prod_r R'(x_r)$$

$$= \frac{\beta(1)}{\beta(2)} R(x),$$

where self-explanatory abbreviations have been introduced. Equation
(5) is meaningless, if both the numerator and denominator of its left-
hand side vanish. If the denominator alone vanishes, the fraction may
properly be regarded as infinite. This will happen; if and only if B_2 is
null, and B_1 is not null given x. That is, it will happen if and only if
$\beta(1) \neq 0$, $\beta(2) = 0$, or if $\beta(1) \neq 0$, and $R(x) = \infty$.

In modern statistical usage, $R'(x_r)$ and $R(x)$ are the **likelihood ratios**
of B_1 to B_2 given x_r and x, respectively, quantities of importance in
many theoretical contexts.

If a person contemplates making the observation **x**, that is, finding
out the value of $x(s)$ for the s that is the true state of the world, it may
properly be asked how probable he considers it that **R** will turn out to
have a particular value. It will be shown, barring two banal exceptions, that, for n sufficiently large, the probability, given B_1, that R is
greater than any preassigned number is almost 1. The possibility
$P(B_1) = 0$ is to be excepted, for then the conditional probability in
question is meaningless. The other exception occurs when $\xi(x_r \mid 1) =$
$\xi(x_r \mid 2)$ for every x_r, that is, when the common distribution of $\mathbf{x}_r$ given
B_1 is the same as it is given B_2; for then observation of $\mathbf{x}_r$ is simply
irrelevant in distinguishing B_1 from B_2, or, a little more technically, $\mathbf{x}_r$
is irrelevant to B_1 given $B_1 \cup B_2$, and

(6) $$P(R'(x_r) = 1 \mid B_1) = 1.$$

Formally, it is to be demonstrated that, unless $P(B_1) = 0$, or (6)
holds,

(7) $$\lim_{n \to \infty} P(R(x) \geq \rho \mid B_1) = 1 \qquad \text{for } 0 \leq \rho < \infty.$$

The problem is quite simple when account is taken of the fact that
$R(\mathbf{x})$ is the product of n random variables, $R'(\mathbf{x}_r)$, that are independent
given B_1. In attacking the problem, two cases are to be distinguished,
according as there are or are not values of x that have positive probability given B_1 but zero probability given B_2.

It is in practice rather fortunate to find instances of the first case,
for then (7) applies with a vengeance. Indeed, suppose that

(8) $$P(R'(x_r) < \infty \mid B_1) = \phi, \qquad \phi < 1.$$

Then

(9) $$P(R = \infty \mid B_1) = 1 - \phi^n,$$

which obviously approaches 1 with increasing n.

The second case, namely $\phi = 1$, is more interesting. Since much is known about *sums* of identically distributed independent random variables, it is natural to investigate

$$(10) \qquad \log R(x) = \sum_r \log R'(x_r),$$

thereby replacing a product by a sum. It is easily seen from the definition of $R'(x_r)$ that $P(R'(x_r) > 0 \mid B_1) = 1$, so, in the case now at hand, the functions $\log R'(x_r)$ are independent real bounded random variables.

Letting

$$(11) \qquad I = E(\log R'(x_r) \mid B_1),$$

the **weak law of large numbers** † implies that, for any $\epsilon > 0$,

$$(12) \qquad \lim_{n \to \infty} P(\log R(x) \geq n(I - \epsilon) \mid B_1) = 1,$$

equivalently,

$$(13) \qquad \lim_{n \to \infty} P(R(x) \geq e^{n(I - \epsilon)} \mid B_1) = 1.$$

The objective will therefore be achieved, if it is demonstrated that $I > 0$ unless (6) holds. But

$$(14) \qquad \begin{aligned} I &= E(\log R'(x_r) \mid B_1) \\ &\geq -\log E(R'^{-1}(x_r) \mid B_1) \\ &= -\log 1 = 0, \end{aligned}$$

as may be argued thus: The inequality in the above calculation is assigned as Exercise 8 in Appendix 2, together with the fact that equality can hold in (14) if and only if $R'^{-1}(\mathbf{x}_r)$ is constant with probability one given B_1. But the expected value of $R'^{-1}(\mathbf{x}_r)$ given B_1 is equal to 1, as (14) asserts and as may be easily verified from the definition of $R'^{-1}(\mathbf{x}_r)$. So, barring the exceptions provided for, $I > 0$, and the demonstration of (7) is complete.

Before the observation, the probability that the probability given x of whichever element of the partition actually obtains will be greater than α is

$$(15) \qquad \sum_i \beta(i) P(P(B_i \mid x) > \alpha \mid B_i),$$

where summation is confined to those i's for which $\beta(i) \neq 0$. Application of (14) (extended to arbitrary pairs of i's) shows that the coefficients

† For the definition of this law, see, if necessary, p. 191 of Feller's book [F1].

of each $\beta(i)$ in the quantity (15), and therefore the quantity itself, approaches 1 as n increases; provided only that no two functions $\xi(\mathbf{x}_r \mid i)$ and $\xi(\mathbf{x}_r \mid i')$ are the same, if $\beta(i)$ and $\beta(i')$ are both different from zero.

To summarize informally, it has now been shown that, with the observation of an abundance of relevant data, the person is almost certain to become highly convinced of the truth, and it has also been shown that he himself knows this to be the case.

It may be remarked, for those familiar with certain theorems, that many refinements of (7) and its consequences could be worked out by application of the strong law of large numbers, the central limit theorem, and the law of the iterated logarithm to $R'(\mathbf{x}_r)$.

The quantity I is coming to be called the **information** of the distribution of $\mathbf{x}_r$ given B_1 with respect to the distribution of $\mathbf{x}_r$ given B_2. More generally, if P and Q are probability measures, confined (for simplicity) to a finite set X with elements x; the information of P with respect to Q is defined by

$$(16) \qquad \sum_x P(x) \log \frac{P(x)}{Q(x)}.$$

This usage stems from work of Claude Shannon in communication engineering, a good account of which is given in [S11]; and also from independent work of Norbert Wiener in a related context [W10]. The ideas of Shannon and of Wiener, though concerned with probability, seem rather far from statistics. It is, therefore, something of an accident that the term "information" coined by them should be not altogether inappropriate in statistics. The situation is still further confused, because, as long ago as 1925, R. A. Fisher emphasized an important notion, which he called "information," in connection with the theory of estimation (Paper 11, *Theory of statistical estimation* in [F6]). At first glance, Fisher's notion seems quite different from that of Shannon and Wiener, but, as a matter of fact, his is a limiting form of theirs. A useful but rather technical exposition relating the several senses of "information" is given by Kullback and Leibler [K15], and I return to the topic in § 15.6.[+]

7 Symmetric sequences of events

A problem often posed by statisticians is to estimate from a sequence of observations the unknown probability p that repeated trials of some sort are successful. On an objectivistic view, this problem is natural and important, for on such a view the probability that a coin falls heads, for example, is a property of the coin that can be determined by experimentation with the coin and in no other way. But on a personalistic

[+] See also (Kullback 1961).

view of probability, strictly interpreted, no probability is unknown to the person concerned, or, at any rate, he can determine a probability only by interrogating himself, not by reference to the external world.

This situation has been interpreted to imply that the personalistic view is wrong, or at any rate inadequate, because it apparently cannot even express one of the most natural and typical problems of statistics. Thus far in this book, I have not argued against the possibility of defining some useful notion of objective probability, but have contented myself with presenting a particular notion of personal probability. Therefore, at this point it might be tempting to seek a dualistic theory admitting both objective and personal probabilities in some kind of articulation with one another. De Finetti [D3] has shown, however, that it is not necessary to do so, that the notion of a coin with unknown probability p can be reinterpreted in terms of personal probability alone.

The present section is devoted to outlining this development due to de Finetti. In the organization of the book as a whole, it plays no logically essential part; it is, rather, a digression intended to give a clearer understanding of the notion of personal probability, especially in relation to objectivistic views. The ideas presented here are but a fragment of those on the same subject in [D2].

Let $\mathbf{x}_r$ be a sequence of random variables taking only the values 0 and 1. The $\mathbf{x}_r$'s are, to all intents and purposes, a sequence of events, the rth of which is the event that $x_r(s) = 1$. To say that these events are independent, each occurring with probability p, is to say that the probability of any finite pattern, $x_1, \cdots, x_n$, initiating the sequence $x_r(s)$ is given by the formula

(1) $$P(x_r(s) = x_r; r = 1, \cdots, n \mid p) = p^y(1 - p)^{n-y},$$

where y is the number of 1's among the x_r's for $r = 1, \cdots, n$.

Mixtures, in a certain sense, of sequences of random variables are often of interest, as they already have been in the preceding section. Suppose, to be explicit, that the world is partitioned by B_i and that, given B_i, the $\mathbf{x}_r$'s are independent with $P(x_r(s) = 1 \mid B_i)$ having some fixed value $p(i)$. Then the unconditional probability of a particular initial sequence is a mixture of the probabilities given by (1) thus:

(2) $$P(x_r(s) = x_r; r = 1, \cdots, n) = \sum_i p(i)^y(1 - p(i))^{n-y}P(B_i).$$

It is natural to generalize (2) formally thus:

(3) $$P(x_r(s) = x_r; r = 1, \cdots, n) = \int p^y(1 - p)^{n-y}\, dM(p),$$

where M is a probability measure on the real numbers in the interval $[0, 1]$.

It is noteworthy that equation (3), understood to apply for every n, is equivalent to the condition that the probability that every n of each prescribed set of n of the x_r's takes the value 1 is

$$(4) \qquad \int p^n \, dM(p).$$

This follows by arithmetic induction from the obvious formula

$$(5) \quad P(x_r(s) = x_r; r = 1, \cdots, n)$$
$$= P(x_r(s) = x_r; r = 1, \cdots, n; x_{n+1}(s) = 0)$$
$$+ P(x_r(s) = x_r; r = 1, \cdots, n; x_{n+1}(s) = 1),$$

which applies to any sequence of random variables taking on only the values 0 and 1.

Equation (3) can very well have an interpretation in such terms that the measure M is not merely an abstract probability measure, but is actually a personal probability. Thus, if $\mathbf{p}$ is a random variable that is (for a given person) distributed according to M, and, if for each p the conditional distribution of the x_r's given p is independent, with $P(x_r(s) = 1) = p$; then (3) obtains. Strictly speaking, the notion of conditional probability as it occurs in the preceding sentence is used in a somewhat wider sense than has been defined in this book, for the probability of any particular p will typically be zero. At least for countably additive measures, the necessary extension of conditional probability and conditional expectation is presented by Kolmogoroff in [K7]; it is a concept of the greatest value in advanced mathematical statistics and in probability generally.

However, in most contexts where objectivists speak of an unknown probability p, there is, so far as an exclusively personalistic view of probability is concerned, no unknown parameter that can play the role of p in (3).

Examination of situations in which "unknown" probability is appealed to, whether justifiably or not, shows that, from the personalistic standpoint, they always refer to symmetric sequences of events in the sense of the following definition. The sequence of random variables x_r, taking only the values 0 and 1, is a *symmetric* † *sequence*, if and only if the probability that any b of the $x_r(s)$'s equal 1 and any c other $x_r(s)$'s equal 0 depends only on the integers b and c.

---

† De Finetti uses the French word for "equivalent."[+]
[+] He and others now prefer "exchangeable." The concept seems to have been first suggested by Jules Haag (1928).

It is easy to verify that any mixture of independent sequences in the sense of (3) is a symmetric sequence. De Finetti has discovered that the converse is also true. These conclusions can be formally summarized thus:

THEOREM 1   A sequence of random variables x_r, taking only the values 0 and 1, is symmetric, if and only if there exists a probability measure M on the interval $[0, 1]$ such that the probability that any prescribed n of the $x_r(s)$'s equal 1 is given by (4). Two such measures, M and M', must be essentially the same,† in the sense that, if B is a subinterval of $[0, 1]$, then $M(B) = M'(B)$.

Considering that de Finetti has published a proof of Theorem 1 in [D2] based on the Fourier integral, that any proof of it must be rather technical, and that the theorem is not the basis of any formal inference later in this book, it seems best not to prove it here.‡

It is Theorem 1 that makes it possible to express propositions referring to unknown probabilities in purely personalistic terms. If, for example, a statistician were to say, "I do not know the p of this coin, but I am sure it is at most one half," that would mean in personalistic terms, "I regard the sequence of tosses of this coin as a symmetric sequence, the measure M of which assigns unit measure to the interval $[0, \frac{1}{2}]$." This condition on M means in turn that for every n the (personal) probability of n consecutive heads is at most 2^{-n}, as is easily verified. I do not insist that propositions couched in terms of a fictitious unknown probability are bad, if understood as suggestive abbreviations, but only that the meaningfulness of such propositions does not constitute an inadequacy of the personalistic view of probability.

The mathematical concept of probability measure or, a trifle more generally, bounded measure is fundamental to mathematics generally. Probability measures, often under other names, are, therefore, employed in many parts of pure and applied mathematics completely unrelated to probability proper. For example, the distribution of mass in a not necessarily rigid body is expressed by a bounded measure that tells how much of the body is in each region of space. We must, therefore, not be surprised if, even in studying probability itself, we come across some probability measures used not to measure probability

† Technical note: If "probability measure" were here understood to mean a countably additive probability measure on the Borel sets of $[0, 1]$, the theorem would remain true, and the essential uniqueness of M would become true uniqueness.

‡ Technical note: Theorem 1 can be proved very quickly and naturally by applying the theory of the Hausdorff moment problem (pp. 8–9 of [S13]) to M, but this method does not seem to generalize readily.+

+ New and general methods are in Hewitt and Savage (1955) and Ryll-Nardzewski (1957). For related work see Bühlmann (1960), Freedman (1962, 1963), Milier-Gruzewska (1949, 1950), and Rényi and Révész (1963).

proper but only for auxiliary purposes. In the event that p is not actually an unknown parameter, the measure M presented by Theorem 1 seems at first sight to be such a purely auxiliary measure, but, as a matter of fact, M does measure certain interesting probabilities, at least approximately. For example, letting

$$(6) \qquad \bar{x}_n = \frac{1}{n} \sum_1^n x_r,$$

it can be shown that

$$(7) \qquad \lim_{n \to \infty} P(\bar{x}_n(s) \leq \delta) = M(p \leq \delta).$$

In words, the person considers the average of any large number of future observations to be distributed approximately the way p is distributed by M. This is an extension of the ordinary weak law of large numbers, proved in [D2] along with a corresponding extension of the strong law.

If the first n terms of a symmetric sequence are observed, how does the rest of the sequence appear to the person in the light of this observation? In the first place, it also is a symmetric sequence but generally of a structure different from that of the original sequence, as may be shown thus: Let

$$(8) \qquad \pi(y, n - y) =_{\mathrm{Df}} P(x_r(s) = x_r; r = 1, \cdots, n),$$

as one may for a symmetric sequence. Then

$$(9) \quad P(x_q(s) = x_q; q = n + 1, \cdots, n + m \mid x_r(s) = x_r, r = 1, \cdots, n)$$

$$= \frac{P(x_p(s) = x_p, p = 1, \cdots, n + m)}{P(x_r(s) = x_r, r = 1, \cdots, n)}$$

$$= \frac{\pi(y + z, (n - y) + (m - z))}{\pi(y, n - y)},$$

where z is the number of 1's among the x_q's, $q = n + 1, \cdots, n + m$. Equation (9) shows that the sequence $\mathbf{x}_q$, $q > n$, given that $x_r(s) = x_r$, $r = 1, \cdots, n$, is a new symmetric sequence characterized by

$$(10) \qquad \pi'(z, m - z) =_{\mathrm{Df}} \frac{\pi(y + z, (n - y) + (m - z))}{\pi(y, n - y)}.$$

The measure M' associated with the new sequence is, according to Theorem 1, essentially determined by the condition that

(11)
$$\int p^m \, dM'(p) = \pi'(m, 0)$$

$$= \frac{\pi(m + y, n - y)}{\pi(y, n - y)}$$

$$= \frac{\displaystyle\int p^{m+y}(1 - p)^{n-y} \, dM(p)}{\pi(y, n - y)}$$

$$= \int p^m \frac{p^y(1 - p)^{n-y}}{\pi(y, n - y)} \, dM(p).$$

Equation (11) makes it plausible that, except for the slight ambiguity permitted by Theorem 1, M' is defined (for Borel sets B) by

(12)
$$M'(B) = \pi^{-1}(y, n - y) \int_B p^y(1 - p)^{n-y} \, dM(p),$$

and this can in fact be demonstrated with some appeal to slightly advanced methods pertaining to the Hausdorff moment problem (pp. 8–9 of [S13]).

It is noteworthy that, if $M(B) = 0$, then $M'(B) = 0$ also. In the event that p really is an unknown parameter, this means that, if the person is virtually certain that the true p is not in B, no amount of evidence can alter that opinion.

Equation (12) shows that M' is generally different from M. Indeed, for fixed $n \geq 1$, M' is clearly the same as M for every y for which $\pi(y, n - y) > 0$, if and only if M assigns the measure 1 to some one value of p. That is, the person regards evidence drawn from a symmetric sequence as irrelevant to the future behavior of the sequence, if and only if at the outset he regards the sequence not merely as symmetric but also as independent.

It can be shown that the person regards it as highly probable that, if he observes a sufficiently long segment of a symmetric sequence, the continuation of the sequence will then be one for which the conditional variance of p,

(13)
$$\int p^2 \, dM'(p) - \left\{ \int p \, dM'(p) \right\}^2,$$

will be small. In the event that p is really an unknown parameter, this implies that the person is very sure that after a long sequence of observations he will assign nearly unit probability to the immediate neighborhood of the value of p that actually obtains—a parallel to the approach to certainty discussed in § 6.

CHAPTER 4

Critical Comments
on Personal Probability

1 Introduction

It is my tentative view that the concept of personal probability introduced and illustrated in the preceding chapter is, except possibly for slight modifications, the only probability concept essential to science and other activities that call upon probability. I propose in this chapter to discuss the shortcomings I see in that particular personalistic view of probability, which, for brevity, shall here be called simply "the personalistic view"; to point out briefly the relationships between it and other views; to criticize other views in the light of it; and to discuss the criticisms holders of other views have raised, or may be expected to raise, against it.

From the standpoint of strict logical organization such critical remarks are somewhat premature, because the personalistic view itself insists that probability is concerned with consistent action in the face of uncertainty. Consequently, until the theory of such action has been completely outlined in later chapters, the view to be criticized cannot even be considered to have been wholly presented. Practically, however, it seems wise not to confine critical comments to the one part of the text that logic may suggest as appropriate, but rather to touch on criticism from time to time, even at the cost of some repetition. Thus, some of what is to be said here has already been said in the introductory chapter and elsewhere, and some of it will be said again.

Views other than the personalistic view are to be discussed here, but it cannot be too distinctly emphasized that the account given of them will be very superficial.† One function of discussing other views is to provide the reader with at least some orientation in the large and diversified body of ideas pertaining to the foundation of statistics that

† Much more extensive comparative material is given by Keynes [K4], by Nagel [N1], and by Carnap [C1]. Koopman [K12] should also be mentioned in this connection.

have been accumulated. A less obvious, but I think no less important and legitimate, function is to cast new light on the personalistic view, especially for those who already hold, or tend to hold, other views.

2 Some shortcomings of the personalistic view

I can answer, to my own satisfaction, some criticisms of the personalistic view that have been brought to my attention. These points are discussed later in the chapter, but in this section I state and discuss as clearly as I can those that I find more difficult and confusing to answer.

According to the personalistic view, the role of the mathematical theory of probability is to enable the person using it to detect inconsistencies in his own real or envisaged behavior. It is also understood that, having detected an inconsistency, he will remove it. An inconsistency is typically removable in many different ways, among which the theory gives no guidance for choosing. Silence on this point does not seem altogether appropriate, so there may be room to improve the theory here. Consider an example: The person finds on interrogating himself about the possible outcome of tossing a particular coin five times that he considers each of the thirty-two possibilities equally probable, so each has for him the numerical probability 1/32. He also finds that he considers it more probable that there will be four or five heads in the five tosses than that the first two tosses will both be heads. Now, reference to the mathematical theory of probability soon shows the person that, if the probability of each of the thirty-two possibilities is 1/32, then the probability of four or five heads out of five is 6/32, and the probability that the first two tosses will be heads is 8/32, so the person has caught himself in an inconsistency. The theory does not tell him how to resolve the inconsistency; there are literally an infinite number of possibilities among which he must choose.

In this particular example, the choice that first comes to my mind, and I imagine to yours, is to hold fast to the position that all thirty-two possibilities are equally likely and to accept the implications of that position, including the implication that four or five heads out of five is less probable than two heads out of two. I do not think that there is any justification for that choice implicit in the example as formally stated, but rather that in the sort of actual situation of which the example is a crude schematization there generally are considerations not incorporated in the example that do justify, or at any rate elicit, the choice.

To approach the matter in a somewhat different way, there seem to be some probability relations about which we feel relatively "sure" as compared with others. When our opinions, as reflected in real or envisaged action, are inconsistent, we sacrifice the unsure opinions to the sure ones. The notion of "sure" and "unsure" introduced here is vague, and my complaint is precisely that neither the theory of personal probability, as it is developed in this book, nor any other device known to me renders the notion less vague.† There is some temptation to introduce probabilities of a second order so that the person would find himself saying such things as "the probability that B is more probable than C is greater than the probability that F is more probable than G." But such a program seems to meet insurmountable difficulties.

The first of these—pointed out to me by Max Woodbury—is this. If the primary probability of an event B were a random variable **b** with respect to secondary probability, then B would have a "composite" probability, by which I mean the (secondary) expectation of **b**. Composite probability would then play the allegedly villainous role that secondary probability was intended to obviate, and nothing would have been accomplished.

Again, once second order probabilities are introduced, the introduction of an endless hierarchy seems inescapable. Such a hierarchy seems very difficult to interpret, and it seems at best to make the theory less realistic, not more.

Finally, the objection concerning composite probability would seem to apply, even if an endless hierarchy of higher order probabilities were introduced. The composite probability of B would here be the limit of a sequence of numbers, $E_n(E_{n-1}(\cdots E_2(P_1(B))\cdots))$, a limit that could scarcely be postulated not to exist in any interpretable theory of this sort. The reader may wish to evaluate for himself the arguments in favor of such a hierarchy put forward by Reichenbach (Chapter 8, [R2]), taking proper account of the differences between Reichenbach's overall view, and his mathematical theory, of probability on one hand and, on the other, the personalistic view and measure-theoretic mathematical theory that are the basis of my critique of higher order probabilities.

The interplay between the "sure" and "unsure" is interestingly expressed by de Finetti (p. 60, [D2]) thus: "The fact that a direct estimate of a probability is not always possible is just the reason that the logical rules of probability are useful. The practical object of these rules is simply to reduce an evaluation, scarcely accessible directly, to others by means of which the determination is rendered easier and more precise."

It may be clarifying, especially for some readers under the sway of the objectivistic tradition, to mention that, if a person is "sure" that

---

† One tempting representation of the unsure is to replace the person's single probability measure P by a set of such measures, especially a convex set. Some explorations of this are Dempster (1968), Good (1962), and Smith (1961).

the probability of heads on the first toss of a certain penny is $\frac{1}{2}$, it does not at all follow that he considers the coin fair. He might, to take an extreme example, be convinced that the penny is a trick one that always falls heads or always falls tails.

Logic, to which the theory of personal probability can be closely paralleled, is similarly incomplete. Thus, if my beliefs are inconsistent with each other, logic insists that I amend them, without telling me how to do so. This is not a derogatory criticism of logic but simply a part of the truism that logic alone is not a complete guide to life. Since the theory of personal probability is more complete than logic in some respects, it may be somewhat disappointing to find that it represents no improvement in the particular direction now in question.

A second difficulty, perhaps closely associated with the first one, stems from the vagueness associated with judgments of the magnitude of personal probability. The postulates of personal probability imply that I can determine, to any degree of accuracy whatsoever, the probability (for me) that the next president will be a Democrat. Now, it is manifest that I cannot really determine that number with great accuracy, but only roughly. Since, as is widely recognized, all the interesting and useful theories of modern science, for example, geometry, relativity, quantum mechanics, Mendelism, and the theory of perfect competition, are inexact; it may not at first sight seem disquieting that the theory of personal probability should also be somewhat inexact. As will immediately be explained, however, the theory of personal probability cannot safely be compared with ordinary scientific theories in this respect.

I am not familiar with any serious analysis of the notion that a theory is only slightly inexact or is almost true, though philosophers of science have perhaps presented some. Even if valid analyses of the notion have been made, or are made in the future, for the ordinary theories of science, it is not to be expected that those analyses will be immediately applicable to the theory of personal probability, normatively interpreted; because that theory is a code of consistency for the person applying it, not a system of predictions about the world around him.

The difficulty experienced in § 2.6 with defining indifference seems closely associated with the difficulty about vagueness raised here.

Another difficulty with the theory of personal probability (or, more properly, with that larger theory of the behavior of a person in the face of uncertainty, of which the theory of personal probability is a part) is that the statement of the theory is not yet necessarily complete. Thus we shall in the next chapter come upon another proposition that demands acceptance as a postulate, and, since even this leaves the person a great deal of freedom, there is no telling when someone will come upon still another postulate that clamors to be adjoined to the others. Strictly speaking, this is not so much an objection to the theory as a warning about what to expect of its future development.

3 Connection with other views

All views of probability are rather intimately connected with one another. For example, any necessary view can be regarded as an extreme personalistic view in which so many criteria of consistency have been invoked that there is no role left for the person's individual judgment. Again, objectivistic views can be regarded as personalistic views according to which comparisons of probability can be made only for very special pairs of events, and then only according to such criteria that all (right-minded) people agree in their comparisons.

From a different standpoint, personalistic views lie not between, but beside, necessary and objectivistic views; for both necessary and objectivistic views may, in contrast to personalistic views, be called objective in that they do not concern individual judgment.

4 Criticism of other views

It will throw some light on the personalistic view to say briefly how some other views seem to compare unfavorably with it.

It is one of my fundamental tenets that any satisfactory account of probability must deal with the problem of action in the face of uncertainty. Indeed, almost everyone who seriously considers probability, especially if he has practical experience with statistics, does sooner or later deal with that problem, though often only tacitly. Even some personalistic views seem to me too remote from the problem of action, or decision. For example, de Finetti in [D2] gives two approaches to personal probability. Of these, one is almost exactly like the view sponsored here, except only that the notion "more probable than" is supposed to be intuitively evident to the person, without reference to any problem of decision. The other is more satisfactory in this respect, being couched in terms of betting behavior, but it seems to me a somewhat less satisfactory approach than the one sponsored here, because it must assume either that the bets are for infinitesimal sums or—anticipating the language of the next chapter—that the utility of money is linear. The theory expressed by Koopman in [K9], [K10], and [K11] and that expressed by Good in [G2] are both personalistic views that tend to ignore decision, or at any rate keep it out of the foreground; but the personalistic view expressed by Ramsey in [R1], like the one sponsored here, takes decision as fundamental. If any necessary view can be formulated at all, it might well be possible to formulate it in terms of decision, but, so far as I know, the notion of decision has not appeared fundamental to the holders of any necessary view. It seems fair to say that objectivistic views, by their very nature, must in principle regard decision as secondary to probability, if relevant at all. Yet, the objectivist A. Wald has done more than anyone else to popularize the notion of decision.

As has already been indicated, from the position of the personalistic view, there is no fundamental objection to the possibility of constructing a necessary view, but it is my impression that that possibility has not yet been realized, and, though unable to verbalize reasons, I conjecture that the possibility is not real. Two of the most prominent enthusiasts of necessary views are Keynes, represented by [K4], and Carnap, who has begun in [C1] to state what he hopes will prove a satisfactory necessary (or nearly necessary) view of probability. Keynes indicated in the closing pages of [K4] that he was not fully satisfied that he had solved his problem and even suggested that some element of objectivistic views might have to be accepted to achieve a satisfactory theory, and Carnap regards [C1] as only a step toward the establishment of a satisfactory necessary view, in the existence of which he declares confidence. That these men express any doubt at all about the possibility of narrowing a personalistic view to the point where it becomes a necessary one, after such extensive and careful labor directed toward proving this possibility, speaks loudly for their integrity; at the same time it indicates that the task they have set themselves, if possible at all, is not a light one.

Keynes, writing in 1921 of what are here called objectivistic views, complained, "The absence of a recent exposition of the logical basis of the frequency theory by any of its adherents has been a great disadvantage to me in criticizing it." (Chap. VIII, Sec. 17, of [K4]). I believe that his complaint applies as aptly to my position today as to his then, though I cannot pretend to have combed the intervening literature with anything like the thoroughness Keynes himself would have employed. Reichenbach, to be sure, presents in great detail an interesting view that must be classified as objectivistic [R2], but it seems far removed from those that dominate modern statistical theory and form the main subject of the following discussion. Whatever objectivistic views may be, they seem, to holders of necessary and personalistic views alike, subject to two major lines of criticism. In the first place, objectivistic views typically attach probability only to very special events. Thus, on no ordinary objectivistic view would it be meaningful, let alone true, to say that on the basis of the available evidence it is very improbable, though not impossible, that France will become a monarchy within the next decade. Many who hold objectivistic views admit that such everyday statements may have a meaning, but they insist, depending on the extremity of their positions, that that meaning is not relevant to mathematical concepts of probability or even to science generally. The personalistic view claims, however, to analyze such statements in terms of mathematical probability, and it considers them important in science and other human activities.

Secondly, objectivistic views are, and I think fairly, charged with circularity. They are generally predicated on the existence in nature of processes that may, to a sufficient degree of approximation, be represented by a purely mathematical object, namely an infinite sequence of independent events. This idealization is said, by the objectivists who rely on it, to be analogous to the treatment of the vague and extended mark of a carpenter's pencil as a geometrical point, which is so fruitful in certain contexts. When it is pointed out to the objectivist that he uses the very theory of probability in determining the quality of the approximation to which he refers, he retorts that the applied geometer—a fictitious character whose reputation for solidity in science is unquestioned—likewise uses geometry in determining the quality of his approximations. Let the geometer then be challenged, and he replies with a threefold reference to experience, saying, "It is a common experience that with sufficient experience one develops good judgment in the use of geometry and thenceforth generally experiences success in the predictions he bases on it." "Now," says the objectivist, "the geometer's answer is my answer." But it seems to critics of objectivistic views that, though the geometer may be entitled to make as many allusions to experience as he pleases, the probabilist is not free to do so, precisely because it is the business of the probabilist to analyze the concept of experience. He, therefore, cannot properly support his position by alluding to experience until he has analyzed that concept, though he can, of course, allude to as many experiences as he wishes.

Two sorts of mixed views call for special comment here.

First, some (among them Carnap [C1]; Koopman [K9], [K10], and [K11]; and Nagel [N1]) hold that two probability concepts play a role in inference, an objectivistic one and a personalistic or a necessary one. This dualism is typically justified as necessary to the analysis of such a concept as that of a coin with unknown probability of falling heads. But, as § 3.7 explains, de Finetti has provided a satisfactory analysis on the basis of personal probability alone.

Second, others—for example, van Dantzig [V1] and Féraud [F2]— finding the conventional objectivistic views circular for the reasons I have cited, try to break the circle by relatively isolated use of subjective ideas. Very crudely, it seems to be their position that in any one context it is allowable for a person to act as though some one event of sufficiently small (objective) probability, chosen at his discretion, were impossible. Quite apart from the relatively technical question of whether any consistent mixed view of this kind can be constructed, holders of personalistic and necessary views alike criticize them as unnecessarily timid, for they embrace subjective ideas, but only gingerly.

5 The role of symmetry in probability

An important and highly controversial question in the foundations of probability is whether and, if so, how symmetry considerations can determine the probabilities of at least some events.

Symmetry considerations have always been important in the study of probability. Indeed, early work in probability was dominated by the notion of symmetry, for it was usually either concerned with, or directly inspired by, symmetrical gambling apparatus such as dice or cards. To illustrate those classical problems, suppose that a gambler is offered several bets concerning the possible outcome of rolling three dice, where it is to be understood that refraining from any bets at all may be among the available "bets." Which of the available bets should the gambler choose? Perhaps I distort history somewhat in insisting that early problems were framed in terms of choice among bets, for many, if not most, of them were framed in terms of equity, that is, they asked which of two players, if either, would have the advantage in a hypothetical bet. But, especially from the point of view of the earlier probabilists, such a question of equity is tantamount to a question of choice among bets, for to ask which of two "equal" betters has the advantage is to ask which of them has the preferable alternative, as was pointed out quite explicitly by D. Bernoulli in [B10].

In effect, the classical workers recommended the following solution to the problem of three dice, with corresponding solutions to other gambling problems:

1. Attach equal mathematical probabilities to each of the 216 ($=6^3$) possible outcomes of rolling the three dice. (There are 6^3 possibilities, because the first, second, and third dice can each show any of six scores, all combinations being possible.)

2. Under the mathematical probability established in Step 1, compute the expected winnings (possibly negative) of the gambler for each available bet.

3. Choose a bet that has the largest expected winnings among those available.

At present it is appropriate to refrain from criticisms of the use made of expected winnings until the next chapter and to concentrate discussion on the notion that the 216 possibilities should be considered equally probable, which can conveniently be done by drastically reducing the class of bets considered to be available. Say, for definiteness, that the only bets to be considered are simply even-money bets of one dollar, that the triple of scores falls in a preassigned subset of the 216 possibilities. When attention is focused on this restricted class of bets, the total recommendation is seen to imply that the probability measure defined in the first step of the recommendation be adopted as the personal probability of the gambler. To put it differently, a gambler who adopts the recommendation will hold the 216 possible outcomes equally probable not only in some abstract sense, but also in the sense of personal probability as defined in § 3.2.

The notion that the 216 possibilities should be regarded as equally probable is familiar to everyone; for it is taken for granted wherever gentlemen gamble as well as in the standard high-school algebra courses, where it serves to illustrate the theory of combinations and permutations.

Traditionally, the equality of the probabilities was supposed to be established by what was called the **principle of insufficient reason**,† thus: Suppose that there is an argument leading to the conclusion that one of the possible combinations of ordered scores, say {1, 2, 3}, is more probable than some other, say {6, 3, 4}. Then the information on which that hypothetical argument is based has such symmetry as to permit a completely parallel, and therefore equally valid, argument leading to the conclusion that {6, 3, 4} is more probable than {1, 2, 3}. Therefore, it was asserted, the probabilities of all combinations must be equal.

The principle of insufficient reason has been and, I think, will continue to be a most fertile idea in the theory of probability; but it is not so simple as it may appear at first sight, and criticism has frequently and justly been brought against it. Holders of necessary views typically attempt to put the principle on a rigorous basis by modifying it in such a way as to take account of such criticism. Holders of personalistic and objectivistic views typically regard the criticism as not altogether refutable, so they do not attempt to establish a formal postulate corresponding to the principle but content themselves—as I shall here —with exhibiting an element of truth in it.

One of the first criticisms is that the principle is not strictly applicable for a person who has had any experience with the apparatus in ques-

† Perhaps what I here call the principle of insufficient reason should be called the principle of cogent reason. See Section 3 of [B15] for the distinction involved.

tion, or even with similar apparatus. Thus, attempts to use the principle, as I have stated it, to prove that there is no such thing as a run
of luck at dice, as actually played, are invalid. The person may have
had relevant experience, directly or vicariously, not only with gambling
apparatus itself, but also with people who make and handle it, including
cheaters.

It is not always obvious what the symmetry of the information is in
a situation in which one wishes to invoke the principle of insufficient
reason. For example, d'Alembert, an otherwise great eighteenth-cen-
tury mathematician, is supposed to have argued seriously that the probability of obtaining at least one head in two tosses of a fair coin is 2/3
rather than 3/4. (Cf. [T3], Art. 464.) Heads, as he said, might appear
on the first toss, or, failing that, it might appear on the second, or,
finally, might not appear on either. D'Alembert considered the three
possibilities equally likely.

It seems reasonable to suppose that, if the principle of insufficient
reason were formulated and applied with sufficient care, the conclusion
of d'Alembert would appear simply as a mistake. There are, however,
more serious examples. Suppose, to take a famous one, that it is known
of an urn only that it contains either two white balls, two black balls,
or a white ball and a black ball. The principle of insufficient reason has
been invoked to conclude that the three possibilities are equally probable, so that in particular the probability of one white and one black
ball is concluded to be 1/3. But the principle has also been applied to
conclude that there are four equally probable possibilities, namely, that
the first ball is white and the second also, that the first is white and the
second black, etc. On that basis, the probability of one white and one
black ball is, of course, 1/2. Personally, I do not try to arbitrate between the two conclusions but consider that the existence of the pair
of them reflects doubt on the notion that a person's knowledge relevant
to any matter admits any full and precise description in terms of
propositions he knows to be true and others about which he knows
nothing.

Most holders of personalistic views do not find the principle of insufficient reason compelling, because they envisage the possibility that
a person may consider one event more probable than another without
having any compelling argument for his attitude. Viewed practically,
this position is closely associated with the first criticism of the principle
of insufficient reason, for the holder of a personalistic view typically
supposes that the person is under the influence of experience, and possibly even biologically determined inheritance, that expresses itself in
his opinions, though not necessarily through compelling argument.

Holders of personalistic views do see some truth in the principle of insufficient reason, because they recognize that there are frequently partitions of the world, associated with symmetrical-looking gambling apparatus and the like, that many and diverse people all consider (very nearly) uniform partitions. As was illustrated in the preceding section, we often feel more "sure" about probabilities derived from the judgment that such partitions are uniform than we do about others. Such partitions are, moreover, very important in that they provide some events the probability of which to diverse people is in agreement. Though the events concerned are often of no importance in themselves, agreement about them can, through the statistical invention of randomization, contribute to agreement about all sorts of issues open to empirical investigation. Widespread though the agreement about the near uniformity of some partitions is, holders of personalistic views typically do not find the contexts in which such agreement obtains sufficiently definable to admit of expression in a postulate.

Holders of purely objectivistic views see no sense at all in the original formulation of the principle of insufficient reason, for it uses "probability" in a manner they consider meaningless. But they too see an element of truth in the principle, which they consider to be established as a part of empirical physics. Thus, for example, they regard it as an experimental fact, admitting some explanation in terms of theoretical physics, that three dice manufactured with reasonable symmetry will exhibit each of the 216 possible patterns with nearly equal frequency, if repeatedly rolled with sufficient violence on a suitable surface.

Holders of personalistic views agree that experiments or, more generally, experiences determine to a large extent when people employ the idea of insufficient reason. Thus, though experiments with gambling apparatus, quite apart from gambling itself, have a fascination that perhaps exceeds their real interest, such experiments are not altogether worthless. On the one hand, they provide strong evidence that a person cannot expect to maintain a symmetrical attitude toward any piece of apparatus with which he has had long experience, unless he is virtually convinced at the outset that the possible states of the apparatus are equally probable and independent from trial to trial. To say it in the more familiar and sometimes more congenial language of objective probability, long experiments with coins, dice, cards, and the like have always shown some bias, and often some dependence from trial to trial. On the other hand (and this has the utmost practical importance), it has been shown that, with skill and experience, gambling apparatus, or its statistical equivalent, can be manufactured in which the bias and the dependence from trial to trial are extremely small. This implies that groups of very diverse people can be brought to agree that repeated trials with certain apparatus are nearly uniform and nearly independent. Thus certain methods of obtaining random numbers and other outcomes of uniform and independent trials, which are vital to many sorts of experimentation, have justifiably found acceptance with the scientific public. A stimulating account of practical methods of obtaining random numbers, and random samples generally, is given by Kendall in Chapter 8 (Vol. I) of [K2].

6 How can science use a personalistic view of probability?

It is often argued by holders of necessary and objectivistic views alike that that ill-defined activity known as science or scientific method consists largely, if not exclusively, in finding out what is probably true, by criteria on which all reasonable men agree. The theory of probability relevant to science, they therefore argue, ought to be a codification of universally acceptable criteria. Holders of necessary views say that, just as there is no room for dispute as to whether one proposition is logically implied by others, there can be no dispute as to the extent to which one proposition is partially implied by others that are thought of as evidence bearing on it, for the exponents of necessary views regard probability as a generalization of implication. Holders of objectivistic views say that, after appropriate observations, two reasonable people can no more disagree about the probability with which trials in a sequence of coin tosses are heads than they can disagree about the length of a stick after measuring it by suitable methods, for they consider probability an objective property of certain physical systems in the same sense that length is generally considered an objective property of other physical systems, small errors of measurement being contemplated in both contexts. Neither the necessary nor the objectivistic outlook leaves any room for personal differences; both, therefore, look on any personalistic view of probability as, at best, an attempt to predict some of the behavior of abnormal, or at any rate unscientific, people.

I would reply that the personalistic view incorporates all the universally acceptable criteria for reasonableness in judgment known to me and that, when any criteria that may have been overlooked are brought forward, they will be welcomed into the personalistic view. The criteria incorporated in the personalistic view do not guarantee agreement on all questions among all honest and freely communicating people, even in principle. That incompleteness, if one will call it such, does not distress me, for I think that at least some of the disagreement we see around us is due neither to dishonesty, to errors in reasoning, nor to friction in communication, though the harmful effects of the latter are almost incapable of exaggeration.

As was mentioned in connection with symmetry, there are partitions that diverse people all consider nearly uniform, though not compelled to that agreement by any postulate of the theory of personal probability. As has also been mentioned and as will be explained later (especially in § 14.8), through the statistical invention of randomization, agreement about partitions pertaining to gambling apparatus of no importance in itself can be made to contribute to agreement in every part of empirical science.

Another mechanism that brings people having some, but not all, opinions in common into more complete agreement was illustrated in §§ 3.6–7. Indeed, it was there shown that in certain contexts any two opinions, provided that neither is extreme in a technical sense, are almost sure to be brought very close to one another by a sufficiently large body of evidence.

It has been countered,[+] I believe, that, if experience systematically leads people with opinions originally different to hold a common opinion, then that common opinion, and it only, is the proper subject of scientific probability theory. There are two inaccuracies in this argument. In the first place, the conclusion of the personalistic view is not that evidence brings holders of different opinions to the same opinions, but rather to similar opinions. In the second place, it is typically true of any observational program, however extensive but prescribed in advance, that there exist pairs of opinions, neither of which can be called extreme in any precisely defined sense, but which cannot be expected, either by their holders or any other person, to be brought into close agreement after the observational program.

I have, at least once, heard it objected against the personalistic view of probability that, according to that view, two people might be of different opinions, according as one is pessimistic and the other optimistic. I am not sure what position I would take in abstract discussion of whether that alleged property of personalistic views would be objectionable, but I think it is clear from the formal definition of qualitative probability that the particular personalistic view sponsored here does not leave room for optimism and pessimism, however these traits be interpreted, to play any role in the person's judgment of probabilities.

+ See (Fisher 1934), p. 287.

CHAPTER 5

Utility

1 Introduction

The postulates P4–6, introduced in Chapter 3, have already led to simplification of the relation $\leq$ in so far as it applies to acts of a special but important form. Indeed, through the introduction of numerical probability, those special comparisons have been reduced to ordinary arithmetic comparison of numbers in such a way that many relations among acts are deducible by simple and systematic arithmetic calculation. In this chapter it will be shown that the arithmetization of comparison among acts can, with the introduction of one mild new postulate, be extended to virtually all pairs of acts.

This far-reaching arithmetization of comparison among acts is achieved by attaching a number $U(f)$ to each consequence f in such a way that $\mathbf{f} \leq \mathbf{g}$ if and only if the expected value of $U(\mathbf{f})$ is numerically less than or equal to that of $U(\mathbf{g})$, provided only that the real-valued functions $U(\mathbf{f})$ and $U(\mathbf{g})$ are essentially bounded. The provision can fail to be met only if there exist acts that are, so to speak, distinctly preferable to any fixed reward or distinctly worse than any fixed punishment.

A function $\mathbf{U}$ that thus arithmetizes the relation of preference among acts will be called a utility. It will be shown that the multiplicity of utilities is not complicated, every utility being simply related to every other. I have chosen to use the name "utility" in preference to any other, in spite of some unfortunate connotations this name has in connection with economic theory, because it was adopted by von Neumann and Morgenstern when in [V4] they revived the concept to which it refers, in a most stimulating way. Their treatment has been of such widespread interest that the introduction of a name other than "utility" at the present time would cause more confusion than it could alleviate.

The next three sections are concerned with the technical exploration of the utility concept. I think readers interested in the details will find it best to read these sections twice as a unit, in the fashion I have been recommending for other material in which definitions and propositions

69

are interlarded with proofs; others will be content with a cursory reading, omitting proofs.

Taking advantage of the simplicity afforded by the introduction of utility, I try in § 5 to make some progress with the problem, pointed out in § 2.5, of specifying criteria for the construction of "small worlds."

Finally, § 6 briefly reports the history of the utility idea. A separate critical section is not necessary, because the criticisms of the theory of utility known to me are incorporated conveniently into the historical section.

2 Gambles

Before discussing utility, it is expedient to establish certain facts, the first being that at least among a rather rich class of acts, namely acts confined with probability one to a finite number of consequences, preference depends only on the probability distribution of the consequences of the acts.

THEOREM 1

HYP.   1. $f_1, \cdots, f_n$ are n elements of F, $n \geq 1$.
    2. $\rho_1, \cdots, \rho_n$ are numbers such that $\Sigma \rho_i = 1$.
    3. $\mathbf{g}$ and $\mathbf{h}$ are acts such that

$$P(g(s) = f_i) = P(h(s) = f_i) = \rho_i, \qquad i = 1, \cdots, n.$$

CONCL. $\mathbf{g} \doteq \mathbf{h}$.

PROOF. The theorem is obvious for $n = 1$. It will be proved by induction, supposing henceforth that $n > 1$.

Let B denote the intersection of the two events that $g(s) = f_n$ and $h(s) \neq f_n$, and let C denote the intersection of the two events that $h(s) = f_n$ and $g(s) \neq f_n$. It is easy to see that $P(B) = P(C)$. C can be partitioned into $C_0, C_1, \cdots, C_{n-1}$, where C_0 is a null event and C_i, $i = 1, \cdots, n - 1$, is the intersection of C with the event that $g(s) = f_i$. By repeated application of Conclusion 7 of Theorem 3.3.3, B can be partitioned into events $B_0, B_1, \cdots, B_{n-1}$ such that $P(B_i) = P(C_i)$, $i = 0, \cdots, n - 1$.

Let $\mathbf{g}_0 = \mathbf{g}$, and define $\mathbf{g}_{i+1}$ step by step for $i = 0, \cdots, n - 2$ thus:

$$(1) \qquad g_{i+1}(s) = f_n \qquad \text{for } s \, \varepsilon \, C_{i+1},$$
$$= f_{i+1} \qquad \text{for } s \, \varepsilon \, B_{i+1},$$
$$= g_i(s) \qquad \text{elsewhere.}$$

It is easily seen from the facts of conditional probability that $\mathbf{g}_{i+1} \doteq \mathbf{g}_i$ given $B_{i+1} \cup C_{i+1}$, and it is even more obvious that $\mathbf{g}_{i+1} \doteq \mathbf{g}_i$ given $\sim(B_{i+1} \cup C_{i+1})$. Therefore $\mathbf{g}_{i+1} \doteq \mathbf{g}_i$, so $\mathbf{g}_{n-1} \doteq \mathbf{g}$. Furthermore,

$P(g_{i+1}(s) = f_j) = P(g_i(s) = f_j) = \rho_j$, so $P(g_{n-1}(s) = f_j) = \rho_j$, $j = 1$, $\cdots$, n. Thus g_{n-1} is not only equivalent to g but also satisfies the hypothesis of the theorem relative to h, so it will suffice to prove the theorem for g_{n-1} and h in place of g and h.

Now g_{n-1} has been constructed to equal f_n in C, except on a null set. Therefore $g_{n-1} \doteq h$ given $C \cup D$, where D is the subset of $\sim C$ on which $g_{n-1} = h = f_n$.

It remains only to show that $g_{n-1} \doteq h$ given $\sim (C \cup D)$. If $\sim (C \cup D)$ is null, that is true automatically; henceforth concentrate on the less trivial situation. If $\sim (C \cup D)$ is not null, then $\leq$ given $\sim (C \cup D)$ satisfies all the postulates assumed thus far, and therefore the consequences $f_1, \cdots, f_{n-1}$; the numbers $\rho_i' = \rho_i/(1 - \rho_n)$, $i = 1, \cdots, n - 1$; the acts g_{n-1} and h; and the relation $\leq$ given $\sim (C \cup D)$ satisfy the hypothesis of the theorem for a case in which it is supposed already to have been proved. ◆

In this chapter the notation $\Sigma \rho_i f_i$ will denote the class of all acts f for which there exist partitions B_i of S such that $P(B_i) = \rho_i$ and $f(s) = f_i$ for $s \in B_i$. Here the f_i's are a finite sequence of consequences (not necessarily distinct), and the ρ_i's a corresponding sequence of non-negative real numbers such that $\Sigma \rho_i = 1$. In view of Conclusion 7 of Theorem 3.3.3, such a class of acts, which will in this chapter be referred to as a *gamble* and denoted by f, g, h, or the like, always has at least one element. Theorem 1 says, in effect, that the person regards all elements of any gamble as equivalent. To put it differently, if the events B_i of a partition have the probabilities ρ_i, and if the act f is such that the consequence f_i will befall the person in case B_i occurs, then the value of f is independent of how the partition B_i is chosen.

Gambles can be mixed, in a sense, to make new gambles, thus: Let f_j be a finite sequence of gambles,

$$(2) \qquad\qquad f_j = \sum_i \rho_{ij} f_{ij},$$

and σ_j a corresponding sequence of non-negative real numbers such that $\Sigma \sigma_j = 1$. The *mixture* of the f_j's with weights σ_j, denoted $\Sigma \sigma_j f_j$, is defined by

$$(3) \qquad\qquad \Sigma \sigma_j f_j = \sum_j \sigma_j \left\{ \sum_i \rho_{ij} f_{ij} \right\}$$

$$= \sum_{i,j} (\sigma_j \rho_{ij}) f_{ij},$$

which is meaningful, the f_{ij}'s being consequences and the $(\sigma_j \rho_{ij})$'s being numbers such that $\Sigma (\sigma_j \rho_{ij}) = 1$. Such mixtures are exemplified by an insurance policy in which the benefit is an annuity payable during the life of the beneficiary, and by a lottery in which the prizes are tickets in other lotteries.

In view of Theorem 1, it is natural to say that $f \leq g$ means that, for every act $\mathbf{f}$ in the class of acts corresponding to f, $\mathbf{f} \leq g$. Corresponding definitions are to be understood for $f \leq \mathbf{g}$, $\mathbf{f} \leq g$, $f \leq g$, etc.

THEOREM 2 If f, g, and h are gambles, and $0 < \rho \leq 1$; then $\rho f + (1 - \rho)h \leq \rho g + (1 - \rho)h$, if and only if $f \leq g$.

PROOF. Let $\mathbf{f}$, $\mathbf{g}$; f_i, g_j; and B_i, C_j be acts, consequences, and partitions such that $\mathbf{f}$ and $\mathbf{g}$ are among the acts represented by f and g, respectively, with $f(s) = f_i$ for $s \varepsilon B_i$ and $g(s) = g_j$ for $s \varepsilon C_j$.

Construct $D_{ij} \subset B_i \cap C_j$ such that $P(D_{ij}) = \rho P(B_i \cap C_j)$, and let $D = \bigcup D_{ij}$. Then $P(D) = \rho$, $P(B_i \mid D) = P(B_i)$, and $P(C_j \mid D) = P(C_j)$.

What is to be proved is, in effect, that $\mathbf{f} \leq \mathbf{g}$ given D, if and only if $\mathbf{f} \leq \mathbf{g}$. In view of Theorem 1 it is clear that whether that is so or not for $\mathbf{f}$ and $\mathbf{g}$ does not depend on the particular choice of D; so, with an obvious temporary extension of terminology, it is to be proved that $\mathbf{f} \leq \mathbf{g}$ given ρ, if and only if $\mathbf{f} \leq \mathbf{g}$.

If $\mathbf{f} \doteq \mathbf{g}$ given α for every $0 < \alpha \leq 1$, there is nothing to prove. Otherwise it can be assumed without loss of generality that, for some α_0, $\mathbf{f} < \mathbf{g}$ given α_0.

In view of Theorem 2.7.2, if $\alpha + \beta \leq 1$, $\mathbf{f} \geq \mathbf{g}$ given α, and $\mathbf{f} \geq \mathbf{g}$ given β; then $\mathbf{f} \geq \mathbf{g}$ given $(\alpha + \beta)$, and similarly $\mathbf{f} \geq \mathbf{g}$ given $\alpha/2$.

Making use of P6 and Theorem 2.7.2, it can easily be shown that, for any α sufficiently close to α_0, $\mathbf{f} < \mathbf{g}$ given α.

The preceding three paragraphs imply that, in the case at hand, $\mathbf{f} < \mathbf{g}$ given α for every α, $0 < \alpha \leq 1$. ◆

THEOREM 3 If $f < g$, and $0 \leq \sigma < \rho \leq 1$, then $\rho f + (1 - \rho)g < \sigma f + (1 - \sigma)g$.

PROOF. In view of the immediately verifiable identities,

$$\rho f + (1 - \rho)g = (\rho - \sigma)f + [1 - (\rho - \sigma)] \times$$

$$\left\{ \frac{\sigma}{1 - (\rho - \sigma)} f + \frac{(1 - \rho)}{1 - (\rho - \sigma)} g \right\},$$

(4)

$$\sigma f + (1 - \sigma)g = (\rho - \sigma)g + [1 - (\rho - \sigma)] \times$$

$$\left\{ \frac{\sigma}{1 - (\rho - \sigma)} f + \frac{(1 - \rho)}{1 - (\rho - \sigma)} g \right\},$$

this theorem is a special case of Theorem 2; unless $\rho = 1$, and $\sigma = 0$, in which case it is trivial. ◆

THEOREM 4   If $\mathbf{f}_1 < \mathbf{f}_2$ and $\mathbf{f}_1 \leq \mathbf{g} \leq \mathbf{f}_2$, then there is one and only one ρ such that $\rho\mathbf{f}_1 + (1 - \rho)\mathbf{f}_2 \doteq \mathbf{g}$.

PROOF.   It follows immediately from Theorem 3 and the principle of the Dedekind cut † that there is one and only one ρ_0 such that

$$
\text{(5)} \quad
\begin{aligned}
\sigma\mathbf{f}_1 + (1 - \sigma)\mathbf{f}_2 < \mathbf{g}, &\quad \text{if } \sigma > \rho_0 \\
\sigma\mathbf{f}_1 + (1 - \sigma)\mathbf{f}_2 > \mathbf{g}, &\quad \text{if } \sigma < \rho_0.
\end{aligned}
$$

According to (5), no number, except possibly ρ_0, can satisfy the equivalence demanded by the theorem.

Finally, using (5) and P6 (much as it was used in the proof of Theorem 2), it follows that ρ_0 does indeed satisfy the equivalence. ◆

3 Utility, and preference among gambles

The idea of utility can most conveniently be introduced in connection with gambles or, equivalently, acts that with probability one are confined to a finite number of consequences, thus: A **utility** is a function **U** associating real numbers with consequences in such a way that, if $\mathbf{f} = \Sigma\rho_i f_i$ and $\mathbf{g} = \Sigma\sigma_j g_j$; then $\mathbf{f} \leq \mathbf{g}$, if and only if $\Sigma\rho_i U(f_i) \leq \Sigma\sigma_j U(g_j)$. Writing $U[\mathbf{f}]$ for $\Sigma\rho_i U(f_i)$, the condition takes the form $U[\mathbf{f}] \leq U[\mathbf{g}]$. Similarly, it is convenient to understand that, for an act $\mathbf{f}$,

$$
\text{(1)} \qquad U[\mathbf{f}] = E(U(\mathbf{f})).
$$

In this notation the following obvious theorem gives a slightly different characterization of utility.

THEOREM 1   A real-valued function of consequences, **U**, is a utility; if and only if $\mathbf{f} \leq \mathbf{g}$ is equivalent to $U[\mathbf{f}] \leq U[\mathbf{g}]$, provided $\mathbf{f}$ and $\mathbf{g}$ are both with probability one confined to a finite set of consequences.

Do the postulates thus far assumed guarantee that any utilities exist at all? Can Theorem 1 be extended to an even wider class of acts? Does a great diversity of utilities exist, or does the relation $\leq$ practically determine the function **U**? These questions, here mentioned in the order in which they most naturally arise, are manifestly of great importance in understanding utility. For technical reasons, they will

† Cf., if necessary, any introduction to the theory of the real numbers for explanation of this principle, e.g., Chapter II of [G3].

be answered in a different order—the third followed by the first in this section, and the second in the next section.

If there is a utility at all, there is surely more than one, because a utility plus a constant and a utility times a positive constant are also obviously utilities; thus:

THEOREM 2   If U is a utility, and ρ, σ are real numbers with $\rho > 0$; then $U' = \rho U + \sigma$ is also a utility.

COROLLARY 1   If there exists a utility, and if $f < g$; then there exists a utility U for which $U(f)$ and $U(g)$ are any preassigned pair of numbers, provided $U(f) < U(g)$.

Theorem 2 says that any increasing linear function of a utility is a utility. The next theorem says that, conversely, any two utilities are necessarily increasing linear functions of one another.

THEOREM 3   If U and U' are utilities, there exist numbers ρ and σ such that $U' = \rho U + \sigma$, $\rho > 0$.

PROOF.   The first step of the proof will be to demonstrate the following identity for the two utilities U and U' and for any three consequences f, g, h.

(2)
$$\begin{vmatrix} 1 & 1 & 1 \\ U(f) & U(g) & U(h) \\ U'(f) & U'(g) & U'(h) \end{vmatrix} = 0.$$

If any two of the consequences f, g, h are equivalent, two columns of the determinant in question are equal, and therefore the determinant vanishes. It can be assumed, then, that no two of f, g, and h are equivalent; and there is no loss in generality, as may be seen by permuting columns, in assuming $f < g < h$. Theorem 2.4 now permits the conclusion that there is a ρ such that $\rho f + (1 - \rho)h \doteq g$. Therefore,

$$1 = \rho 1 \quad + (1 - \rho)1$$

(3)
$$U(g) = \rho U(f) + (1 - \rho)U(h)$$

$$U'(g) = \rho U'(f) + (1 - \rho)U'(h).$$

Thus the middle column of the determinant is linearly dependent on the other two, so the determinant vanishes, as was asserted.

Now let g and h be any fixed pair of consequences such that $g < h$, the existence of such a pair being assured by P5. Equation (2) can be successively rewritten, where f is an arbitrary consequence, thus:

(4) $1[U(g)U'(h) - U(h)U'(g)] - U(f)[U'(h) - U'(g)]$

$$+ U'(f)[U(h) - U(g)] = 0,$$

(5) $U'(f) = \dfrac{U'(h) - U'(g)}{U(h) - U(g)} U(f) - \dfrac{U(g)U'(h) - U(h)U'(g)}{U(h) - U(g)},$

which proves the theorem; for $U'(h) - U'(g)$ and $U(h) - U(g)$ are both positive. ◆

COROLLARY 2   If **U** and **U'** are utilities such that, for some $g < h$, $U(g) = U'(g)$ and $U(h) = U'(h)$; then **U** and **U'** are the same, that is, for every f, $U(f) = U'(f)$.

To summarize, if there is a utility at all, there are an infinite number, but the array of utilities is not complicated; for all can be generated from any one by increasing linear transformations.

Turn now to the question of existence.

THEOREM 4   There exists a utility.

PROOF. Von Neumann and Morgenstern prove essentially this theorem, as well as the preceding one, in the appendix of [V4]. The following proof is theirs, expressed, as the teacher used to say, in my own words.

For this proof only, certain special nomenclature is introduced. A set of gambles **F** is *convex*; if and only if, for every **f**, **g** ε **F** and ρ, $0 \le \rho \le 1$, ρ**f** $+ (1 - \rho)$**g** ε **F**. An *interval* **I** of gambles is the set of all gambles **f** such that, for some fixed g and h (which determine the interval), $g \le$ **f** $\le h$. A *hyper-utility* **V** on a convex set **F** is a real-valued function of the gambles of **F**, such that **f** $\le$ **g**, if and only if $V($**f**$) \le V($**g**$)$, and such that $V(\rho$**f** $+ (1 - \rho)$**g**$) = \rho V($**f**$) + (1 - \rho)V($**g**$)$.

The following remarks about this special nomenclature are obvious and will be repeatedly used in the proof, without explicit reference. The set of all gambles is convex. The intersection of two convex sets is convex. Every interval is convex. There is an interval containing any finite set of gambles. If there is a hyper-utility on the set of all gambles, it is a utility when confined to consequences.

By the same method that led to the proofs of Theorems 2 and 3, if there is a hyper-utility on **F** containing **g** and **h**, with **g** < **h**, then there is one and only one hyper-utility **V** on **F** such that $V($**g**$) = 0$ and $V($**h**$) = 1$.

If I is the interval determined by $g < h$, then, according to Theorem 2.4, there is for every f in I a unique number, call it $V(f)$, such that

(6) $$f \doteq (1 - V(f))g + V(f)h.$$

By repeated use of Theorem 2.2, it follows for any f, f' ε I that

$$
\begin{aligned}
(7) \quad \rho f + (1 - \rho)f' &\doteq \rho\{(1 - V(f))g + V(f)h\} \\
&\quad + (1 - \rho)\{(1 - V(f'))g + V(f')h\} \\
&= \{1 - [\rho V(f) + (1 - \rho)V(f')]\}g \\
&\quad + [\rho V(f) + (1 - \rho)V(f')]h,
\end{aligned}
$$

so V is a hyper-utility on the convex set I.

From here on in this proof, let g, h be a fixed pair of consequences with $g < h$. Making use of the preceding two paragraphs, there is a unique hyper-utility assigning the values 0 and 1 to g and h, respectively, on any one interval containing g and h. The intersection of two such intervals is a convex set containing g and h, and on the intersection the hyper-utilities associated with the two intervals are both hyper-utilities attaching 0 and 1 to g and h, respectively; they must, therefore, be equal to one another on the intersection.

Any gamble f is an element of some interval containing g and h. Let $V(f)$ be the common value assigned to f by all the hyper-utilities that are defined on intervals containing f, g, and h and that assign the values 0 and 1 to g and h, respectively. Since there is always at least one such interval for any gamble f, the function V is defined for all gambles.

The proof will be complete when it is shown that V is a hyper-utility for the convex set of all gambles. Let f and f' be any two gambles and ρ a number, $0 \le \rho \le 1$. There is an interval containing f, f', g, h, and $\rho f + (1 - \rho)f'$. In that interval the function V is a hyper-utility. Therefore $V(\rho f + (1 - \rho)f') = \rho V(f) + (1 - \rho)V(f')$ and $V(f) \le V(f')$, if and only if $f \le f'$. ◆

4 The extension of utility to more general acts

The requirement that an act have only a finite number of consequences may seem, from a practical point of view, almost no requirement at all. To illustrate, the number of time intervals that might possibly be the duration of a human life can be regarded as finite, if you agree that the duration may as well be rounded to the nearest minute, or second, or microsecond, and that there is almost no possibility of its exceeding a thousand years. More generally, it is plausible that, no matter what set of consequences is envisaged, each consequence can be practically identified with some element of a suitably chosen finite, though possibly enormous, subset. It might therefore seem of little or no importance to extend the concept of utility to acts having an infinite number of consequences. If that argument were valid, it could easily be extended to reach the conclusion that infinite sets are irrelevant to all practical affairs, and therefore to all parts of applied mathematics. But it is one of the most profound lessons of mathematical experience that infinite sets, tactfully handled, can lead to great simplification of situations that could, in principle, but only with enormous difficulty, be treated in terms of finite sets. How difficult it would be to study geometry if one made at the outset the "simplifying assumption" that to all intents and purposes at most $10^{1,000}$ points in space can be discriminated from one another! Again, it is generally more convenient and fruitful to think of the annual cash income of an individual or firm as a continuous variable with an infinite number of possible values than as a discrete variable confined to some large finite number of values, even if it is known that the income must be some integral number of cents less, say, than 10^{10}.

One way to extend the concept of utility to acts with an infinite number of consequences would be to postulate: If $U[\mathbf{f}]$ and $U[\mathbf{g}]$ both exist (the values $+\infty$ and $-\infty$ being regarded as possible); $\mathbf{f} \leq \mathbf{g}$, if and only if $U[\mathbf{f}] \leq U[\mathbf{g}]$. I see no serious objection to making this assumption outright, though it might be complained that the assumption is motivated more by general mathematical intuition and experience than by intuitive standards of consistency among decisions, which I have tried to take as my sole guide thus far. A statement almost as strong as the one in question can, however, be derived on adjoining a new postulate, P7, more in the spirit of P1–6. That rather technical program will be carried out in the next several paragraphs. Those not interested can safely skip to the paragraph following Corollary 1 on page 80.

Suppose that every possible consequence of the act $\mathbf{g}$ is at least as attractive to the person as the act $\mathbf{f}$ considered as a whole; then it seems to me within the spirit of the sure-thing principle to conclude that $\mathbf{f} \leq \mathbf{g}$; the same might as fairly have been said for the relations $\geq$, and also for the two relations $\leq$ given B and $\geq$ given B. This idea is formalized in the following postulate, which, according to the conventions of mathematical double-talk, is to be interpreted as two propositions—one having $\leq$ and the other $\geq$ throughout.

P7   If $\mathbf{f} \leq (\geq) g(s)$ given B for every $s \, \varepsilon \, B$, then $\mathbf{f} \leq (\geq) \mathbf{g}$ given B.

Attention has been called to the mathematically useful fact that, if P1–6 apply to a relation $\leq$, then they also apply to any relation $\leq$ given B, provided B is not null. It is obvious that the same is true for P1–7, a fact that will be used often. It is also noteworthy that P1–7 obviously imply the propositions that arise if in them every instance of the sign $\leq$ is replaced by $\geq$ and every instance of $\geq$ is replaced by $\leq$. Therefore in any deduction from P1–7 every instance of the signs $\leq$ and $\geq$ can be reversed to produce a deduction that may be called the *symmetric dual* of the original deduction. This remark, a legitimate child of the principle of insufficient reason, has not been important heretofore, because almost all deductions thus far made have been their own symmetric duals. Since that will not be so of some of the lemmas in the present section, much needless writing and thinking can be saved by agreeing at the outset that, once a result is proved, it and its symmetric dual may be used as if both had been explicitly proved.

Before going to work with P7, some may wish to see an example of a mathematical structure satisfying P1–6 but not satisfying P7. Moreover, understanding of such an example will do much to clarify the uses to be made of P7. To construct the example, begin by letting S be a set carrying a finitely additive probability measure P under which S can be partitioned into subsets of arbitrarily small probability. Let the set of consequences be the half-open interval of numbers $0 \leq f < 1$. Let $U(f) = f$, $U[\mathbf{f}] = E(\mathbf{f})$, and

$$(1) \qquad\qquad V[\mathbf{f}] = \lim_{\epsilon \to 0} P\{f(s) \geq 1 - \epsilon\}.$$

Since the probability in (1) decreases with ϵ, there is no question about the existence of the limit. Now let $W[\mathbf{f}] = U[\mathbf{f}] + V[\mathbf{f}]$, and define $\mathbf{f} \leq \mathbf{g}$ to mean that $W[\mathbf{f}] \leq W[\mathbf{g}]$. Checking postulates P1–6, it will be found that the $\leq$ thus defined satisfies them all, and that what has here been called $U(f)$ is indeed a utility for $\leq$. But if, for example, there is an $\mathbf{f}$ such that $U[\mathbf{f}] = V[\mathbf{f}] = \frac{1}{2}$, P7 is violated, as can be seen by comparing $\mathbf{f}$ to the act that, for each s, takes as value the maximum of $\frac{3}{4}$ and $f(s)$. Whether there can be such an $\mathbf{f}$, may, so far as I know, depend on the choice of S and P. But, if the positive integers are taken as S, and P is so chosen that though the probability of any one integer is 0 the probability of the set of even integers is $1/2$, a possibility assured by the note to Section 3 of Chapter II on p. 231 of [B4], the function equal to 0 at the odd integers and equal to $(1 - 1/n)$ at each even n is such an $\mathbf{f}$. Finite, as opposed to countable, additivity seems to be essential to this example; perhaps, if the theory were worked out in a countably additive spirit from the start, little or no counterpart of P7 would be necessary.[+]

+ Fishburn (1970, Exercise 21, p. 213) has suggested an appropriate weakening of P7.

Several lemmas depending on P7 are now to be proved preparatory to proving that $U[\mathbf{f}]$ governs preference for a very large class of acts. It is to be understood throughout the section that $\mathbf{U}$ is any fixed utility. The truth of each lemma is intuitively clear, in the sense that each could justifiably be accepted as a postulate if need be. Since they are also easy to prove and of secondary interest, condensed proofs will suffice.

LEMMA 1   If, for every consequence h, $\mathbf{f} \leq h$, and $\mathbf{g} \leq h$; then $\mathbf{f} \doteq \mathbf{g}$.

PROOF.   Consider in the light of P7 that $\mathbf{f} \leq g(s)$ and $\mathbf{g} \leq f(s)$ for every s. ◆

LEMMA 2   If there exists a consequence f_0 such that $\mathbf{f} \leq f_0$, and if $U(f(s)) \leq U_0$ for every s, then there exists a gamble $\mathbf{g}$ such that $\mathbf{f} \leq \mathbf{g}$ and $U[\mathbf{g}] \leq U_0$.

PROOF.   If $U(f_0) \leq U_0$, then $\mathbf{g}$ can be taken to consist of f_0 alone. Otherwise, let f_1 be any consequence such that $U(f_1) \leq U_0$ and let $\mathbf{g}$ be the unique mixture of f_0 and f_1 such that $U(\mathbf{g}) = U_0$. ◆

LEMMA 3

HYP.   1. The B_i's, $i = 1, \cdots, n$, are a partition, and the U_i's are corresponding numbers.
2. $\mathbf{f}$ is an act such that $U(f(s)) \leq U_i$ for $s \in B_i$.
3. $\mathbf{f}$ is a gamble such that $\mathbf{f} \leq \mathbf{f}$.

CONCL. $U[\mathbf{f}] \leq \Sigma U_i P(B_i)$.

PROOF.   If the lemma were false, it would be false even for some $\mathbf{f} < \mathbf{f}$. Then it may be assumed, modifying $\mathbf{f}$ if need be by means of P6 and Lemma 1, that there exists for each i an f_i such that $\mathbf{f} \leq f_i$ given B_i. Now, in view of Lemma 2, there exists for each i a $\mathbf{g}_i$ such that $\mathbf{f} \leq \mathbf{g}_i$ given B_i and $U[\mathbf{g}_i] \leq U_i$. Let $\mathbf{g} = \Sigma P(B_i)\mathbf{g}_i$, and observe that $\mathbf{f} \leq \mathbf{f} \leq \mathbf{g}$. Therefore, $U[\mathbf{f}] \leq U[\mathbf{g}] = \Sigma P(B_i)U(\mathbf{g}_i) \leq \Sigma P(B_i)U_i$. ◆

An act will be called bounded if its utility is, according to ordinary mathematical usage, an essentially bounded random variable; the notion is put in a more formal and self-contained way as follows: A *bounded act* is an act $\mathbf{f}$ such that, for some two numbers U_0 and U_1, $P\{U_0 \leq U(f(s)) \leq U_1\} = 1$. The definition is clearly not dependent on the choice of $\mathbf{U}$.

THEOREM 1   If $\mathbf{f}$ and $\mathbf{g}$ are bounded, then $\mathbf{f} \leq \mathbf{g}$, if and only if $U[\mathbf{f}] \leq U[\mathbf{g}]$.

PROOF.   If there exist g and h such that $g \leq \mathbf{f} \leq h$, then there is, by Theorem 2.4, a mixture $\mathbf{f}$ of g and h such that $\mathbf{f} \doteq \mathbf{f}$. The null event on which $U(f(s))$ is not between U_0 and U_1 may as well be disregarded; the rest can be partitioned into $n + 1$ events B_i defined by the condition that $s \varepsilon B_i$ if and only if $V_{i-1} \leq U(f(s)) < V_i$, $i = 1, \cdots, n + 1$, where

$$(2) \qquad V_i = \left\{ \left(1 - \frac{i}{n} \right) U_0 + \frac{i}{n} U_1 \right\}, \qquad i = 0, \cdots, n + 1.$$

Applying Lemma 3 and its symmetric dual,

$$(3) \qquad \Sigma V_{i-1} P(B_i) \leq U[\mathbf{f}] \leq \Sigma V_i P(B_i).$$

Similarly, according to Exercise 3 of Appendix 1,

$$(4) \qquad \Sigma V_{i-1} P(B_i) \leq U[\mathit{f}] \leq \Sigma V_i P(B_i).$$

Therefore

$$(5) \qquad \big| U[\mathbf{f}] - U[\mathit{f}] \big| \leq \Sigma(V_i - V_{i-1}) P(B_i) = (U_1 - U_0)/n,$$

whence $U(\mathbf{f}) = U(\mathit{f})$.

To consider the remaining case, suppose that the bounded act $\mathbf{f}$ exceeds (is exceeded by) every consequence; call it for the moment *big* (*little*). According to Lemma 1, all big (and, dually, all little) acts are equivalent to one another. Furthermore, it is, for example, easily seen that, if an act is big, then for $\epsilon > 0$,

$$(6) \qquad P\big\{ U(f(s)) \geq \sup_f U(f) - \epsilon \big\} = 1.$$

(Some may be more familiar with the notation "LUB" and "GLB," read "least upper bound" and "greatest lower bound," than with the corresponding "sup" and "inf," read "supremum" and "infimum." If even these older terms are not familiar, see Exercise 4 of Appendix 2.) Therefore, if there are big (little) acts, they all have the same expected utility, namely $\sup U(f)$ ($\inf U(f)$).

Suppose now that $\mathbf{f} \leq \mathbf{g}$. It is possible that $\mathbf{f}$ and $\mathbf{g}$ are both little; that $\mathbf{f}$ is little, and $\mathbf{g}$ is equivalent to some gamble; that $\mathbf{f}$ is little and $\mathbf{g}$ big; that $\mathbf{f}$ and $\mathbf{g}$ are each equivalent to some gamble; that $\mathbf{f}$ is equivalent to some gamble, and $\mathbf{g}$ is big; or, finally, that they are both big. In each of these cases, a simple argument shows that $U[\mathbf{f}] \leq U[\mathbf{g}]$. The converse arguments are similar. ◆

COROLLARY 1   If $\mathbf{f}$ and $\mathbf{g}$ are bounded, and $P(B) > 0$, then $\mathbf{f} \leq \mathbf{g}$ given B, if and only if $E(U(\mathbf{f}) - U(\mathbf{g}) \mid B) \leq 0$.

It would be possible to explore unbounded acts for which expected utility exists to see whether expected utility governs preferences among even such acts under postulates P1–7 or under some extension of them.[+]

_____

[+] Peter Fishburn (1970, pp. 194, 206–207) and I have since discovered to my surprise that these postulates imply bounded utility, which puts the next several paragraphs in a new light.

I do not think, however, that the question is sufficiently interesting to warrant attention here, especially since there is some reason, first stated by Gabriel Cramer in a letter partially reproduced in [B10], to postulate that there are upper and lower bounds to utility, in which case all acts would necessarily be bounded.

Even without P7, the postulates imply, in the following sense, that no gamble has infinite or minus infinite utility.

An act f has *infinite (minus infinite) utility*; if and only if, for some $g < (>) h$ and for every $\epsilon > 0$, there is a B with $P(B) \leq \epsilon$ and such that the act equal to f on B and to g on $\sim B$ exceeds (is exceeded by) h. A gamble or a consequence would be said to have *infinite (minus infinite) utility*, if one of the acts corresponding to it had infinite (minus infinite) utility.

Indeed, Theorem 2.4, a deduction from P1–6, obviously implies that there are no infinite or minus infinite gambles or consequences. It may, however, be mentioned that Pascal held that, in just the sense at hand, salvation is an infinite consequence ([P2], pp. 189–191). Again, it is often said, in effect, that the utility to a person of immediate death is a consequence of minus infinite utility, but casual observation shows that this is not true of anyone—at least not of anyone who would cross the street to greet a friend. In the same vein, medicine often gives lip service to the idea that the death of a patient is of minus infinite utility, and, of course, doctors do go to great lengths to keep their patients alive; but a doctor who took the idea too seriously would make a nuisance of himself and soon find himself with no patients to treasure.

If the utility of consequences is unbounded, say from above,† then, even in the presence of P1–7, acts (though not gambles) of infinite utility can easily be constructed. My personal feeling is that, theological questions aside, there are no acts of infinite or minus infinite utility, and that one might reasonably so postulate, which would amount to assuming utility to be bounded.

Justifiable though it might be, that assumption would entail a certain mathematical awkwardness in many practical contexts. For example, as will be discussed at greater length in Chapter 15, it sometimes seems reasonable to suppose that the penalty for acting as though a particular unknown number were $\hat\mu$ instead of its true value, μ, is proportional to $\delta^2 = (\mu - \hat\mu)^2$. But, if the possible values of μ are unbounded, then so are the possible values of δ, so utility is here taken to be unbounded. On close scrutiny of such an example one always finds that

† That is, if, for every V, there is a consequence f such that $V \leq U(f)$. This manner of speaking is permissible; because in view of Theorem 3.3, if one utility is bounded, all are.

it is not really reasonable to assume the penalty even roughly proportional to δ^2 for large values of δ^2, but rather that large values are so improbable that the error made in misappraising the penalty associated with them is negligible compared to the saving in simplicity resulting from the misappraisal. If the assumption of bounded utility were made part of the theory of personal probability, then any example in which unbounded utility is used for mathematical simplicity would be in contradiction to the postulates. I propose, therefore, not to assume bounded utility formally, but to remember that problems involving unbounded utility are to be handled cautiously.

To take stock of the chapter thus far, utility having been established, it is now superfluous to consider that consequences may be of all sorts, since the postulates imply that in virtually every context a consequence is adequately characterized by its utility, some one utility function having been chosen from the linear family of possibilities. Therefore, unless the contrary is clearly indicated, f, g, and h will henceforth mean not exactly consequences in the sense used to date, but rather real numbers measuring utility in units to be called **utiles**. Correspondingly, an act $\mathbf{f}$ will henceforth be understood to be a real-valued random variable. The entire theory of preference, at least for bounded acts, can now be summarized by the following résumé:

R $\mathbf{f} \leq \mathbf{g}$ given B, if and only if $P(B) = 0$, or $E(\mathbf{f} - \mathbf{g} \mid B) \leq 0$.

From now on, though not formulated as a postulate, it is to be assumed without further quibbling that R holds, provided only that $E(\mathbf{f})$ and $E(\mathbf{g})$ exist and are finite; no attempt will be made to compare acts for which the expected value does not exist or is infinite.

If a person is free to decide among a set $\mathbf{F}$ of acts, he will presumably choose one the expectation of which is $v(\mathbf{F})$, where

(7) $$v(\mathbf{F}) = \sup_{\mathbf{f} \,\varepsilon\, \mathbf{F}} E(\mathbf{f}),$$

provided that such a one exists. This provision must be mentioned, even though a set $\mathbf{F}$ for which $v(\mathbf{F}) = \infty$ will, by convention, not be considered to give rise to a valid decision problem; for, if $\mathbf{F}$ is infinite in number, there may be no act in $\mathbf{F}$ with expectation quite as great as $v(\mathbf{F})$. Nonetheless, $v(\mathbf{F})$ may, in a sense, be regarded as the value or utility of the set of acts $\mathbf{F}$, as is discussed in the penultimate paragraph of § 6.5.

5 Small worlds

Allusion was made in the penultimate paragraph of § 2.5 to the practical necessity of confining attention to, or isolating, relatively simple situations in almost all applications of the theory of decision developed in this book. As was mentioned there, I find it difficult to say with any completeness how such isolated situations are actually arrived at and justified. The purpose of the present section is to take some steps toward the solution of that problem or, at any rate, to set the problem forth as clearly as I can. This section, though important for a critical evaluation of the thesis of this book, is not essential to a casual reading.

Making an extreme idealization, which has in principle guided the whole argument of this book thus far, a person has only one decision to make in his whole life. He must, namely, decide how to live, and this he might in principle do once and for all. Though many, like myself, have found the concept of overall decision stimulating, it is certainly highly unrealistic and in many contexts unwieldy.† Any claim to realism made by this book—or indeed by almost any theory of personal decision of which I know—is predicated on the idea that some of the individual decision situations into which actual people tend to subdivide the single grand decision do recapitulate in microcosm the mechanism of the idealized grand decision. One application of the theory of utility to overall decisions has, however, been attempted by Milton Friedman in [F11].

The problem of this section is to say as clearly as possible what constitutes a satisfactory isolated decision situation. The general method of attack I propose to follow, for want of a better one, is to talk in terms of the grand situation—tongue in cheek—and in those terms to analyze and discuss isolated decision situations. I hope you will be able to agree, as the discussion proceeds, that I do not lean too heavily on the concept of the grand decision situation.

Consider a simple example. Jones is faced with the decision whether to buy a certain sedan for a thousand dollars, a certain convertible also for a thousand dollars, or to buy neither and continue carless. The simplest analysis, and the one generally assumed, is that Jones is deciding between three definite and sure enjoyments, that of the sedan, the convertible, or the thousand dollars. Chance and uncertainty are considered to have nothing to do with the situation. This simple analysis may well be appropriate in some contexts; however, it is not difficult to recognize that Jones must in fact take account of many uncertain future possibilities in actually making his choice. The relative

---

† Unrealistic though the concept is, it would be a mistake, arising out of elliptical presentation, to suppose that the concept predicates the choice of a complete lifelong policy by new-born babies. If a person ever reached such a level of maturity as to be able to make a lifelong choice for his life from that time on, he would then become a person to whom the concept could be literally applied.

fragility of the convertible will be compensated only if Jones's hope to arrange a long vacation in a warm and scenic part of the country actually materializes; Jones would not buy a car at all if he thought it likely that he would immediately be faced by a financial emergency arising out of the sickness of himself or of some member of his family; he would be glad to put the money into a car, or almost any durable goods, if he feared extensive inflation. This brings out the fact that what are often thought of as consequences (that is, sure experiences of the deciding person) in isolated decision situations typically are in reality highly uncertain. Indeed, in the final analysis, a consequence is an idealization that can perhaps never be well approximated. I therefore suggest that we must expect acts with actually uncertain consequences to play the role of sure consequences in typical isolated decision situations.

Suppose now, to elaborate the example, that Jones is presented with a choice between tickets in several different lotteries such that, whichever he chooses and whatever tickets are drawn, he will win either nothing, the sedan, the convertible, or a thousand dollars. None of these four consequences—not even "nothing"—is actually a sure consequence in the strict sense, as I think you will now understand. I propose to analyze Jones's present decision situation in terms of a "small world." The more colloquial Greek word, microcosm, will be reserved for a special kind of small world to be described later. To describe the state of the small world is to say which prize is associated with each of the tickets offered to Jones. The small-world acts actually available to Jones are acceptance of one or another of the tickets. The generic small-world act is an arbitrary function taking as its value one of the four small-world consequences according to which small-world state obtains.

It will be noticed that the small-world states are in fact events in the grand world, that indeed they constitute a partition of the grand world. If there are an infinite number of small-world states, as indeed there must be, if the small world is to satisfy the postulates P1–7, then the partitio₋ in question becomes an infinite partition.† These considerations lead to the following technical definitions.

Let the *grand world S* be, as always, a set with elements s, s', $\cdots$. The *grand-world consequences F* may as well be taken to be a bounded

---

† Technical note: It is mathematically more general and elegant not to insist that the small world have states at all, but rather to speak of a special class of events as small-world events. This class should be closed under complements and finite unions. In short, the small-world events, and thereby the small world itself, constitute a Boolean subalgebra of the Boolean algebra of the grand-world events.

set of real numbers. The *grand-world acts* are then real-valued functions **f**, **g**, **h**, $\cdots$. The preference ordering between acts is determined by the condition that **f** $\leq$ **g** if and only if

(1) $E(\mathbf{f} - \mathbf{g}) \leq 0,$

where the expected value indicated in (1) is derived from a probability measure P characteristic of the grand world or, to be more exact, of the person's attitude toward the grand world.

The construction of a *small world* $\bar{S}$ from the grand world S begins with the partition of S into subsets, or *small-world states* $\bar{s}$, $\bar{s}'$, $\cdots$ (not necessarily finite in number). Throughout this technical discussion, it will be necessary to bear in mind certain double interpretations such as that $\bar{s}$ is both an element of $\bar{S}$ and a subset of S. Strictly speaking, a *small-world event* $\bar{B}$ in $\bar{S}$ is a collection of subsets of S and not itself a subset of S. However, the union of all the elements of $\bar{B}$, regarded as subsets of S, is an event in S; call it $[\bar{B}]$.

The small world, as I mean to define it, is determined not only by the definition of a state, but also by the definition of small-world consequences. A *small-world consequence* is a grand-world act. A set $\bar{F}$ of grand-world acts, regarded as small-world consequences, is thus part of the definition of any given small world. It will be mathematically simplest, and cost little if anything in insight, to suppose that the elements of $\bar{F}$ are finite in number. They will be denoted $\bar{f}$, $\bar{g}$, $\bar{h}$, $\cdots$; and, when the small-world consequence $\bar{f}$ is recognized as a grand-world act, $f(s)$ will denote the grand-world consequence of $\bar{f}$ at the grand-world state s.

A small-world act $\bar{\mathbf{f}}$ is, of course, a function from small-world states $\bar{s}$ to small-world consequences $\bar{f}$. In this isolated technical discussion, we will hobble along with the notations $\bar{f}(\bar{s})$ for the small-world consequence attached to $\bar{s}$ by $\bar{\mathbf{f}}$, and $f(s; \bar{s})$ for the grand-world consequence attached to s by $\bar{f}(\bar{s})$ recognized as a grand-world act. Each small-world act $\bar{\mathbf{f}}$ gives rise to a unique grand-world act $\hat{\mathbf{f}}$, defined thus:

(2) $\hat{f}(s) = _{\mathrm{Df}} f(s; \bar{s}(s)),$

where $\bar{s}(s)$ means that small-world state $\bar{s}$ of which the grand-world state s is an element.

The distinction between $\bar{\mathbf{f}}$ and $\hat{\mathbf{f}}$, like some other distinctions I have thought it worth while to make in the present complicated context, is perhaps pedantic. At any rate, it is to be understood as part of the definition of a small world that $\bar{\mathbf{f}} \leq \bar{\mathbf{g}}$ if and only if $\hat{\mathbf{f}} \leq \hat{\mathbf{g}}$, that is, in view of (1), if and only if $E(\hat{f}) \leq E(\hat{g})$. In this connection, it is useful to note that

$$(3) \qquad E(\hat{f}) = \sum_{\bar{k}\,\varepsilon\,\bar{F}} E(\hat{f} \mid \bar{f}(\bar{s}(s)) = \bar{k})P(\bar{f}(\bar{s}(s)) = \bar{k})$$

$$= \sum_{\bar{k}} E(\bar{k} \mid \bar{f}(\bar{s}(s)) = \bar{k})P(\bar{f}(\bar{s}(s)) = \bar{k}).$$

It may be advantageous to review (3), and thereby the whole technical definition of a small world, in terms of an example. A small-world act, typified by the purchase of a lottery ticket, amounts to accepting the consequences of one of several ordinary grand-world acts according to which element of a partition does in fact obtain. For example, the participant in a lottery may drive away a car, lead away a goat, face a firing squad, or remain in the status quo, according to the terms of the lottery and according to which ticket he has in fact drawn. Letting the example of the lottery stand for the general situation, the expected utility of a lottery ticket can be computed by the partition formula (3.5.3) from the conditional expectation associated with each ticket, which is what (3) does.

It may fairly be said that a lottery prize is not an act, but rather the opportunity to choose from a number of acts. Thus a cash prize puts its possessor in a position to choose among many purchases he could not otherwise afford. I believe that analysis to be more nearly correct, but it is more complicated; and, if one thinks of each set of acts made available by a lottery prize as represented by a best act of that set, the more complicated analysis seems superfluous, at least in a first attack.

A small world is completely satisfactory for the use to which I mean to put it, if and only if it itself satisfies the seven postulates and leads to—more technically, agrees with—a probability $\bar{P}$ such that

$$(4) \qquad\qquad \bar{P}(\bar{B}) = P([\bar{B}])$$

for all $\bar{B} \subset \bar{S}$ and has a utility $\bar{U}$ such that

$$(5) \qquad\qquad \bar{U}(\bar{f}) = E(\bar{f})$$

for all $\bar{f}\,\varepsilon\,\bar{F}$. For the present context, call such a completely satisfactory small world a *microcosm;* if the small world satisfies the postulates, but does not necessarily admit $\bar{P}$ as its probability nor $\bar{U}$ as a utility, call it a *pseudo-microcosm.*

To display the circumstances under which a small world is a pseudo-microcosm, I shall briefly comment on each of the postulates in the form given on the end papers of this book, referring to them here as

$\bar{P}$1–7, as opposed to P1–7, to emphasize that they are here being considered with respect to $\bar{S}$ and $\bar{F}$.

$\bar{P}$1   *Simple ordering.*

Automatically satisfied. Indeed it is directly implied by P1.

$\bar{P}$2   *Conditional preference well defined.*

Automatic.

$\bar{P}$3   *Conditional preference does not effect consequences.*

Requires exactly that, for every $\bar{f}$, $\bar{g}$ ε $\bar{F}$, and $\bar{B} \subset \bar{S}$, either:

a. $\bar{f} \leq \bar{g}$ given $[\bar{B}]$,   if and only if $\bar{f} \leq \bar{g}$,   or

b. $\bar{h} \leq \bar{k}$ given $[\bar{B}]$,   for every $\bar{h}$, $\bar{k}$ ε $\bar{F}$.

In these inequalities the elements of $\bar{F}$ are of course interpreted as grand-world acts.

$\bar{P}$4   *Qualitative personal probability well defined.*

Requires exactly that, if $\bar{f} < \bar{g}$ and $\mathbf{h}_{\bar{B}} \leq \mathbf{h}_{\bar{C}}$, where

$$h_{\bar{B}}(s) = \bar{g} \qquad \text{for } s \text{ ε } [\bar{B}]$$
$$= \bar{f} \qquad \text{for } s \text{ ε } \sim[\bar{B}] \cdot$$

(6)

$$h_{\bar{C}}(s) = \bar{g} \qquad \text{for } s \text{ ε } [\bar{C}]$$
$$= \bar{f} \qquad \text{for } s \text{ ε } \sim[\bar{C}];$$

then $h'_{\bar{B}} \leq h'_{\bar{C}}$, where $h'_{\bar{B}}$ and $h'_{\bar{C}}$ are defined in terms of $\bar{f}'$, $\bar{g}'$, $\bar{f}' < \bar{g}'$, in analogy with (6).

This postulate is automatic in case $\bar{F}$ has at most two elements.

$\bar{P}$5   *The person has some definite preference.*

Requires $\bar{f} < \bar{g}$ for some $\bar{f}$, $\bar{g}$ ε $\bar{F}$.

$\bar{P}$6   *Partition of worlds into tiny events.*

It is clear that this postulate is not automatic, that is, it is not implied by the validity of P1–7 for the grand world. It is not even implied by P1–7 together with $\bar{P}$1–5, though in the presence of all these $\bar{P}$6 could undoubtedly be weakened. There seems to be little to gain in the present context by reducing $\bar{P}$6 to such minimal terms, nor by expressing it, as $\bar{P}$1–5 have been expressed, in grand-world terms alone; for $\bar{P}$6 does not lend itself easily to such treatment, though it would be

easy to decide in any instance whether $\bar{P}6$ obtained without undue reference to the grand world.

$\bar{P}7$ *Strong form of sure-thing principle.*

Automatic, in view of the explicit assumption that $\bar{F}$ has only a finite number of elements.

To summarize, a small world is a pseudo-microcosm, if and only if it satisfies $\bar{P}3$–6. The possibility of enlarging an arbitrary small world in such a way as to satisfy those conditions has already been implicitly discussed in connection with P3–6. To recall the arguments that were adduced, one might review the example about the egg in § 3.1, and the further discussion of that example in the opening paragraph of § 3.2; the remark in § 3.2, introducing P5; and the example about the coin following P6′ in § 3.3.

It is encouraging to possess the arguments just cited tending to show that any small world can without overwhelming difficulty be embedded in a somewhat larger small world that is a pseudo-microcosm. A pseudo-microcosm is, however, completely satisfactory, only if it is actually a microcosm, that is, only if it leads to a probability measure and a utility well articulated with those of the grand world. The problem of deciding under what circumstances that occurs is much facilitated by the fact that the probability measure and a utility of a pseudo-microcosm can be written down explicitly, as the next few paragraphs show.

To study the problem, suppose the small world is a pseudo-microcosm. Then, in view of $\bar{P}5$, let $\bar{g}$, $\bar{h}$ be elements of $\bar{F}$ such that $\bar{g} < \bar{h}$, and let

$$(7) \qquad Q(\bar{B}) =_{\mathrm{Df}} \frac{E(\bar{h} - \bar{g} \mid [\bar{B}])}{E(\bar{h} - \bar{g})} P([\bar{B}])$$

$$= E^{-1}(\bar{h} - \bar{g}) \int_{[\bar{B}]} \{h(s) - g(s)\} \, dP(s).$$

By using $\bar{P}3$ to check the positivity, it is easily verified that Q is a probability measure on $\bar{S}$. The probability measure Q agrees with the relation $\leq$ between small-world events, which is easily verified on rewriting (3) for the special small-world act $\bar{f}_{\bar{B}}$ that takes the value $\bar{h}$ for $\bar{s} \in \bar{B}$ and $\bar{g}$ for $\bar{s} \in \sim\bar{B}$ thus:

$$(8) \qquad E(\hat{f}_{\bar{B}}) = E(\bar{h} \mid [\bar{B}])P([\bar{B}]) + E(\bar{g} \mid \sim[\bar{B}])P(\sim[\bar{B}])$$

$$= E(\bar{h} - \bar{g} \mid [\bar{B}])P([\bar{B}]) + E(\bar{g})$$

$$= E(\bar{h} - \bar{g})Q(\bar{B}) + E(\bar{g}).$$

Since $\bar{g}$ and $\bar{h}$ are essentially arbitrary, there are many ways to construct a probability measure that agrees with the relation $\leq$ between small-world events, but, in the presence of $\bar{P}$1–6, all of them must (in view of Corollary 3.3.1) be the same as Q. That consideration leads to the formula

$$(9) \qquad E(\bar{f} - \bar{f}' \mid [\bar{B}])P([\bar{B}]) = E(\bar{f} - \bar{f}')Q(\bar{B})$$

for all $\bar{f}, \bar{f}' \in \bar{F}$ and $\bar{B} \subset \bar{S}$.

Using (9) and recalling that $U(\bar{f})$ has been defined as $E(\bar{f})$, (3) can be rewritten thus:

$$(10) \qquad E(\hat{\mathbf{f}}) = E(\bar{g}) + \sum_k E(\bar{k} - \bar{g} \mid \bar{f}(\bar{s}(s)) = \bar{k})P(\bar{f}(\bar{s}(s)) = \bar{k})$$

$$= \sum_k U(\bar{k})Q(\bar{f}(\bar{s}) = \bar{k}).$$

The question whether a given pseudo-microcosm is really a microcosm is the question whether $Q(\bar{B}) = P([\bar{B}])$ and whether U is a utility for the pseudo-microcosm. The answer to the second part is immediate and, I think, somewhat surprising, for (10) shows that for any pseudo-microcosm U is indeed a utility.

Unfortunately, the condition $Q(\bar{B}) = P([\bar{B}])$ is not also automatic. The possibility of its failing to be satisfied is illustrated by the following simple mathematical example. Let S be the unit square $0 \leq x, y \leq 1$, and let

$$(11) \qquad E(\mathbf{f}) = \int_0^1 \int_0^1 f(x, y) \, dx \, dy.$$

It is of no real moment that the integral in (11), if understood in the Lebesgue or Riemann sense, is not defined for all bounded functions. Let the elements of $\bar{S}$ be the vertical line segments, $x = \text{constant}$. Finally, suppose that the elements of $\bar{F}$ consist of the function zero and any finite number of non-negative multiples of a fixed positive function $\mathbf{h} = \bar{h}$. It is easy to verify that $\bar{S}$ as thus defined is a pseudo-microcosm and that

$$(12) \qquad Q(\bar{B}) = \int_{x' \in \bar{B}} q(x') \, dx'$$

where

$$(13) \qquad q(x') = \frac{\displaystyle\int_0^1 h(x', y) \, dy}{\displaystyle\int_0^1 \int_0^1 h(x, y) \, dx \, dy}.$$

Unless $\mathbf{q}$ is 1 for every x', which will not at all typically be the case, $\bar{S}$ is not really a microcosm.

The general condition that a pseudo-microcosm be a microcosm—i.e., that $Q(\bar{B}) = P([\bar{B}])$—is evidently, in view of (9),

$$(14) \qquad E(\bar{f} - \bar{f}' \mid [\bar{B}]) = E(\bar{f} - \bar{f}')$$

for every $\bar{f}$, $\bar{f}' \in \bar{F}$ and every $\bar{B}$ for which $P([\bar{B}]) > 0$. Incidentally, that condition alone practically implies that a small world $\bar{S}$, not necessarily assumed to be a pseudo-microcosm, is a real microcosm. More exactly, it implies all the postulates $\bar{P}$1–7, except $\bar{P}$6; and it implies that the probability measure P agrees with the relation $\leq$ between small-world events. Also, if a small world is a pseudo-microcosm, it is enough that (14) should hold for *some* pair of functions for which the right-hand side of the equation does not vanish.

Equation (14) is, however, unsatisfactory in that it seems incapable of verification without taking the grand world much too seriously. Some consolation may derive from the fact that if $\bar{f}$ and $\bar{f}'$ are constants they automatically satisfy (14). Two such absolute, or grand-world, consequences would suffice, for, as has just been remarked, it is sufficient that (14) be satisfied for two materially different small-world consequences, in the presence of $\bar{P}$1–7 (which are verifiable without any detailed knowledge of the grand world). It must, however, be admitted, as has already been mentioned, that the very idea of a grand-world consequence takes the grand world pretty seriously—a point forced into my reluctant mind by a conversation with Francesco Brambilla.

I feel, if I may be allowed to say so, that the possibility of being taken in by a pseudo-microcosm that is not a real microcosm is remote, but the difficulty I find in defining an operationally applicable criterion is, to say the least, ground for caution.

There certainly seem to be cases in which one could confidently assume (14), though thus far formal analysis of the source of such security escapes me. Consider, for example, a lottery in which numbered tickets are drawn from a drum. It seems clear that for an ordinary person the outcome of the lottery is utterly irrelevant to his life, except through the rules of the lottery itself. In other terms equally loose, the value of a thousand dollars, or of a car, to a person would not ordinarily depend at all on what numbers were drawn in a lottery, unless the person himself (or perhaps some other person or organization with whom he had some degree of contact) held tickets in the lottery. A more precise formulation, which does indeed imply (14), is that the events that represent the outcome of the lottery are all statistically

independent of the grand-world acts, or functions, that typically enter as prizes in a lottery. This suggests once more that it would be desirable, if possible, to find a simple qualitative personal description of independence between events. (Compare the first paragraph after (3.5.2).)

6   Historical and critical comments on utility

A casual historical sketch of the concept of utility will perhaps have some interest as history. At any rate, most of the critical ideas pertaining to utility that I wish to discuss find their places in such a sketch as conveniently as in any other organization I can devise. Much more detailed material on the history of utility, especially in so far as the economics of risk bearing is concerned, is to be found in Arrow's review article [A6]. Stigler's historical study [S18] emphasizes the history of the now almost obsolete economic notion of utility in riskless situations, a notion still sometimes confused with the one under discussion.

As was mentioned in § 4.5, the earliest mathematical studies of probability were largely concerned with gambling, particularly with the question of which of several available cash gambles is most advantageous. Early probabilists advanced the maxim that the gamble with the highest expected winnings is best or, in terms of utility, that wealth measured in cash is a utility function. Some sense can be seen in that maxim, which will here be called by its traditional though misleading name, the **principle of mathematical expectation.** First, it has often been argued that the principle follows for the long run from the weak law of large numbers, applied to large numbers of independent bets, in each of which only sums that the gambler considers small are to be won or lost. Second, Daniel Bernoulli, who, in [B10], was one of the first to introduce a general idea of utility corresponding to that developed in the preceding three sections, made the following analysis of the principle, which justifies its application in limited but important contexts. If the consequences f to be considered are all quantities of cash, it is reasonable to suppose that $U(f)$ will change smoothly with changes in f. Therefore, if a person's present wealth is f_0, and he contemplates various gambles, none of which can greatly change his wealth, the utility function can, for his particular purpose, be approximated by its tangent at f_0, that is,

(1) $$U(f) \simeq U(f_0) + (f - f_0)U'(f_0),$$

a linear function of f. Since a constant term is irrelevant to any comparison of expected values, the approximation amounts to regarding utility as proportional to wealth, that is, to following the principle of

mathematical expectation. So far as I know, the only other argument for the principle that has ever been advanced is one concerning equity between two players. As Bernoulli says, that argument is irrelevant at best; and neither of the relevant arguments justifies categorial acceptance of the principle. None the less, the principle was at first so categorically accepted that it seemed paradoxical to mathematicians of the early eighteenth century that presumably prudent individuals reject the principle in certain real and hypothetical decision situations.

Daniel Bernoulli (1700–1782), in the paper [B10], seems to have been the first to point out that the principle is at best a rule of thumb, and he there suggested the maximization of expected utility as a more valid principle. Daniel Bernoulli's paper reproduces portions of a letter from Gabriel Cramer to Nicholas Bernoulli, which establishes Cramer's chronological priority to the idea of utility and most of the other main ideas of Bernoulli's paper. But it is Bernoulli's formulation together with some of the ideas that were specifically his that became popular and have had widespread influence to the present day. It is therefore appropriate to review Bernoulli's paper in some detail.

Being unable to read Latin, I follow the German edition [B11].

Bernoulli begins by reminding his readers that the principle of mathematical expectation, though but weakly supported, had theretofore dominated the theory of behavior in the face of uncertainty. He says that, though many arguments had been given for the principle, they were all based on the irrelevant idea of equity among players. It seems hard to believe that he had never heard the argument justifying the principle for the long run, even though the weak law of large numbers was then only in its mathematical infancy. *Ars Conjectandi* [B12], then a fairly up-to-date and most eminent treatise on probability, does seem to give only the argument about equity, and that in countless forms. This treatise by Daniel's uncle, Jacob (= James) Bernoulli (1654–1705), incidentally, contains the first mathematical advance toward the weak law, proving it for the special case of repeated trials.

Many examples show that the principle of mathematical expectation is not universally applicable. Daniel Bernoulli promptly presents one: "To justify these remarks, let us suppose a pauper happens to acquire a lottery ticket by which he may with equal probability win either nothing or 20,000 ducats. Will he have to evaluate the worth of the ticket as 10,000 ducats; and would he be acting foolishly, if he sold it for 9,000 ducats? "

Other examples occur later in the paper as illustrations of the use of the utility concept. Thus a prudent merchant may insure his ship against loss at sea, though he understands perfectly well that he is

thereby increasing the insurance company's expected wealth, and to the same extent decreasing his own. Such behavior is in flagrant violation of the principle of mathematical expectation, and to one who held that principle categorically it would be as absurd to insure as to throw money away outright. But the principle is neither obvious nor deduced from other principles regarded as obvious; so it may be challenged, and must be, because everyone agrees that it is not really insane to insure.

Bernoulli cites a third, now very famous, example illustrating that men of prudence do not invariably obey the principle of mathematical expectation. This example, known as the St. Petersburg paradox (because of the journal in which Bernoulli's paper was published) had earlier been publicized by Nicholas Bernoulli,† and Daniel acknowledges it as the stimulus that led to his investigation of utility. Suppose, to state the St. Petersburg paradox succinctly, that a person could choose between an act leaving his wealth fixed at its present magnitude or one that would change his wealth at random, increasing it by $(2^n - f)$ dollars with probability 2^{-n} for every positive integer n. No matter how large the admission fee f may be, the expected income of the random act is infinite, as may easily be verified. Therefore, according to the principle of mathematical expectation, the random act is to be preferred to the status quo. Numerical examples, however, soon convince any sincere person that he would prefer the status quo if f is at all large. If f is \$128, for example, there is only 1 chance in 64 that a person choosing the random act will so much as break even, and he will otherwise lose at least \$64, a jeopardy for which he can seek compensation only in the prodigiously improbable winning of a prodigiously high prize.

Appealing to intuition, Bernoulli says that the cash value of a person's wealth is not its true, or moral, worth to him. Thus, according to Bernoulli, the dollar that might be precious to a pauper would be nearly worthless to a millionaire—or, better, to the pauper himself were he to become a millionaire. Bernoulli then postulates that people do seek to maximize the expected value of moral worth, or what has been called moral expectation.

Operationally, the moral worth of a person's wealth, so far as it concerns behavior in the face of uncertainty, is just what I would call the utility of the wealth, and moral expectation is expectation of utility.

† Daniel refers to this Nicholas Bernoulli as his uncle, but, in view of dates mentioned in the last section of Daniel's paper and the genealogy in Chapter 8 of [B9], I think he must have meant his elder cousin (1687–1759), perhaps using "uncle" as a term of deference.

It seems mystical, however, to talk about moral worth apart from probability and, having done so, doubly mystical to postulate that this undefined quantity serves as a utility. These obvious criticisms have naturally led many to discredit the very idea of utility, but §§ 2–4 show (following von Neumann and Morgenstern) that there is a more cogent, though not altogether unobjectionable, path to that concept.

Bernoulli argued, elaborating the example of the pauper and the millionaire, that a fixed increment of cash wealth typically results in an ever smaller increment of moral wealth as the basic cash wealth to which the increment applies is increased. He admitted the possibility of examples in which this **law of diminishing marginal utility,** as it has come to be called in the literature of economics, might fail. For example, a relatively small sum might be precious to a wealthy prisoner who required it to complete his ransom. But Bernoulli insisted that such examples are unusual and that as a general rule the law may be assumed. In mathematical terms, the law says that utility as a function of money is a **concave** (i.e., the negative of a convex) function.†
It follows from the basic inequality concerning convex functions (Theorem 1 of Appendix 2) that a person to whom the law of diminishing marginal utility applies will always prefer the status quo to any fair gamble, that is, to any random act for which the change in his expected wealth is zero, and that he will always be willing to pay something in addition to its actuarial, or expected, value for insurance against any loss to himself. The law of diminishing marginal utility has been very popular, and few who have considered utility since Bernoulli have discarded it, or even realized that it was not necessarily part and parcel of the utility idea. Of course, the law has been embraced eagerly and uncritically by those who have a moral aversion to gambling.

Bernoulli went further than the law of diminishing marginal utility and suggested that the slope of utility as a function of wealth might, at least as a rule of thumb, be supposed, not only to decrease with, but to be inversely proportional to, the cash value of wealth. This, he pointed out, is equivalent to postulating that utility is equal to the logarithm (to any base) of the cash value of wealth. To this day, no other function has been suggested as a better prototype for Everyman's utility function. None the less, as Cramer pointed out in his aforementioned letter, the logarithm has a serious disadvantage; for, if the logarithm were the utility of wealth, the St. Petersburg paradox could be

---

† Often the meanings of "convex" and "concave" as applied to functions are interchanged. A function is here called convex if it appears convex, in the ordinary sense of the word, when viewed from below. Such a function is, of course, also concave from above, whence the confusion. Cf. Appendix 2.

amended to produce a random act with an infinite expected utility (i.e., an infinite expected logarithm of income) that, again, no one would really prefer to the status quo. To take a less elaborate example, suppose that a man's total wealth, including an appraisal of his future earning power, were a million dollars. If the logarithm of wealth were actually his utility, he would as soon as not flip a coin to decide whether his wealth should be changed to ten thousand dollars—roughly $500 per year—or a hundred million dollars. This seems preposterous to me. At any rate, I am sure you can construct an example along the same lines that will seem preposterous to you. Cramer therefore concluded, and I think rightly, that the utility of cash must be bounded, at least from above. It seems to me that a good argument can also be adduced for supposing utility to be bounded from below, for, however wealth may be interpreted, we all subject our total wealth to slight jeopardy daily for the sake of a large probability of avoiding more moderate losses. But the logarithm is unbounded both from above and from below; so, though it might be a reasonable approximation to a person's utility in a moderate range of wealth, it cannot be taken seriously over extreme ranges.

Bernoulli's ideas were accepted wholeheartedly by Laplace [L1], who was very enthusiastic about the applications of probability to all sorts of decision problems. It is my casual impression, however, that from the time of Laplace until quite recently the idea of utility did not strongly influence either mathematical or practical probabilists.

For a long period economists accepted Bernoulli's idea of moral wealth as the measurement of a person's well-being apart from any consideration of probability. Though "utility" rather than "moral worth" has been the popular name for this concept among English-speaking economists, it is my impression that Bernoulli's paper is the principal, if not the sole, source of the notion for all economists, though the paper itself may often have been lost sight of. Economists were for a time enthusiastic about the principle of diminishing marginal utility, and they saw what they believed to be reflections of it in many aspects of everyday life. Why else, to paraphrase Alfred Marshall ·(pp. 19, 95 of [M2]), does a poor man walk in a rain that induces a rich man to take a cab?

During the period when the probability-less idea of utility was popular with economists, they referred not only to the utility of money, but also to the utility of other consequences such as commodities (and services) and combinations (or, better, patterns of consumption) of commodities. The theory of choice among consequences was expressed by the idea that, among the available consequences, a person prefers those that have the highest utility for him. Also, the idea of diminishing marginal utility was extended from money to other commodities.

The probability-less idea of utility in economics has been completely discredited in the eyes of almost all economists, the following argument against it—originally advanced by Pareto in pp. 158–159 and the Mathematical Appendix of [P1]—being widely accepted. If utility is regarded as controlling only consequences, rather than acts, it is not true—as it is when acts, or at least gambles, are considered and the formal definition in § 3, is applied—that utility is determined except for a linear transformation. Indeed, confining attention to consequences, any strictly monotonically increasing function of one utility is another utility. Under these circumstances there is little, if any, value in talking about utility at all, unless, of course, special economic considerations should render one utility, or say a linear family of utilities, of particular interest. That possibility remains academic to date, though one attempt to exploit it was made by Irving Fisher, as is briefly discussed in the paragraph leading to Footnote 155 of [S18]. In particular, utility as a function of wealth can have any shape whatsoever in the probability-less context, provided only that the function in question is increasing with increasing wealth, the provision following from the casual observation that almost nobody throws money away. The history of probability-less utility has been thoroughly reported by Stigler [S18].

What, then, becomes of the intuitive arguments that led to the notion of diminishing marginal utility? To illustrate, consider the poor man and the rich man in the rain. Those of us who consider diminishing marginal utility nonsensical in this context think it sufficient to say simply that it is a common observation that rich men spend money freely to avoid moderate physical suffering whereas poor men suffer freely rather than make corresponding expenditures of money; in other terms, that the rate of exchange between circumstances producing physical discomfort and money depends on the wealth of the person involved.

In recent years there has been revived interest in Bernoulli's ideas of utility in the technical sense of §§ 2–4, that is, as a function that, so to speak, controls decisions among acts, or at least gambles. Ramsey's essays in [R1], which in spirit closely resemble the first five chapters of this book, present a relatively early example of this revival of interest. Ramsey improves on Bernoulli in that he defines utility operationally in terms of the behavior of a person constrained by certain postulates. Ramsey's essays, though now much appreciated, seem to have had relatively little influence.

Between the time of Ramsey and that of von Neumann and Morgenstern there was interest in breaking away from the idea of maximizing expected utility, at least so far as economic theory was concerned (cf. [T1a]). This trend was supported by those who said that Bernoulli gives no reason for supposing that preferences correspond to the expected value of some function, and that therefore much more general possibilities must be considered. Why should not the range, the variance, and the skewness, not to mention countless other features, of the distribution of some function join with the expected value in determining preference? The question was answered by the construction of Ramsey and again by that of von Neumann and Morgenstern, which has been slightly extended in §§ 2–4; it is simply a mathematical fact that, almost any theory of probability having been adopted and the sure-thing principle having been suitably extended, the existence of a function whose expected value controls choices can be deduced. That does not mean that as a theory of actual economic behavior the theory of utility is absolutely established and cannot be overthrown. Quite the contrary, it is a theory that makes factual predictions many of which can easily be observed to be false, but the theory may have some value in making economic predictions in certain contexts where the departures from it happen not to be devastating. Moreover, as I have been arguing, it may have value as a normative theory.

Von Neumann and Morgenstern initiated among economists and, to a lesser extent, also among statisticians an intense revival of interest in the technical utility concept by their treatment of utility, which appears as a digression in [V4].

The von Neumann-Morgenstern theory of utility has produced this reaction, because it gives strong intuitive grounds for accepting the Bernoullian utility hypothesis as a consequence of well-accepted maxims of behavior. To give readers of this book some idea of the von Neumann-Morgenstern theory, I may repeat that the treatment of utility as applied to gambles presented in § 3 is virtually copied from their book [V4]. Indeed, their ideas on this subject are responsible for almost all of my own. One idea now held by me that I think von Neumann and Morgenstern do not explicitly support, and that so far as I know they might not wish to have attributed to them, is the normative interpretation of the theory.

Of course, much of the new interest in utility takes the form of criticism and controversy. The greater part of this discussion that has come to my attention has not yet been published. A list of references leading to most of that which has is [B7], [W14], [S1], [C4], [F13], [A2].

I shall successively discuss each of the recent major criticisms of the modern theory of utility known to me. My method in each case will be first to state the criticism in a form resembling those in which it is typically put forward, regardless of whether I consider that form well chosen. I will then discuss the criticism, elaborating its meaning and indicating its rebuttal, when there seems to me to be one.

(a) Modern economic theorists have rigorously shown that there is no meaningful measure of utility. More specifically, if any function **U** fulfills the role of a utility, then so does any strictly monotonically increasing function of **U**. It must, therefore, be an error to conclude that every utility is a linear function of every other.

This argument has been advanced with a seriousness that is surprising, considering that it concedes little intelligence or learning to the proponents of the utility theory under discussion and considering that it results, as will immediately be explained, from the baldest sort of a terminological confusion. To be fair, I must go on to say that I have never known the argument to be defended long in the presence of the explanation I am about to give.

In ordinary economic usage, especially prior to the work of von Neumann and Morgenstern, a utility associated with gambles would presumably be simply a function **U** associating numbers with gambles in such a way that $f \leq g$, if and only if $U(f) \leq U(g)$; though economic discussion of utility was, prior to von Neumann and Morgenstern, almost exclusively confined to consequences rather than to gambles or to acts. It is unequivocally true, as I have already brought out, that any monotonic function of a utility in this wide classical sense is itself a utility. What von Neumann and Morgenstern have shown, and what has been recapitulated in § 3, is that, granting certain hypotheses, there exists at least one classical utility **V** satisfying the very special condition

$$(2) \qquad\qquad V(\alpha f + \beta g) = \alpha V(f) + \beta V(g),$$

where f and g are any gambles and α, β are non-negative numbers such that $\alpha + \beta = 1$. Furthermore, if I may for the moment call a classical utility satisfying (2) a von Neumann-Morgenstern utility, every von Neumann-Morgenstern utility is an increasing linear function of every other. To put the point differently, the essential conclusion of the von Neumann-Morgenstern utility theory is that (2) can be satisfied by a classical utility, but not by very many. The confusion arises only because von Neumann and Morgenstern use the already pre-empted word "utility" for what I here call "von Neumann-Morgenstern utility." In retrospect, that seems to have been a mistake in tactics, but one of no long-range importance.

(b) The postulates leading to the von Neumann-Morgenstern concept of utility are arbitrary and gratuitous.

Such a view can, of course, always be held without the slightest fear of rigorous refutation, but a critic holding it might perhaps be persuaded away from it by a reformulation of the postulates that he might find more appealing than the original set, or by illuminating examples. In particular, P1–7 are quite different from, but imply, the postulates of von Neumann and Morgenstern. Incidentally, the main function of the von Neumann-Morgenstern postulates themselves is to put the essential content of Daniel Bernoulli's "postulate" into a form that is less gratuitous in appearance. At least one serious critic, who had at first found the system of von Neumann and Morgenstern gratuitous, changed his mind when the possibility of deriving certain aspects of that system from the sure-thing principle was pointed out to him.

(c) The sure-thing principle goes too far. For example, if two lotteries with cash prizes (not necessarily positive) are based on the same set of lottery tickets and so arranged that the prize that will be assigned to any ticket by the second lottery is at least as great as the prize assigned to that ticket by the first lottery, then there is no doubt that virtually any person would find a ticket in the first lottery not preferable to the same ticket in the second lottery. If, however, the prizes in each lottery are themselves lottery tickets, such that the prize associated with any ticket in the first lottery is not preferred by the person under study to the prize associated with the same ticket by the second lottery, the conclusion that the person will not prefer a ticket in the first lottery to the same ticket in the second is no longer compelling.

This point resembles the preceding one in that the intuitive appeal of an assumption can at most be indicated, not proved. I do think it cogent, however, to stress in connection with this particular point that a cash prize is to a large extent a lottery ticket in that the uncertainty as to what will become of a person if he has a gift of a thousand dollars is not in principle different from the uncertainty about what will become of him if he holds a lottery ticket of considerable actuarial value.

Perhaps an adherent to the criticism in question would think it relevant to reply thus: Though cash sums are indeed essentially lottery tickets, a sum of money is worth at least as much to a person as a smaller sum, in a peculiarly definite and objective sense, because money can, if one desires, always be quickly and quietly thrown away, thereby making any sum available to a person who already has a larger sum. But I have never heard that reply made, nor do I here plead its cogency.

(d) An actual systematic deviation from the sure-thing principle and, with it, from the von Neumann-Morgenstern theory of utility, can be exhibited. For example, a person might perfectly reasonably prefer to subsist on a packet of Army K rations per meal than on two ounces of the best caviar per meal. It is then to be expected, according to the sure-thing principle, that the person would prefer the K rations to a lottery ticket yielding the K rations with probability 9/10 and the caviar diet with probability 1/10. That expectation is no doubt fulfilled, if the lottery is understood to determine the person's year-long diet once and for all. But, if the person is able to have at each meal a lottery ticket offering him the K rations or the caviar with the indicated probabilities, it is not at all unlikely, granting that he likes caviar and has some storage facilities, that he will prefer this "lottery diet." This conclusion is in defiance of the principle that "the theory of consumer demand is a static theory." (Cf. [W14].)

I admit that the theory of utility is not static in the indicated sense, as the foregoing example conclusively shows. But there is not the slightest reason to think of a lottery producing either a steady diet of caviar or a steady diet of K rations as being the same lottery as one having a multitude of different prizes almost all of which are mixed chronological programs of caviar and K rations. The fact that a theory of consumer behavior in riskless situations happens to be static in the required sense (under certain special assumptions about storability and the linearity of prices) is no argument at all that the theory of consumer behavior in risky circumstances should be static in the same sense (as I mention in a note appended to [W14]).

(e) If the von Neumann-Morgenstern theory of utility is not static, it is not subject to repeated empirical observation and is therefore vacuous. (Cf. [W14].)

I think the discussion in § 3.1 of how to determine the preferences of a hot man for a swim, a shower, and a glass of beer, and the discussion in § 5 of the practicality of identifying pseudo-microcosms are steps toward showing how the theory can be put to empirical test without making repeated trials on any one person.

(f) Casual observation shows that real people frequently and flagrantly behave in disaccord with the utility theory, and that in fact behavior of that sort is not at all typically considered abnormal or irrational.

Two different topics call for discussion under this heading. In the first place, it is undoubtedly true that the behavior of people does often flagrantly depart from the theory. None the less, all the world knows from the lessons of modern physics that a theory is not to be altogether rejected because it is not absolutely true. It seems not unreasonable to suppose, and examples could easily be cited to confirm, that in the extremely complicated subject of the behavior of people very crude theory can play a useful role in certain contexts.

Second, many apparent exceptions to the theory can be so reinterpreted as not to be exceptions at all. For example, a flier may be observed doing a stunt that risks his life, apparently for nothing. That seems to be in complete violation of the theory; but, if in addition it is known that the flier has a real and practical need to convince certain colleagues of his courage, then he is simply paying for advertising with the risk of his life, which is not in itself in contradiction to the theory. Or, suppose that it were known more or less objectively that the flier has a need to demonstrate his own courage to himself. The theory would again be rescued, but this time perhaps not so convincingly as before. In general, the reinterpretation needed to reconcile various sorts of behavior with the utility theory is sometimes quite acceptable and sometimes so strained as to lay whoever proposes it open to the charge of trying to save the theory by rendering it tautological. The same sort of thing arises in connection with many theories, and I think there is general agreement that no hard-and-fast rule can be laid down as to when it becomes inappropriate to make the necessary reinterpretation. For example, the law of the conservation of energy (or its atomic age variant, the law of the conservation of mass *and* energy) owes its success largely to its being an expression of remarkable and reliable facts of nature, but to some extent also to certain conventions by which new sorts of energy are so defined as to keep the law true. A stimulating discussion of this delicate point in connection with the theory of utility is given by Samuelson in [S1].

(g) Introspection about certain hypothetical decision situations suggests that the sure-thing principle and, with it, the theory of utility are normatively unsatisfactory. Consider an example based on two decision situations each involving two gambles.†

Situation 1.   Choose between

Gamble 1. $500,000 with probability 1; and
Gamble 2. $2,500,000 with probability 0.1,
             $500,000 with probability 0.89,
             status quo with probability 0.01.

† This particular example is due to Allais [A2]. Another interesting example was presented somewhat earlier by Georges Morlat [C4].

Situation 2.   Choose between

Gamble 3. $500,000 with probability 0.11,
            status quo with probability 0.89; and
Gamble 4. $2,500,000 with probability 0.1,
            status quo with probability 0.9.

Many people prefer Gamble 1 to Gamble 2, because, speaking qualitatively, they do not find the chance of winning a *very* large fortune in
place of receiving a large fortune outright adequate compensation for
even a small risk of being left in the status quo.   Many of the same
people prefer Gamble 4 to Gamble 3; because, speaking qualitatively,
the chance of winning is nearly the same in both gambles, so the one
with the much larger prize seems preferable.   But the intuitively acceptable pair of preferences, Gamble 1 preferred to Gamble 2 and Gamble 4 to Gamble 3, is not compatible with the utility concept or, equivalently, the sure-thing principle.   Indeed that pair of preferences implies
the following inequalities for any hypothetical utility function.

$$U (\$500,000) > 0.1U (\$2,500,000) + 0.89U (\$500,000) + 0.1U (\$0),$$

(3)

$$0.1U (\$2,500,000) + 0.9U (\$0) > 0.11U (\$500,000) + 0.89U (\$0);$$

and these are obviously incompatible.

Examples † like the one cited do have a strong intuitive appeal; even
if you do not personally feel a tendency to prefer Gamble 1 to Gamble 2
and simultaneously Gamble 4 to Gamble 3, I think that a few trials
with other prizes and probabilities will provide you with an example
appropriate to yourself.

If, after thorough deliberation, anyone maintains a pair of distinct
preferences that are in conflict with the sure-thing principle, he must
abandon, or modify, the principle; for that kind of discrepancy seems
intolerable in a normative theory.   Analogous circumstances forced
D. Bernoulli to abandon the theory of mathematical expectation for
that of utility [B10].   In general, a person who has tentatively accepted
a normative theory must conscientiously study situations in which the
theory seems to lead him astray; he must decide for each by reflection
—deduction will typically be of little relevance—whether to retain his
initial impression of the situation or to accept the implications of the
theory for it.

To illustrate, let me record my own reactions to the example with

† Allais has announced (but not yet published) an empirical investigation of the
responses of prudent, educated people to such examples [A2].

which this heading was introduced. When the two situations were first presented, I immediately expressed preference for Gamble 1 as opposed to Gamble 2 and for Gamble 4 as opposed to Gamble 3, and I still feel an intuitive attraction to those preferences. But I have since accepted the following way of looking at the two situations, which amounts to repeated use of the sure-thing principle.

One way in which Gambles 1–4 could be realized is by a lottery with a hundred numbered tickets and with prizes according to the schedule shown in Table 1.

TABLE 1.   PRIZES IN UNITS OF $100,000 IN A LOTTERY REALIZING
GAMBLES 1–4

		Ticket Number		
		1	2–11	12–100
Situation 1	Gamble 1	5	5	5
	Gamble 2	0	25	5
Situation 2	Gamble 3	5	5	0
	Gamble 4	0	25	0

Now, if one of the tickets numbered from 12 through 100 is drawn, it will not matter, in either situation, which gamble I choose. I therefore focus on the possibility that one of the tickets numbered from 1 through 11 will be drawn, in which case Situations 1 and 2 are exactly parallel. The subsidiary decision depends in both situations on whether I would sell an outright gift of $500,000 for a 10-to-1 chance to win $2,500,000— a conclusion that I think has a claim to universality, or objectivity. Finally, consulting my purely personal taste, I find that I would prefer the gift of $500,000 and, accordingly, that I prefer Gamble 1 to Gamble 2 and (contrary to my initial reaction) Gamble 3 to Gamble 4.

It seems to me that in reversing my preference between Gambles 3 and 4 I have corrected an error. There is, of course, an important sense in which preferences, being entirely subjective, cannot be in error; but in a different, more subtle sense they can be. Let me illustrate by a simple example containing no reference to uncertainty. A man buying a car for $2,134.56 is tempted to order it with a radio installed, which will bring the total price to $2,228.41, feeling that the difference is trifling. But, when he reflects that, if he already had the car, he certainly would not spend $93.85 for a radio for it, he realizes that he has made an error.

One thing that should be mentioned before this chapter is closed is that the law of diminishing marginal utility plays no fundamental role in the von Neumann-Morgenstern theory of utility, viewed either empirically or normatively. Therefore the possibility is left open that utility as a function of wealth may not be concave, at least in some intervals of wealth. Some economic-theoretical consequences of recognition of the possibility of non-concave segments of the utility function have been worked out by Friedman and myself [F12], and by Friedman alone [F11]. The work of Friedman and myself on this point is criticized by Markowitz [M1].+

+ See also Archibald (1959) and Hakansson (1970).

CHAPTER 6

Observation

1 Introduction

With the construction of utility, the theory of decision in the face of uncertainty is, in a sense, complete. I have no further postulates to propose, and those I have proposed have been shown to be equivalent to the assumption that the person always decides in favor of an act the expected utility of which is as large as possible, supposing for simplicity that only a finite number of acts are open to him. At the level of generality that has led to this conclusion there seems to be little or nothing left to say. To go further now means to go into more detail, to investigate special types of decision problems. One type of decision problem of central importance is that in which the person is called upon to make an observation and then to choose some act in the light of the outcome of the observation.

The consideration of such observational decision problems is a step toward those problems of great interest for statistics in which the person must decide what observation to make, that is, of course, what to look at, not what to see. They are the problems of designing experiments and other observational programs.

Some remarks on observation were made in Chapter 3, but only now that the theory of utility is established is it possible to give a relatively complete analysis of the concept.

Observation is a concept essential to the study of statistics proper, most of what has been said thus far being preliminary to, but not really part of, statistics; even after this chapter and the next one, on observation, there will still remain a major transition. One important feature of much of what is ordinarily called statistics is, according to my analysis, concerned with the behavior not of an isolated person, but of a group of persons acting, for example, in concert. In later chapters I will deal, so far as I am able, with the problem of group action, but preliminary considerations bearing on it will be made and pointed out from time to time in this chapter and the next.

Though the details of these two chapters may seem mathematically forbidding, drastic simplifying assumptions are made in them to keep extraneous difficulties to a minimum. These typically take the form of assuming that certain sets of acts, events, and values of random variables are finite. Even in elementary applications of the theory, these simplifying assumptions seldom actually hold. In some contexts, it is quite elementary to relax them sufficiently; in others, serious mathematical effort has been required; and some are still at the frontier of research. Relaxations of the assumptions will be touched on from time to time, sometimes explicitly but sometimes only implicitly in the choice of suggestive notation and nomenclature.

Beyond this introduction, the present chapter is divided into four sections: § 2 analyzes informally and then formally the notion of a cost-free observation; §§ 3 and 4 discuss certain obvious but important conditions under which one observation, and similarly one set of acts, is more valuable than another; § 5 abstractly discusses problems of designing experiments or, perhaps more generally, observational programs.

2   What an observation is

To begin with an informal survey of observation, consider a decision problem, that is, a person faced with a decision among several acts. Calling it the basic decision problem and the acts associated with it the basic acts, a new decision problem would arise, if the person were informed before he made his decision that a particular event, say B, obtained. The new decision problem is related to the basic decision problem in a simple way; for the acts associated with it are also the basic acts, and the decision is to be made by computing the expected utility given B of the basic acts and deciding on one that maximizes the conditional expected utility. The basic problem may be modified in still another, though closely related, way. Let the person say in advance, for each possible B_i, which of the basic acts he will decide on when he is informed, as he is to be, which element B_i of a given partition obtains. This will be called the derived decision problem arising from the basic decision problem and the observation of i, and its acts will be called derived acts. Technically speaking, the derived acts are determined by arbitrarily assigning one basic act to each element of the partition. For any state s, the consequence of a derived act is the consequence for s of the basic act associated with the particular B_i in which s lies. The terms informally introduced in this paragraph are defined formally later in the section.

A derived decision problem is not necessarily different in kind from the basic problem; indeed it is quite possible that the basic problem can itself be viewed as derived from some other basic problem and observation.

Formidable though the description of a derived problem may seem at first reading, its solution is, in a sense, easy and has already almost been given; for it is clear that, if $P(B_i) > 0$, the person will decide to associate with B_i a basic act the expected utility of which given B_i is as high as possible, and, if $P(B_i) = 0$, it is immaterial to the person which basic act is associated with B_i.

It is almost obvious that the value of a derived problem cannot be less, and typically is greater, than the value of the basic problem from which it is derived. After all, any basic act is among the derived acts, so that any expected utility that can be attained by deciding on a basic act can be attained by deciding on the same basic act considered as a derived act. In short, the person is free to ignore the observation. That obvious fact is the theory's expression of the commonplace that knowledge is not disadvantageous.

It sometimes happens that a real person avoids finding something out or that his friends feel duty bound to keep something from him,' saying that what he doesn't know can't hurt him; the jealous spouse and the hypochondriac are familiar tragic examples. Such apparent exceptions to the principle that forewarned is forearmed call for analysis. At first sight, one might be inclined to say that the person who refuses freely proffered information is behaving irrationally and in violation of the postulates. But perhaps it is better to admit that information that *seems* free may prove expensive by doing psychological harm to its recipient. Consider, for example, a sick person who is certain that he has the best of medical care and is in a position to find out whether his sickness is mortal. He may decide that his own personality is such that, though he can continue with some cheer to live in the fear that he may possibly die soon, what is left of his life would be agony, if he knew that death were imminent. Under such circumstances, far from calling him irrational, we might extol the person's rationality, if he abstained from the information. On the other hand, such an interpretation may seem forced. (Cf. Criticism (f) of § 5.6.)

Examples of decisions based on observation are on every hand, but it will be worth while to examine one in some detail before undertaking an abstract mathematical analysis of such decisions. Any example would have to be highly idealized for simplicity, because the complexity of any real decision problem defies complete explicit description, but particular simplicity is in order here.

The person in the example is considering whether to buy some of the grapes he sees in a grocery store and, if so, in what quantity. To his taste, the grapes may be of any of three qualities, poor, fair, and excellent. Call the qualities Q generically and 1, 2, and 3 individually. From what the person knows at the moment, including of course the appearance of the grapes, he cannot be certain of their quality, but he attaches personal probability to each of the three possibilities according to Table 1.

TABLE 1. $P(Q)$

Q(uality)	1	2	3
P(robability)	1/4	1/2	1/4

The person can decide to buy 0, 1, 2, or 3 pounds of grapes; these are the basic acts of the example. Taking one consideration with another, he finds the consequences of each act, measured in utiles, in each of the three possible events to be those given in the body of Table 2. The expected utilities in the right margin of Table 2 follow, of course, from Table 1 and the body of Table 2.

TABLE 2. UTILITY $f(Q)$ FOR EACH f AND EACH Q

f	1	2	3	$E(\mathbf{f})$
0	0	0	0	0
1	−1	1	3	1
2	−3	0	5	1/2
3	−6	−2	6	−1

The entries in Table 2 have not been chosen haphazardly, but with an attempt at verisimilitude. Thus it is supposed that if the person buys grapes of poor quality his dissatisfaction with the bargain will accelerate rapidly with the amount bought, which seems reasonable, especially if the keeping quality of poor grapes is low. He is, of course, unaffected by the quality if he buys none. Again, buying a few fair grapes may be mildly desirable, but overbuying is not. Finally, excellent grapes are worth buying, even in large quantities, but the utility of the purchase increases less than proportionally to the amount bought.

The correct solution of the basic decision problem is to buy 1 pound of grapes; for that act has, according to the right margin of Table 2, an expected utility of 1, which is the largest that can be attained.

Now, suppose the person is free to make an observation, that is, a new observation in addition to those that may have contributed to the determination of the probabilities in the basic problem. It may be, for example, that the grocer invites him to eat a few of the grapes or that the person is going to ask the woman beside him how they look to her. Let there be five possible outcomes of his observation; call them x

generically and 1, 2, 3, 4, and 5 individually. I assume, though this feature is rather incidental to the example, that low values of x tend to be suggestive of low quality. The joint distribution of x and Q, that is, the probability that x and Q simultaneously have any given pair of values, is of central technical importance. Those probabilities, each multiplied by 128 for simplicity of presentation, are given in the body of Table 3. The right-hand and bottom margins of the table give,

TABLE 3. $128P(x \cap Q)$

x	Q 1	2	3	$128P(x)$
1	15	5	1	21
2	10	15	2	27
3	4	24	4	32
4	2	15	10	27
5	1	5	15	21
	32	64	32	128
		$128P(Q)$		

also multiplied by 128, the probability of each value of x and of each value of Q. The marginal entries are, of course, obtained by adding rows and columns. As indicated in the lower right-hand corner of the table, the probabilities assumed do indeed add up to 1, and the bottom margin recapitulates Table 1.

Conditional probabilities can easily be read from Table 3. Thus, for example, the conditional probability that x is 2, given that Q is 3, is $2/32$, and the conditional probability that Q is 2, given that x is 4, is $15/27$. It will be seen in later sections that the distribution of x given Q is, in a sense, even more fundamental than the joint distribution of x and Q.

There are $4^5 = 1{,}024$ derived acts, since one of the four basic acts can be assigned arbitrarily to each of the five possible outcomes of the observation. It is an easy exercise, using Tables 2 and 3, to verify Table 4, which shows the conditional expectation of the utility of each

TABLE 4. $E(\mathbf{f} \mid x)$

$\mathbf{f}$	x 1	2	3	4	5
0	0/21	0/27	0/32	0/27	0/21
1	−7/21	11/27	32/32	43/27	49/21
2	−40/21	−20/27	8/32	44/27	72/21
3	−94/21	−78/21	−48/32	18/27	74/21

basic act given each possible outcome of the observation. For each x, the highest expected utility, given that value of x, has been italicized. Thus, for example, only if x is 1 will the person refrain from buying grapes altogether, and only if x is 5 will he risk buying 3 pounds. In full, the best derived act, call it g, is to buy 0, 1, 1, **2**, or 3 pounds, if x is 1, 2, 3, 4, or 5, respectively. The value of the derived problem is the expected value of g, which is computed thus:

(1)
$$E(g) = \sum_x E(g \mid x) P(x)$$

$$= (0 + 11 + 32 + 44 + 74)/128$$

$$= 161/128 \simeq 1.26 \text{ utiles.}$$

Since the value of the basic problem is 1 utile, the envisaged observation is worth 0.26 utile; that is, the person would if necessary pay up to 0.26 utile for the observation.

Exercise

1. Suppose that the person could directly observe the quality of the grapes. Show that his best derived act would then yield 2 utiles, and show that it could not possibly lead him to buy 2 pounds of the grapes.

The notion of a decision problem based on an observation will now be formally described, with special reference to mathematical notation and other technical details.

1. There is a set of **basic acts, F** with elements **f, f′**, etc.

In the example of the grapes **F** consisted of the four envisaged acts of buying 0, 1, 2, or 3 pounds of grapes.

The convention laid down at the end of § 5.4, requiring that the consequences of acts be measured in utiles, will be adhered to, and it will be supposed that $v(\mathbf{F})$ is finite.

2. The **observation** is a (not necessarily real) random variable **x** associating with each state s an **observed value** $x(s)$ in some set X of possible observed values x, $x′$, etc.

In the example of the grapes, the states s (of which the postulates require that there be an infinite number) were never fully described, and consequently the random variable **x** was not fully described either. In the same sense it may be said that the basic acts, which are also really random variables, were not fully described either. All that is really important, however, is to know the simultaneous distribution of the consequences of the acts in **F** and of the values of **x**. In the example of the grapes that information was implicit in Tables 2 and 3.

For mathematical simplicity in the formal work to follow, it will generally be assumed that X has only a finite number of elements, though the assumption can and must be relaxed in many practical situations. When X is assumed finite, the random variable $\mathbf{x}$ is, for all purposes of the present context, simply a partition of S, namely, the partition into the sets on which $\mathbf{x}$ is constant. Indeed, earlier in this section, the notion of observation was described in terms of a partition, but the description in terms of a random variable is more familiar in statistics and may have technical advantages, especially when the restriction that X be finite is relaxed.

3. The set of *strategy functions* is the set of all functions associating an element of $\mathbf{F}$ with each element x of X. Let the values of the generic strategy function be denoted by $\mathbf{f}(x)$ and the function itself by $\mathbf{f}(\mathbf{x})$.

The notion of strategy function was not introduced in the informal description of observation, nor in the example of the grapes, because it is but a mathematical intermediary to the definition of derived acts and did not seem to call for explicit expression in the less formal contexts.

4. To each strategy function $\mathbf{f}(\mathbf{x})$ corresponds a **derived act g,** in the set of all derived acts $\mathbf{F}(\mathbf{x})$, defined by

$$(2) \qquad g(s) = f(s; x(s)) \qquad \text{for all } s \in S.$$

It was explained that in the example of the grapes there are 4^5 derived acts. In the same way, it can be seen in general that if X has ξ and $\mathbf{F}$ has ϕ elements there are ϕ^ξ derived acts.

5. The **value of F given** x,

$$(3) \qquad v(\mathbf{F} \mid x) =_{\mathrm{Df}} \sup_{\mathbf{f} \in \mathbf{F}} E(\mathbf{f} \mid x).$$

This is the function of x indicated, for the example of the grapes, by italics in Table 4.

3 Multiple observations, and extensions of observations and of sets of acts

If several random variables $\mathbf{x}_1, \cdots, \mathbf{x}_n$, associating elements of S with elements of sets $X_1, \cdots, X_n$, are simultaneously under discussion, it is natural to form the new random variable, denoted $\mathbf{x} = \{\mathbf{x}_1, \cdots, \mathbf{x}_n\}$, that associates with each element of S an ordered n-tuple of elements of $X_1, \cdots, X_n$, respectively. If the context is such that $\mathbf{x}_1, \cdots, \mathbf{x}_n$ are thought of as observations, then $\mathbf{x}$ can also be thought of as an observation and will sometimes be called a **multiple observation**—to emphasize the manner of its formation. To illustrate, any item such as profession or body temperature that might be entered on a patient's history can be thought of as an observation; but the whole history, or a filing cabinet of histories, can also be thought of as an observation, the history being a multiple observation of items, and the cabinet a multiple observation of histories.

Consider two observations **x** and **y**. It is an interesting possibility that **x** and **y** are so related to each other that knowledge of the value of **x** would (almost certainly) imply (almost certain) knowledge of **y**. In that case, observation of **x** implies essentially the observation of **y** and generally something besides, which suggests the following three definitions.

If and only if **x** and **y** are observations such that, for all s and s' in some B of probability one, $x(s) = x(s')$ implies $y(s) = y(s')$; then **x** is an **extension** of **y**, and **y** is a **contraction** of **x**. If **x** is an extension of **y**, and **y** is an extension of **x**, then **x** and **y** are **equivalent**.

Strictly speaking, one should say not that **x** and **y** are equivalent, but rather that they are equivalent regarded as observations, for this would not be a good concept of equivalence to apply to random variables regarded as such. For example, a pair of equivalent observations can obviously be a pair of real random variables with different expected values. Some properties of the relations of extension, contraction, and equivalence between observations are given by the following easy but important exercises. Throughout this set of exercises it is unnecessary to suppose the observations confined to a finite set of values; in the case of Exercise 3b, it is impossible to do so.

Exercises

1. **x** and **y** are equivalent, if and only if **x** is both an extension and a contraction of **y**.

2a. If $P\{x(s) = y(s)\} = 1$, **x** and **y** are equivalent.

2b. Any observation **x** is equivalent to itself.

3a. If there is a value y_0 such that $P\{y(s) = y_0\} = 1$, then every **x** is an extension of **y**, and any two such observations are equivalent. Such an observation, of course, amounts to observing nothing at all and will therefore be called a **null observation**.

3b. If $x(s) = s$ for almost all $s \in S$, then **x** extends every **y**.

4. If **x** is an extension of **y**, and **y** is an extension of **z**, then **x** is an extension of **z**. State and verify the analogous fact about equivalence.

5a. If **y'** is a function associating an element of Y with each element of X, and **x** is an observation, then the observation **y** such that $y = y'(x)$ is a contraction of **x**.

5b. If **y** is a contraction of **x**, then there is a function **y′** such that $P\{y(s) = y'(x(s))\} = 1$. What freedom is there in the choice of the function **y′**?

5c. What are the implications of Exercises 5a and 5b for equivalence between observations?

6. If **x** and **y** are observations and $z = \{x, y\}$ is the corresponding double observation, then **z** is an extension of **x** and of **y**. (This exercise seems to call for a converse saying that every extension can be regarded as a double observation, but no really neat one suggests itself to me. None the less, in thinking about extensions and contractions, the sort brought out by the exercise is a typical and stimulating example.)

7. $\{x, y\}$ is equivalent to **x**, if and only if **x** extends **y**.

The relations of extension, contraction, and equivalence have parallels for sets of acts, defined thus:

If **F** and **G** are (non-vacuous) sets of acts such that, for some B of probability one, there is for each **g** ε **G** an **f** ε **F** with $f(s) = g(s)$ for all s ε B; then **F** is an **extension** of **G**, and **G** is a **contraction** of **F**. If **F** is an extension of **G**, and **G** is an extension of **F**, then **F** and **G** are **equivalent**.

More exercises

8. If **F** is an extension of (equivalent to) **G**, then $v(\mathbf{F}) \geq (=) v(\mathbf{G})$.

9. Discuss the analogues of Exercises 1, 2b, and 4 for sets of acts.

10. If $\mathbf{F} \supset \mathbf{G}$, then **F** extends **G**.

11. If **F(x)** is derived from **F** on observation of **x**, then **F(x)** extends **F**.

12. HYP.

> **F(x)** is derived from **F** on observation of **x**;
> **F(y)** is derived from **F** on observation of **y**;
> **F(x, y)** is derived from **F** on observation of $\{x, y\}$;
> **F(x; y)** is derived from **F(x)** on observation of **y**.

CONCL.

1. **F(x, y)** is equivalent to **F(x; y)**.

2. **F(x, y)** extends **F(x)** and **F(y)**.

3. If **x** is equivalent to **y**, then **F(x)** is equivalent to **F(y)**.

4. If **y** extends **x**; then **F(x, y)** is equivalent to **F(y)**, **F(y)** is equivalent to **F(x; y)**, and **F(y)** extends **F(x)**.

13a. Under the hypothesis of 12, the equivalences and relations of extension among the sets of acts arising out of two observations can, with evident conventions, be diagrammed thus:

$$\boxed{\begin{array}{ccc} \mathbf{x};\mathbf{y} & \mathbf{x},\mathbf{y} & \mathbf{y};\mathbf{x} \end{array}}$$

$$\downarrow \qquad\qquad\qquad \downarrow$$

$$\mathbf{x} \longrightarrow 0 \longleftarrow \mathbf{y}$$

13b. If $\mathbf{y}$ extends $\mathbf{x}$, the diagram becomes

$$\boxed{\begin{array}{cccc} \mathbf{x};\mathbf{y} & \mathbf{x},\mathbf{y} & \mathbf{y};\mathbf{x} & \mathbf{y} \end{array}} \rightarrow \mathbf{x} \rightarrow 0.$$

13c. If $\mathbf{x}$ and $\mathbf{y}$ are equivalent, the diagram becomes

$$\boxed{\begin{array}{ccc} \mathbf{x};\mathbf{y} & \mathbf{x},\mathbf{y} & \mathbf{y};\mathbf{x} \\ \mathbf{x} & & \mathbf{y} \end{array}} \rightarrow 0.$$

14. If $\mathbf{F}(\mathbf{x})$ and $\mathbf{G}(\mathbf{x})$ are derived from $\mathbf{F}$ and $\mathbf{G}$, respectively, and if $\mathbf{F}$ extends $\mathbf{G}$, then $\mathbf{F}(\mathbf{x})$ extends $\mathbf{G}(\mathbf{x})$.

15. $v(\mathbf{F}(\mathbf{x})) = E[v(\mathbf{F} \mid \mathbf{x})] = \int v(\mathbf{F} \mid x(s))\, dP(s) \geq v(\mathbf{F}).$

4 Dominance and admissibility

According to Exercise 3.14, if one set of acts, regarded as basic, extends another, the first is at least as valuable as the second in the light of any observation whatever. This section explores a relation, dominance, which has the same property but is not so strict as extension. Dominance is of some importance for the theory of personal probability as it has been developed thus far. But its ·importance will be even greater in the study of statistics proper, where interpersonal agreement is of particular interest; for, as the definition shortly to be given will make clear, two people having different personal probabilities will agree as to whether one of two sets of acts dominates another, if only they agree which events have probability zero—a condition generally met in practice, and one that could if desired be dispensed with by a slight change in the definition of dominance.

It will be seen that dominance and notions related to it are intimately associated with the sure-thing principle. Indeed, probability being taken for granted, the basic facts about dominance seem to give a complete expression of the sure-thing principle. Dominance and related concepts were much stressed by Wald, in [W3] for example.

Two or three notions, the logical connections among them, and those between them and extension, are to be treated. The logical connections being many but simple, I think that the material lends itself better to formal than to expository treatment, for in such a context the reader who looks for the motivating ideas sees them himself more easily than he comprehends someone else's verbalization of them. This section will therefore consist primarily of a group of formal definitions and several exercises.

If and only if $P(f(s) \geq g(s)) = 1$, **f dominates g.** If and only if some (every) element of **F** dominates (is dominated by) **g, F dominates (is dominated by) g.** If and only if **F** dominates every element of **G, F dominates G.** If and only if **f** dominates **g,** but **g** does not dominate **f, f strictly dominates g.** If and only if **f** ε **F**, and **f** is not strictly dominated by any element of **F, f** is **admissible** (with respect to **F**).

Involving as they do acts as well as sets of acts, the definitions, strictly speaking, introduce four different kinds of dominance. However, this complexity can be alleviated, with a slight lapse of logic, by identifying each act **f** with the set of acts of which **f** is the only element, for it is easily seen that this identification is in such harmony with the definition that, once it is made, the four kinds of dominance collapse into one.

Exercises

1a. Consider analogues of Exercises 3.2b and 3.4.

1b. When can two acts dominate each other?

2a. If **F** extends **G,** then **F** dominates **G.** Discuss the converse.

2b. **F(x)** dominates **F.**

2c. If **F** ⊃ **G,** then **F** dominates **G.**

3a. If **F** ⊂ **G,** and **F** dominates **G,** then each admissible element of **G** dominates and is dominated by an element of **F.**

3b. After any finite number of non-admissible elements is deleted from **F,** what remains of any subset of **F** that dominated **F** continues to dominate **F.**

3c. Though the set of admissible elements of **F** may in some instances dominate **F,** no proper subset of the set of admissible elements can ever do so; but, if any other subset dominates **F,** some proper subset of it also does so.

3d. If **F** is finite, the set of admissible elements of **F** dominates **F.**

3e. Discuss the role of "finite" in 3b and 3d.

4a. If the set of admissible elements of **F** dominates **G**, and **G** dominates **F**, then the set of admissible elements of **F** is equivalent to the set of admissible elements of **G**.

4b. If **F** and **G** dominate each other, and either is finite, then the sets of admissible elements of **F** and **G**, respectively, are equivalent to each other, and each dominates both **F** and **G**.

5. If **F** dominates **G**, then $v(\mathbf{F}) \geq v(\mathbf{G})$.

6. If **F** dominates **G**, then, for any observation **x**, **F(x)** dominates **G(x)**.

5   Outline of the design of experiments

Often, especially in statistics, a decision problem can be seen as the problem of deciding which of several experiments—or which of several observational programs, if that is really a more general term—to undertake.

In this section the notion of the decision problem derived from a basic decision problem and an observation must be elaborated a little, because, as derived acts have been treated thus far, they correspond to the possibility of making an observation free of charge.   Though observations are sometimes free, there is typically a cost associated with making them; information must typically be bought either from other people or, more often from nature, so to speak.   The cost of information may be money, trouble, one's own life, that of another, or any of innumerable possibilities, but all can in principle be measured in terms of utility.   The cost of an observation in utility may be negative as well as zero or positive; witness the cook that tastes the broth.

In principle, if a number of experiments are available to a person, he has but to choose one whose set of derived acts has the greatest value to him, due account being taken of the cost of observation.   That simple formulation, like some others in this book, is, in a sense, oversimple; it abstracts from the enormous variety of considerations that enter into the careful design of any experiment.   The possibility of so abstracting from variety does not remove the ultimate necessity of studying some aspects of that variety in detail.   R. A. Fisher's *The Design of Experiments* [F4], for example, is concerned almost exclusively with experiments based on a special technique called the analysis of variance, and it is but an introduction to even that important facet of statistics.   Again, there is a growing literature (in which the work of A. Wald is outstanding) on sequential analysis, which is concerned in principle with all experiments in which later parts of the experiment are conducted in the light of what happens in earlier parts; but this literature has, by necessity, been confined to a relatively tiny part of that domain.

Before turning to a more formal recapitulation of the outline of the design of experiments, this may be a good place for a few speculative words about the difference, if any, between experiment and observation.

Some sciences are commonly called experimental as opposed to others that are called observational. Aerodynamics, the psychology of rote learning, and the genetics of fruit flies would typically be called experimental sciences; and, to take parallel examples, meteorology, the psychology of dreams, and human genetics would be called observational. But it is widely agreed, and the most casual consideration makes it clear, that any basic difference that may really be present resides not in the sciences themselves but in the methods typical of each. To illustrate the role of observation in sciences ordinarily considered experimental and vice versa, observations of wild populations of fruit flies have been useful in the study of the genetics of fruit flies; the effects of fatigue, for example, on dream content may well be the subject of an experiment; and, except for the atom, no topic in science is more popular today than experimental rain making. The illustrations could be extended indefinitely, and there is also a less direct sort exemplified by the discipline called experimental medicine, which typically studies experiments on animals with the hope, often justified, that the findings thus obtained can be extrapolated to humans.

The problem, then, is to distinguish *an* experiment from *an* observation. Except for brevity, it might be better to say mere observation, for, in general usage, an experiment would be considered a special sort of observation.

The first apparent contrast that comes to mind is that experimentation is generally thought of as active and observation as passive. But, upon examination, it is seen that observation is also active, for observations are typically made by going somewhere to observe, or waiting attentively till something happens. Often it is not only the observer himself who must be transported and put in readiness to make an observation, but also a considerable body of apparatus. What demands more activity than the modern observation of a solar eclipse?

Another apparent contrast is that the experimenter acts on the thing he observes, whereas the observer acts only on himself and on instruments of observation that may be regarded as extensions of his own sense organs. If this criterion were accepted altogether naively, there would be no such thing as a physiological experiment on one's self; even sophisticated interpretations might find it difficult to embrace psychological experiments on one's self.

Finally, experiments as opposed to observations are commonly supposed to be characterized by reproducibility and repeatability. But the observation of the angle between two stars is easily repeatable and with highly reproducible results in double contrast to an experiment to determine the effect of exploding an atomic bomb near a battleship. All in all, however useful the distinction between observation and experiment may be in ordinary practice, I do not yet see that it admits of any solid analysis. At any rate, no formal use of the distinction will be attempted in this book.

Return now to the notion of observation subject to cost. It may be that the value of the random variable x is observable but only at a cost c, a real-valued random variable measured in utiles. If, as heretofore, $F(x)$ denotes the set of acts derived from F on cost-free observation of x, let $F(x) - c$ denote the set of derived acts subject to the random cost c. This notation is interpreted to mean that, if f is the generic element of $F(x)$, then $f - c$ (which, being a utility-valued function of s, is an act) is the generic act of the set $F(x) - c$. Very often the cost of an observation is independent of s, but not, for example, for him that tests the sharpness of a thorn with his finger. Since observations are typically paid for before, or simultaneously with, making the observation, the cost is typically observed along with the observation proper. Put differently, the cost c is typically a contraction of the observation x. Thus, if in some special context any advantage were to be gained by so doing, it would not be drastic to assume the cost of observing x to be a function of the form $c'(x)$; but, as a matter of fact, no such advantage has come to my attention. It is not difficult to think of experiments to which the assumption does not apply. For example, in the present state of uncertainty about the long-term effects of x-rays, anyone conducting a short-term experiment in which young human beings were subjected to large doses of x-radiation would risk costs that might not overtly manifest themselves for half a century, or even for generations.

Much that would ordinarily be called observation cannot be described by saying that the random cost is simply to be subtracted from each derived act of the corresponding observation thought of as free of cost. Allowing that it may be legendary, the form of trial by ordeal in which the guilty floated safely to be hanged and the innocent drowned to be exonerated epitomizes such a situation; except in point of absurdity, ordinary industrial destructive testing of electric fuses and other products is much the same. Strictly speaking, discrepancy occurs even in the ordinary context in which the cost of observation is a fixed sum of money; for the utility of money is not strictly linear, so the cost of observation typically affects different derived acts somewhat differently. This sort of situation is indeed so common as to introduce at least a slight error into almost every application of the notion of cost as a sub-tractive term. It would therefore be desirable to extend considerably the notion of cost of observation, but, thus far, I see no way to do so that does not destroy the mathematical advantage of singling problems of observation out of the class of decision problems generally.

It is convenient now to analyze the appropriateness of regarding the number $v(\mathbf{F})$ as a measure of the value of $\mathbf{F}$. As must already be clear to the reader, if a person is to make a preliminary decision limiting his next decision to one or another of several sets of acts, say, $\mathbf{F}$, $\mathbf{G}$, and $\mathbf{H}$, then his preliminary decision will select a set that has the highest value of v, and the preliminary and secondary decisions, regarded as a single grand decision, amount to the problem of deciding on an act from $\mathbf{F} \cup \mathbf{G} \cup \mathbf{H}$. So far as this use of v is concerned, any increasing mono-tonic function of v such as v^3 or 3^v would be equally satisfactory, but v has an advantage in arithmetic simplicity when costs of observation are involved. Consider, for example, the problem of whether to make a particular observation at the random cost $\mathbf{c}$ or to make no observation at all. The two sets of acts involved may then be symbolized by $(\mathbf{F}(\mathbf{x}) - \mathbf{c})$ and $\mathbf{F}$, respectively. The peculiar simplicity of v as a meas-ure of the value of a set of acts, in this context, is exhibited by the almost obvious fact that $v(\mathbf{F}(\mathbf{x}) - \mathbf{c}) = v(\mathbf{F}(\mathbf{x})) - E(\mathbf{c})$. It may be remarked in passing that v is a particularly good measure in any problem where $\mathbf{F}$, $\mathbf{G}$, or $\mathbf{H}$ is, so to speak, made available by lot, a possibility realized in (7.3.2), for example.

Finally, if one among several observations is to be chosen, each with its own random cost (possibly including the null observation), the per-son will choose an observation for which $v(\mathbf{F}(\mathbf{x})) - E(\mathbf{c})$ is as large as possible. If the number of observations among which decision is to be made is infinite, that function may not attain a maximum value, but the value of the situation to the person can reasonably be regarded as the supremum of the function; there are, of course, observations among those available for which the supremum is arbitrarily nearly attained.

CHAPTER 7

Partition Problems

1 Introduction

In the introduction of the preceding chapter it was explained that the treatment of decision problems in general had been carried to a logical conclusion, and that to study decision problems further it had become necessary to specialize. The notion of observation was accordingly chosen as the subject of specialization. The situation now repeats itself at a new level, for I have now covered the main points that occur to me about observation in general, though I see considerably more to say about a certain type of observation.

The type of observation problem to which the present chapter is devoted, though relatively special, is still very general. Indeed, its generality is suggested by the fact that no other type of problem is systematically treated in modern statistics. In objectivistic terms, it would be described as the type of decision problem in which the consequence of each basic act depends only on which of several (possibly infinitely many) probability distributions does in fact apply to the random variable to be observed.

Modern statistics has no name for this type of problem, because it recognizes no other type; and no particularly suggestive name occurs to me. I am therefore tentatively adopting the noncommittal name "partition problem." Such motivation as there is for that name will be apparent when the concept is defined.

In non-objectivistic terms, a partition problem has the following structure. There are, of course, basic acts **F** and an observation x. The peculiar feature is a random variable **b**, which is typically not subject to observation, with the property that every **f** in **F** is constant given that **b** has any particular value b.

In many practical problems **b** takes on an infinity, even a non-denumerable infinity, of values, but systematic consideration of such problems would involve those advanced mathematical techniques that are explicitly being avoided in this book. Glossing over such questions of technique for the moment, the state of the world, which is itself a

120

random variable, might play the role of **b**; with respect to this **b**, any observational decision problem would presumably be a partition problem. It may, therefore, be inaccurate to call partition problems special, but they are special whenever **b** is not equivalent to the state of the world.

As has just been mentioned, the general policy of this book with respect to mathematical technique restricts formal treatment of partition problems here to those in which **b** assumes only a finite number of different values, that is to say, those in which **b** is to all intents and purposes a partition B_i, whence the name "partition problem." For the reader who is not familiar with the elements of the geometry of n-dimensional convex bodies, there will be a distinct expository advantage in confining the formal treatment still further to twofold partitions. At the same time, by explicit statements and by the use of suggestive notation, all readers will be given at least some idea of the extension of the theory to n-fold partitions; indeed, a reader familiar, for example, with Sections 16.1–2 of [V4], or with [B20] will find the extension as plain as if it had been made explicitly. Thus the restriction to twofold as opposed to n-fold partitions will be to the advantage of some and to the disadvantage of none.

Partition problems are even closer than are observational problems generally to the subject matter of statistics proper. In particular, in the course of this chapter, multipersonal considerations will from time to time be pointed out in connection with partition problems.

2   Structure of (twofold) partition problems

A central feature of a twofold partition problem is, of course, a twofold partition, or **dichotomy**, B_i, $i = 1, 2$. By way of abbreviation let $\beta(i) = P(B_i)$, and $\beta = \{\beta(1), \beta(2)\}$. The $\beta(i)$'s can be any two numbers such that $\beta(i) > 0$ and $\Sigma\beta(i) = \beta(1) + \beta(2) = 1$. Since $\beta(2) = 1 - \beta(1)$, it might seem superfluous to have a special notation for $\beta(2)$; but this redundancy more than pays for itself in symmetry, especially in the extension of the theory to n-fold partitions. The possibility that one of the $\beta(i)$'s vanishes has been ruled out, for it is neither typical nor interesting, and its retention would mar the exposition of the theory.

Each basic act $\mathbf{f} \in \mathbf{F}$ is characterized by a pair of numbers f_i such that

(1)
$$P(f(s) = f_i \mid B_i) = 1$$

for each i. The technical assumption will be made that as $\mathbf{f}$ ranges over $\mathbf{F}$ the numbers f_i are bounded from above for each i, which is a little more stringent than the now familiar assumption that $v(\mathbf{F}) < \infty$.

The assumption expressed by (1) is made for definiteness and simplicity, though its full force will seldom be used. The possibility of relaxing (1) in certain contexts will be mentioned from time to time, especially since this possibility is of some interest even in the exploitation of (1) itself. In particular, for several pages now it will scarcely ever be necessary to assume anything about the structure of $\mathbf{F}$ relative to B_i, except that $E(\mathbf{f} \mid B_i)$ is bounded from above for each i; for making the abbreviation $f_i = E(\mathbf{f} \mid B_i)$, almost everything from here through Exercise 1 applies verbatim.

The expected utility of any $\mathbf{f} \, \varepsilon \, \mathbf{F}$ can be computed in several forms thus:

$$
\begin{aligned}
(2) \qquad E(\mathbf{f}) &= E(\mathbf{f} \mid B_1)P(B_1) + E(\mathbf{f} \mid B_2)P(B_2) \\[2mm]
&= f_1\beta(1) + f_2\beta(2) \\[2mm]
&= \Sigma f_i\beta(i) \\[2mm]
&= f_2 + (f_1 - f_2)\beta(1).
\end{aligned}
$$

The first of these forms expresses the expected value in general terms; the second utilizes abbreviations; the third is an obvious mathematical transcription of the second, particularly suggestive of extension to the n-fold situation; the fourth sacrifices the symmetry exhibited by the preceding three in order to take advantage of the relation between $\beta(1)$ and $\beta(2)$. From the fourth form of (2), it is clear that, for fixed $\mathbf{f}$, $E(\mathbf{f})$ is a linear function of $\beta(1)$. Henceforth that fact, for example, would be expressed in symmetric form by saying that $E(\mathbf{f})$ is linear in β, and the dependence of $E(\mathbf{f})$ on β might be explicitly indicated by writing $E(\mathbf{f} \mid \beta)$.

Since in any one decision problem β is constant, it might seem pointless to emphasize that $E(\mathbf{f} \mid \beta)$ is linear in β. But there are, in fact, two different reasons for being interested in variation of β. In the first place, once the observation $\mathbf{x}$ has been observed to have the value x, the basic, or a priori, decision problem is replaced by an a posteriori problem in which $P(B_i \mid x)$ plays the role originally played by $P(B_i) = \beta(i)$. Second, interest in comparing different people is becoming increasingly more explicit as the book proceeds. In particular, it is of interest to compare people who have available the same set of basic acts and who, at least so far as the distribution of $\mathbf{x}$ and the acts in $\mathbf{F}$ are concerned, have the same conditional personal probability given B_i, but who attach different probabilities $\beta(i)$ to the elements of the partition.

To emphasize its dependence on β, $v(\mathbf{F})$ will sometimes be written $v(\mathbf{F} \mid \beta)$; its computation in the following fashion is fundamental to the theory of partition problems.

$$
(3) \qquad v(\mathbf{F} \mid \beta) = \sup_{\mathbf{f} \,\varepsilon\, \mathbf{F}} E(\mathbf{f} \mid \beta)
$$

$$
= \sup_{\mathbf{f} \,\varepsilon\, \mathbf{F}} [f_1 \beta(1) + f_2 \beta(2)]
$$

$$
= k(\beta),
$$

where $k(\beta)$ is defined by the equation in which it occurs. According to Exercise 4 of Appendix 2, the function $\mathbf{k}$ is convex in β, that is, $\mathbf{k}$ is convex when recognized as a function of $\beta(1)$ alone. Interpreted as a pair of a priori probabilities, β is confined to the open interval defined by $\Sigma\beta(j) = 1$, $\beta(i) > 0$, but it is valuable to recognize that $\mathbf{k}$ is defined, convex, and continuous on the closed interval $\Sigma\beta(j) = 1$, $\beta(i) \geq 0$.

Many typical features of the relationship between $\mathbf{F}$ and B_i are illustrated graphically by Figure 1. The abscissa of that graph represents

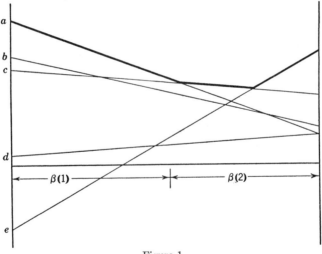

Figure 1

both $\beta(1)$ and $\beta(2)$, as indicated, and the ordinate is measured in utiles. The straight lines, the left ends of which are marked a, b, c, d, and e, graph as functions of β the expected values of the five basic acts of the particular problem represented. The ordinates at their right and left ends, respectively, are the corresponding values of the f_1's and f_2's. The graph of $\mathbf{k}$ is marked by heavy line segments. It is seen that the lines a, c, and e, and they alone, touch the graph of $\mathbf{k}$, for they represent the only acts that are optimal for some value of β. The act represented by d is inadmissible (if (1) is taken literally), being in fact strictly dominated by every other act except e, and it is therefore superfluous to the person, no matter what the value of β; b is obviously equally superfluous, but for a different reason.

In many typical problems in which $\mathbf{F}$ has an infinity of elements, $\mathbf{k}$ is, unlike the $\mathbf{k}$ in Figure 1, strictly convex; that is, its only intervals of linearity are point intervals.

Exercise

1. Compute and graph $\mathbf{k}$ for the set $\mathbf{F}$ of dichotomous acts of the form

$$f_1(\phi) = 1 - (1 + \phi)^2;$$
$$-2 \le \phi \le +2$$
$$f_2(\phi) = 1 - (1 - \phi)^2;$$

Answer. $k(\beta) = [\beta(1) - \beta(2)]^2 = [2\beta(1) - 1]^2.$

Turn now to the relations between an observation $\mathbf{x}$ and the dichotomy B_i. As before, it will be assumed for mathematical simplicity that the values of $\mathbf{x}$ are confined to a finite set X. The probability that $\mathbf{x}$ attains the value x given B_i, written $P(x \mid B_i)$, is fundamental in connection with partition problems. For one thing, as has already been indicated, there is interest in considering people who, though differing with respect to β, agree with respect to $P(x \mid B_i)$. The probability $P(x, B_i)$ that $\mathbf{x}$ attains the value x and that B_i simultaneously obtains, the probability $P(x)$ that $\mathbf{x}$ attains the value x, and the probability $\beta(i \mid x)$ of B_i given that $x(s) = x$ are derived from $P(x \mid B_i)$ and β by means of Bayes' rule (3.5.4) and the partition rule (3.5.3) thus:

(4) $$P(x, B_i) = P(x \mid B_i)\beta(i).$$

(5) $$P(x) = \sum_i P(x, B_i).$$

(6) $$\beta(i \mid x) = P(x, B_i)/P(x),$$

if $P(x) \neq 0$; and $\beta(i \mid x)$ is meaningless otherwise. It must be remembered that $P(x, B_i)$, $P(x)$, and $\beta(i \mid x)$ depend on the value of β and that a really complete notation would show that dependence. On the other hand, the condition that $P(x) \neq 0$ is independent of the value of β.

When a second observation $\mathbf{y}$ is to be discussed, $\beta(i \mid y)$ is, in defiance of strict logic, to be understood as the analogue of $\beta(i \mid x)$; that is, as the conditional probability of B_i given that $y(s) = y$, not as the same function as $\beta(i \mid x)$ with y substituted for x. Corresponding conventions apply to $P(y)$, $P(y \mid B_i)$, and $P(y, B_i)$. Finally, free use will be made of such contractions as $\beta(x)$ for $\{\beta(1 \mid x), \beta(2 \mid x)\}$.

Equation (1) implies that

$$(7) \qquad E(\mathbf{f} \mid B_i, x) = E(\mathbf{f} \mid B_i)$$

for all $\mathbf{f} \; \varepsilon \; \mathbf{F}$ and for all x such that $P(x \mid B_i) > 0$. Equation (7) is the mathematical essence of the concept of a partition problem, and virtually all that is to be said about partition problems applies verbatim, if (7), even without (1), applies to such observations as may be under discussion.

In view of (7),

$$(8) \qquad E(\mathbf{f} \mid \beta, x) = \sum_i E(\mathbf{f} \mid B_i, x) P(B_i \mid x)$$

$$= \sum_i f_i \beta(i \mid x),$$

if $P(x) > 0$.

3　The value of observation

If the observation $\mathbf{x}$ is made, and it is found that $\tilde{x}(s) = x$, then the a posteriori value of the set of basic acts, written $v(\mathbf{F} \mid x)$, or more fully $v(\mathbf{F} \mid \beta, x)$, will typically be different from the a priori value $v(\mathbf{F} \mid \beta)$. Indeed, in view of (2.8),

$$(1) \qquad v(\mathbf{F} \mid \beta, x) = \sup_{\mathbf{f} \, \varepsilon \, \mathbf{F}} E(\mathbf{f} \mid \beta, x)$$

$$= v(\mathbf{F} \mid \beta(x))$$

$$= k(\beta(x)).$$

This is the first illustration of the technical convenience of the function $\mathbf{k}$.

It is known on general principles that $v(\mathbf{F}(\mathbf{x})) \geq v(\mathbf{F})$, but there is some interest in reverifying the inequality in the present context; in particular, it is possible here to say in interesting terms just when equality can obtain.

$$(2) \qquad v(\mathbf{F}(\mathbf{x}) \mid \beta) = E(v(\mathbf{F} \mid \beta(\mathbf{x})) \mid \beta)$$

$$= E(k(\beta(\mathbf{x})) \mid \beta)$$

$$\geq k(E(\beta(\mathbf{x}) \mid \beta)),$$

where the terminal inequality is an application of Theorem 1 of Appendix 2. To appreciate the inequality (2), it is necessary to calculate $E(\beta(i \mid \mathbf{x}))$ explicitly. This calculation, typical of many the reader must henceforth be expected to make for himself, runs as follows, where it is to be understood that the summation with respect to x applies only to those terms for which $P(x)$ is different from 0.

$$(3) \qquad E(\beta(i \mid \mathbf{x}) \mid \beta) = \sum_x \beta(i \mid x)P(x)$$

$$= \sum_x \frac{P(x, B_i)}{P(x)} P(x)$$

$$= \sum_x P(x, B_i)$$

$$= P(B_i) = \beta(i).$$

Substituting (3) into (2) leads to the anticipated conclusion that

$$(4) \qquad v(\mathbf{F(x)} \mid \beta) \geq k(\beta) = v(\mathbf{F} \mid \beta).$$

According to Theorem 1 of Appendix 2, $v(\mathbf{F(x)} \mid \beta)$ is definitely greater than $v(\mathbf{F} \mid \beta)$ unless $\beta(\mathbf{x})$ is confined with probability one to some interval of linearity of $\mathbf{k}$, in which case the observation $\mathbf{x}$ may fairly be called **irrelevant** to the basic decision problem at hand. If $\mathbf{x}$ is irrelevant, the interval of linearity to which $\beta(\mathbf{x})$ is confined must, in view of (3), contain β. In the particularly interesting case—and the only possible one, if $k(\beta)$ is strictly convex—in which $\beta(\mathbf{x})$ is with probability one equal to a constant value, that value must therefore be β. An observation for which $\beta(\mathbf{x})$ is with probability one equal to β may fairly be called **utterly irrelevant,** because it is irrelevant no matter what set $\mathbf{F}$ of basic acts is associated with the dichotomy.

To say that $\mathbf{x}$ is utterly irrelevant is to say that, with probability one,

$$(5) \qquad \beta(i \mid x) = \frac{P(x \mid B_i)\beta(i)}{P(x)}$$

$$= \beta(i).$$

Since $\beta(i) > 0$, (5) is equivalent to the condition that

$$(6) \qquad P(x \mid B_i) = P(x),$$

at least when $P(x) > 0$. Furthermore, it is obvious from (2.5), again noting that $\beta(i) > 0$, that, if $P(x) = 0$, then $P(x \mid B_i) = 0$. Therefore $\mathbf{x}$ is utterly irrelevant, if and only if (6) holds for all x and i; that is, if and only if the distribution of $\mathbf{x}$ given B_i is independent of i. This form of the condition is intuitively evoked by the words "utterly irrelevant" and has the advantage of not involving β.

It is noteworthy that whether an observation is utterly irrelevant depends neither on the particular set of basic acts, nor on the value of β, so people will agree on what is utterly irrelevant independent of their personal a priori probabilities and the acts among which they are free to choose.

The greatest lower bound in $\mathbf{x}$ of $v(F(\mathbf{x}) \mid \beta)$, namely $v(\mathbf{F} \mid \beta)$, and the circumstances under which this bound is attained having been established, it is natural to turn to a parallel investigation of the least upper bound. A foothold for that investigation is found in the remark that the chord joining the ends of the graph of $\mathbf{k}$ never lies below the graph. Analytically,

$$(7) \qquad k(\beta) \leq \beta(1)k(1, 0) + \beta(2)k(0, 1) = l(\beta),$$

where $l(\beta)$ is defined by the context. Unless one of the $\beta(i)$'s vanishes, equality holds in (7), if and only if $k(\beta)$ is a linear function. In view of (7) and (3),

$$(8) \qquad v(\mathbf{F}(\mathbf{x}) \mid \beta) = E(k(\beta(\mathbf{x})) \mid \beta) \leq E(l(\beta(\mathbf{x})) \mid \beta) = l(\beta).$$

The inequality (8) gives an upper bound for $v(\mathbf{F}(\mathbf{x}))$. In graphical terms it says that, for any β, no observation can add more to the value $k(\beta)$ of $\mathbf{F}$ than the vertical distance at β between the graph of $\mathbf{k}$ and the graph of the chord joining the ends of $\mathbf{k}$.

Equality obtains in (8), if $\mathbf{k}$ is linear, in which case the upper and lower bounds are equal to each other irrespective of the value of β and the nature of the observation. If $\mathbf{F}$ is dominated by a single $\mathbf{f}$, that is, if there is a single $\mathbf{f}$ optimal given B_i for both values of i, then $\mathbf{k}$ is linear. It can easily be verified that, provided $\mathbf{F}$ is finite and (1) actually obtains, this is indeed the only circumstance under which $\mathbf{k}$ is linear, and, even if these provisions are not satisfied, the possibilities are not much more interesting.

Suppose, then, that $\mathbf{k}$ is not linear; equality can hold in (8), if and only if $\beta(\mathbf{x})$ is with probability confined to the ends of the interval, a condition that does not depend at all on $\mathbf{F}$. By simple considerations, which have by now been rendered familiar, this condition on $\mathbf{x}$ is equivalent to the condition that

$$(9) \qquad P(x \mid B_1)P(x \mid B_2) = 0,$$

for all x. An observation satisfying (9) may fairly be called **definitive,** because, if (1) obtains, such an observation removes all uncertainty about the outcome of each $\mathbf{f} \; \varepsilon \; \mathbf{F}$, no matter what β may be.

Perhaps many of the observations made in everyday life are definitive, or practically so. Once Old Mother Hubbard looked in the cupboard, her doubts were reduced to the vanishing point. None the less, definitive observations do not play an important part in statistical theory, precisely because statistics is mainly concerned with uncertainty, and there is no uncertainty once an observation definitive for the context at hand has been made.

4 Extension of observations, and sufficient statistics

It was shown in § 6.4 that a **statistic**, or contraction, **y** of an observation **x** is never worth more than **x** and is typically worth less. The purpose of the present section is to explore the relation between an observation and a contraction of itself in the case of a partition problem, especially to explore the special conditions in that case under which the statistic is as valuable as the observation itself.

Let **x** and **y** be two observations such that **y** is a statistic of **x**, that is, such that, for some function **y**′, $y(s) = y'(x(s))$ with probability one. The values of $F(x)$ and $F(y)$ can be compared by the following calculation, which in the light of the preceding section will need but little explanation.

$$(1) \qquad v(\mathbf{F}(\mathbf{x})) = E(k(\beta(\mathbf{x})) \mid \beta)$$

$$= \sum_y E(k(\beta(\mathbf{x})) \mid \beta, y) P(y).$$

$$(2) \qquad E(k(\beta(\mathbf{x}) \mid \beta, y) \geq k(E(\beta(\mathbf{x})) \mid \beta, y)),$$

if $P(y) > 0$.

$$(3) \qquad E(\beta(i \mid \mathbf{x}) \mid \beta, y) = \sum_x \beta(i \mid x) P(x \mid y)$$

$$= \sum_x \frac{\beta(i \mid x) P(x, y)}{P(y)},$$

if $P(y) > 0$.

Because of the special relationship between **x** and **y**, $P(x, y) = 0$ unless $y'(x) = y$, in which case $P(x, y) = P(x)$. Understanding that the summation indicated by Σ' in (4) below extends only over those values of x for which $y'(x) = y$, the calculation is continued thus:

$$(4) \qquad E(\beta(i \mid \mathbf{x}) \mid \beta, y) = \Sigma' \frac{P(x, B_i)}{P(x)} \frac{P(x)}{P(y)}$$

$$= \Sigma' \frac{P(x, B_i)}{P(y)}$$

$$= \frac{P(y, B_i)}{P(y)}$$

$$= \beta(i \mid y).$$

Therefore,

$$(5) \qquad v(\mathbf{F}(\mathbf{x}) \mid \beta) \geq \sum_y k(\beta(y)) P(y) = v(\mathbf{F}(\mathbf{y}) \mid \beta).$$

After the preceding section, it seems almost superfluous to explain that the point of the calculation above is not to obtain the inequality (5), which has already been derived with less labor and greater generality in Exercises 6.3.8 and 6.3.13b, but to be able to discuss when equality holds in (5). The calculation makes it clear that equality holds in (5), if and only if equality holds in (2) for every y of positive probability. This in turn is equivalent to the condition that, given y, $\beta(\mathbf{x})$ is confined with probability one to an interval of linearity of $\mathbf{k}$. A sufficient condition for that is that, given y, $\beta(\mathbf{x})$ be confined with probability one to a single value, which cannot be other than $\beta(y)$; if $\mathbf{k}$ is strictly convex, the almost certain confinement of $\beta(\mathbf{x})$ to $\beta(y)$ is also necessary. Now, if, for every y of positive probability, $P(\beta(x(s)) = \beta(y) \mid y) = 1$, then it is true that $\beta(x) = \beta(y)$ with unconditional probability one, that is,

(6) $$P(\beta(x(s)) = \beta(y(s))) = 1.$$

The condition (6) clearly does not depend on $\mathbf{F}$, and the following calculation so expresses it as to make clear that it does not depend on β either. Equation (6) is satisfied, if and only if

(7) $$\frac{P(x \mid B_i)\beta(i)}{P(x)} = \frac{P(y'(x) \mid B_i)\beta(i)}{P(y'(x))},$$

when $P(x) > 0$; or, if and only if

(8) $$\frac{P(x \mid B_i)}{P(y \mid B_i)} = \frac{P(x)}{P(y)},$$

when $P(x \mid B_i) > 0$; or, again, if and only if

(9) $$P(x \mid B_i, y) = P(x \mid y),$$

when $P(y \mid B_i) > 0$; or finally if and only if $P(x \mid B_i, y)$ is independent of i for those values of i for which it is defined. In this form, and yet another to be derived in connection with (10), the condition is widely studied in modern statistical theory and a statistic satisfying the condition is there called a **sufficient statistic**. The name is well justified; for, as has just been shown, it is sufficient, for any purpose to which $\mathbf{x}$ might be put, to know $\mathbf{y}$, if and only if $\mathbf{y}$ is a sufficient statistic for $\mathbf{x}$.

A different, and perhaps more congenial, approach to sufficient statistics is the following. If the person observes the particular value y of $\mathbf{y}$, his original basic decision problem is replaced by a new one with the same basic acts, but with β replaced by $\beta(y)$. Strictly speaking, this will fail to be a partition problem, in case $\beta(y)$ is $(0, 1)$ or $(1, 0)$, or, for brevity, if $\beta(y)$ is *extreme*. To see whether $v(\mathbf{F}(\mathbf{x}) \mid \beta)$ is really greater than $v(\mathbf{F}(\mathbf{y}) \mid \beta)$, it is enough to investigate whether, for some y of positive probability for which $\beta(y)$ is not extreme, $\mathbf{x}$ is relevant to the partition problem based on $\beta(y)$, for if $\beta(y)$ is extreme there can be no value in following the observation that y has occurred by the observation of $\mathbf{x}$. Therefore, $\mathbf{x}$ will be a worthless addition to $\mathbf{y}$, if, for every y for which $\beta(y)$ is not extreme, $\mathbf{x}$ is utterly irrelevant, that is, if $\mathbf{y}$ is sufficient for $\mathbf{x}$. If $\mathbf{k}$ is strictly convex, the condition is also necessary.

The recognition of sufficient statistics in explicit problems is often facilitated by the following **factorability criterion**. A statistic $\mathbf{y}$ is sufficient for $\mathbf{x}$ if and only if there exists at least one pair of functions $\mathbf{R}$ and $\mathbf{S}$ such that

$$(10) \qquad P(x \mid B_i) = R(y'(x); i)S(x).$$

The necessity of the condition follows from the exhibition of a particular $\mathbf{R}$ and $\mathbf{S}$ for a sufficient statistic thus:

$$(11) \qquad \begin{aligned} P(x \mid B_i) &= \sum_y P(x \mid B_i, y)P(y \mid B_i) \\ &= \sum_y P(x \mid y)P(y \mid B_i) \\ &= P(y'(x) \mid B_i)P(x \mid y'(x)). \end{aligned}$$

On the other hand, if $P(x \mid B_i)$ can be expressed in the form (10), $\mathbf{y}$ can be seen to be sufficient for $\mathbf{x}$ thus: If $P(x \mid B_i, y)$ is meaningful, it is given by

$$(12) \qquad \begin{aligned} P(x \mid B_i, y) &= \frac{P(x, y \mid B_i)}{P(y \mid B_i)} \\ &= 0, \qquad\qquad \text{if } y'(x) \neq y, \\ &= \frac{P(x \mid B_i)}{P(y \mid B_i)}, \qquad \text{if } y'(x) = y, \\ &= \frac{S(x)}{\displaystyle\sum_{y(x')=y} S(x')}, \end{aligned}$$

which is independent of i. The reader may be interested in asking himself, as an exercise, what freedom there is in choosing $\mathbf{R}$ and $\mathbf{S}$ when at least one such pair of factors exists.

Interest in sufficient statistics is not confined, of course, to twofold, or even finite, partitions. With that in mind, the various criteria for sufficient statistics have been given in such terms as to be valid for any finite partition and the usual infinite ones. They require some modification if the observations are not confined to a finite, or at any rate denumerable, set of values, but formal details of that important extension will not be given here. Elementary treatments are given in most textbooks of mathematical statistics; more advanced and general treatments are given in [B2], [L6], and [H3].

There are several examples of sufficient statistics in the exercises below, others are given in almost any fairly advanced textbook on statistics (in particular, in [C9]), and one other general example of extraordinary importance is treated in the next section.

Exercises

In these exercises, let $\mathbf{x}$ denote a multiple observation $\mathbf{x} = \{\mathbf{x}_1, \cdots, \mathbf{x}_n\}$, where, given B_i, the $\mathbf{x}_r$'s are independent and identically distributed. There will be no real advantage here in thinking of the partition as twofold, or even finite, and for some of the exercises it will be impractical to do so.

1. Let $P(x_r \mid B_i) = p_i,$ if $x_r = 1,$

           $= q_i,$ if $x_r = 0,$

           $= 0,$ otherwise,

where $p_i + q_i = 1$; and let $y'(x) = \sum_r x_r$.

Show that:

(a) $P(x \mid B_i) = p_i^y q_i^{n-y}$;

(b) $\mathbf{y}$ is sufficient for $\mathbf{x}$, using the factorability criterion;

(c) $P(y \mid B_i) = \binom{n}{y} p_i^y q_i^{n-y}$, where, as always, $\binom{n}{y} = n!/y!(n-y)!$;

(d) $P(x \mid y'(x)) = \binom{n}{y'(x)}^{-1}.$

2. For each positive integer i, let

$$P(x_r \mid B_i) = i^{-1}, \qquad \text{if } x_r \le i,$$

$$= 0, \qquad \text{otherwise,}$$

where the values of $\mathbf{x}_r$ are confined to the positive integers; and let $y'(x) = \max_r x_r$. Show that:

(a) $P(x \mid B_i) = i^{-n},$ if $y \le i,$

           $= 0,$ otherwise;

(b) $\mathbf{y}$ is sufficient for $\mathbf{x}$.

3. In the two exercises above it has been possible to choose the factor S identically equal to 1. To exhibit a more typical example, let i, x_r, and y be confined to the positive integers with $y'(x) = \max x_r$, as in the preceding exercise, and let

$$P(x_r \mid B_i) = \frac{2x_r}{i(i + 1)}, \qquad \text{if } x_r \le i,$$

$$= 0, \qquad \text{otherwise.}$$

Show that:

(a) $P(x \mid B_i) = \left(\frac{2}{i(i + 1)}\right)^n \prod_r x_r, \qquad \text{if } y \le i,$

$$= 0, \qquad \text{otherwise.}$$

(b) $\mathbf{y}$ is sufficient for $\mathbf{x}$.

4. Put no restriction on the conditional distributions $P(x_r \mid B_i)$, except that $\mathbf{x}_r$ be confined with probability one to some fixed finite set. Say, for the moment, that two values x and x' of $\mathbf{x}$ are *team mates*, if one arises from the other by permutation of the component observations. This divides the possible values of $\mathbf{x}$ into *teams*, and, academic though it may seem, the team to which x belongs can be taken as $y'(x)$. Show that the probability of x given $y'(x)$ and B_i is independent of i (if it is defined at all), so that the statistic $y'(\mathbf{x})$ is sufficient for $\mathbf{x}$.

If the values of the $\mathbf{x}_r$'s happen to be real numbers, then for any x it is possible to permute the component observations to obtain a non-decreasing sequence of n (not necessarily distinct) numbers, and only one such non-decreasing sequence can be so obtained from each x. The sequence thus attached through $\mathbf{x}$ to each s is called in statistical usage the sequence of **order statistics** corresponding to $\mathbf{x}$. Since team mates, and only team mates, have the same order statistics, the set of order statistics regarded as a single statistic is equivalent to the team statistic $y'(\mathbf{x})$ defined more generally in the paragraph above and is therefore sufficient.

5. Let $\mathbf{x}_r$ given B_i be subject to the **normal probability density with mean μ_i, and variance σ_i^2**, that is,

$$(13) \qquad \phi(x_r \mid B_i) = (2\pi\sigma_i^2)^{-\frac{1}{2}} \exp\left\{-(x_r - \mu_i)^2/2\sigma_i^2\right\}.$$

This situation, though elementary, does not fall within the technical scope of this book, because $\mathbf{x}_r$ is not confined to a finite set of values. The reader familiar with probability densities will see, however, that the density of $\mathbf{x}$ is

$$(14) \quad \phi(x_1, \cdots, x_n \mid B_i) = (2\pi\sigma_i^2)^{-n/2} \exp\left\{-\frac{\Sigma x_r^2}{2\sigma_i^2} + \frac{\mu_i \Sigma x_r}{\sigma_i^2} - n\frac{\mu_i^2}{2\sigma_i^2}\right\},$$

which suggests that $\mathbf{y}$, defined by

(15) $y'(x) = \{\Sigma x_r^2, \Sigma x_r\},$

may fairly be called a sufficient statistic for $\mathbf{x}$.

Show in the same heuristic way that, if σ_i is independent of i, then $y'(x) = \Sigma x_r$ defines a sufficient statistic; and that, if μ_i is independent of i, then $y'(x) = n\Sigma x_r^2 - (\Sigma x_r)^2$ does so.

6. If $\mathbf{w}$ and $\mathbf{z}$ are observations independent of each other given B_i, under what conditions can $\mathbf{w}$ be sufficient for $\{\mathbf{w}, \mathbf{z}\}$?

7. To break away from independent observations, suppose that, in the event B_i, n cards are dealt from a thoroughly shuffled deck of $n + i$ cards each bearing a different serial number from 1 through $n + i$. Let $\mathbf{w}_r$ be the number on the rth card dealt and $\mathbf{w} = \{\mathbf{w}_1, \cdots, \mathbf{w}_n\}$. Show that $\max_r \mathbf{w}_r$ defines a sufficient statistic for $\mathbf{w}$ and that the $\mathbf{w}_r$'s are not independent.

8. If $\mathbf{z}$ extends $\mathbf{w}$, and $\mathbf{w}$ is sufficient for $\mathbf{y}$, then $\mathbf{z}$ is also sufficient for $\mathbf{y}$.

9. If $\mathbf{z}$ is sufficient for $\mathbf{w}$, and $\mathbf{y}$ is independent of both $\mathbf{z}$ and $\mathbf{w}$, then $\{\mathbf{z}, \mathbf{y}\}$ is sufficient for $\{\mathbf{w}, \mathbf{y}\}$.

10. Every definitive statistic is sufficient.

In virtually all statistics texts it would be said that the $\mathbf{y}$ defined by (15) constitutes not one statistic, but two; similarly, the set of order statistics would ordinarily be referred to as n statistics rather than as one. There are contexts in which it is appropriate to try to count statistics in that fashion, but, so far as the theory of sufficient statistics is concerned, it often seems fruitless, if not positively detrimental, to do so.

The concept of sufficient statistics has proved of great value in statistical theory and practice. The reason for this does not seem to me altogether easy to analyze, but, as the exercises above illustrate, the families of distributions most frequently studied in statistics are generally rich in sufficient statistics. It is hard to separate cause from effect here; for the distributions that are most studied tend to be those having the greatest mathematical simplicity, and the presence of striking sufficient statistics, such as those exhibited by Exercises 1, 2, 3, 5, and 7, are among the sources of mathematical simplicity most often met in the study of particular families of distributions.

It must be emphasized that sufficient statistics often provide a significant saving in the mechanical labor of storing and presenting data. Thus, in any experiment faithfully represented by Exercise 1, it is sufficient, in both the technical and ordinary senses of the word, to record a single integer y in place of the list of x_r's, which might well be very long. Several of the other exercises would in principle also lead to great savings of this sort, but Exercise 5 is the only other that arises frequently in practice.

The concept of sufficient statistics was introduced, together with much of the theory associated with it, by R. A. Fisher (cf. index, [F6]). The subject has been one of continuing interest and has been explored in several directions; key references are [B2], [E1], [L6], [H3], [K15], and [M5], and (LeCam 1964).

5   Likelihood ratios

The random variable $\beta(\mathbf{x})$ has played so important a role in preceding sections that the reader will probably not be surprised to find that $\beta(\mathbf{x})$ is a sufficient statistic for $\mathbf{x}$, a conclusion that, in the light of the factorability criterion (4.10), can be seen thus:

$$
(1) \qquad P(x \mid B_i) = \frac{P(B_i \mid x)}{\beta(i)} P(x)
$$

$$
= \frac{\beta(i \mid x)}{\beta(i)} P(x).
$$

If a statistic is sufficient, it is sufficient irrespective of the value of β; moreover, any multiple of it by a non-zero constant is also sufficient. Therefore, (1) implies that for any numbers $\alpha(i)$, such that $\alpha(i) > 0$, the multiple observation $\mathbf{r}(\alpha)$ defined by

$$
r_i(x; \alpha) =_{\mathrm{Df}} \frac{P(x \mid B_i)}{\Sigma \alpha(j) P(x \mid B_j)}
$$

$$
(2)
$$

$$
r(x; \alpha) =_{\mathrm{Df}} \{r_1(x, \alpha), r_2(x, \alpha)\}
$$

is a sufficient statistic for $\mathbf{x}$. Since

$$
(3) \qquad \sum_j \alpha(j) r_j(x; \alpha) = 1
$$

there is some redundancy in retaining both components, but this redundancy is more than compensated by the advantage of retaining symmetry, especially when n-fold partitions are contemplated.

Formally, the $\mathbf{r}(\alpha)$'s are an infinite family of sufficient statistics, one for each α; but to all intents and purposes they represent but one sufficient statistic, for any $\mathbf{r}(\alpha)$ is equivalent to any other, say $\mathbf{r}(\alpha')$, as can be demonstrated thus:

$$(4) \qquad r_i(x, \alpha) = \frac{P(x \mid B_i)/\Sigma\alpha'(k)P(x \mid B_k)}{\Sigma\alpha(j)\{P(x \mid B_j)/\Sigma\alpha'(k)P(x \mid B_k)\}}$$

$$= \frac{r_i(x, \alpha')}{\Sigma\alpha(j)r_j(x, \alpha')}$$

Having such a multiplicity of forms for what is essentially one important statistic is rather embarrassing, so there is some incentive to pick a standard form. Setting each $\alpha(j) = 1$ recommends itself as convenient and leads to the particular statistic $\mathbf{r} = \{\mathbf{r}_1, \mathbf{r}_2\}$, where

$$(5) \qquad \mathbf{r}_i(x) = \frac{P(x \mid B_i)}{\sum_j P(x \mid B_j)}.$$

This form is indeed convenient for twofold and, more generally, for n-fold partitions, but, where infinite partitions are to be dealt with, its apparent naturalness is misleading, for the sum in the denominator of (5) is then typically divergent. In the case of twofold partitions, a convenient form for the statistic is that of a likelihood ratio, in the sense introduced in § 3.6, for it is easy to see that, infinite numbers being admitted, $P(x \mid B_1)/P(x \mid B_2)$ is equivalent to $\mathbf{r}$. Henceforth, any statistic equivalent to $\mathbf{r}$ will be called a **likelihood ratio** of $\mathbf{x}$ with respect to the partition B_i—a definition that does not seriously conflict with ordinary statistical usage of the term.

Figure 1 illustrates a geometric interpretation of likelihood ratios that is sometimes valuable. The figure can best be described by telling how to draw it. First draw a pair of cartesian coordinate axes for variables u_1 and u_2. Next draw the two line segments represented by $u_1 + u_2 = 1$ and $(u_1/\alpha(1)) + (u_2/\alpha(2)) = 1$ with the u_i's non-negative. The left ends of these segments are indicated in Figure 1 by a and b, respectively, the particular value $\alpha = \{1/3, 2/3\}$ being used for illustration. Now plot the point $\{P(x \mid B_1), P(x \mid B_2)\}$. If x has positive probability (for any, and therefore for all, β); this point will be different from the origin O, so it will be possible to draw the (dashed) line connecting the origin with the point $\{P(x \mid B_1), P(x \mid B_2)\}$. This line (or ray through the origin, as it is often called) must necessarily pierce the line segments a and b. The important geometrical fact, which the reader will have no difficulty in verifying, is that these intersections occur at the points $\{r_1(x), r_2(x)\}$ and $\{r_1(x, \alpha), r_2(x, \alpha)\}$, respectively.

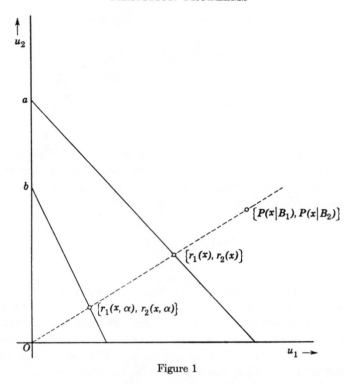

Figure 1

It is also obvious that the ratio $P(x \mid B_1)/P(x \mid B_2)$ is the reciprocal of the slope of the ray.

Since, to each x that occurs with positive probability, there corresponds a ray through the origin, the ray can be taken as a statistic; according to the geometrical construction of the preceding paragraph, this statistic is equivalent to $\mathbf{r}$ and is therefore a likelihood ratio of $\mathbf{x}$ with respect to the partition B_i.

The ray connecting the origin with a point $\{u_1, u_2\}$ can conveniently be represented by the suggestive notation $u_1 : u_2$, though, of course, different pairs of numbers can represent the same ray. More explicitly, if λ is any number different from 0, $\lambda u_1 : \lambda u_2$ represents the same ray as $u_1 : u_2$. In analytical projective geometry any pair of numbers representing a ray in this fashion is called a set of **homogeneous coordinates** of the ray. The redundancy of the notation $u_1 : u_2$ may be removed by, for example, characterizing the ray by the reciprocal of its slope u_1/u_2. Such non-homogeneous coordinatization entails a sacrifice in symmetry and the necessity of admitting infinity as a meaningful value of the quotient; both losses are quite troublesome in extension of these geometric concepts to cartesian space of n dimensions, which is necessary in connection with n-fold partitions. In homogeneous coordinates the likelihood ratio can conveniently be represented by any of the equally good sets of homogeneous coordinates, $P(x \mid B_1):P(x \mid B_2)$, $r_1(x):r_2(x)$, and $r_1(x, \alpha):r_2(x, \alpha)$. Finally, it may be remarked that $P(x \mid B_1)/P(x \mid B_2)$ is a non-homogeneous coordinate. Thus the many equivalent forms in which the likelihood ratio statistics can be naturally expressed corresponds to the many different notations by which a ray through the origin can be naturally designated.

The most remarkable fact about the likelihood ratio considered as a statistic is that it is *necessary*, so to speak, as well as sufficient. By that I mean that to have the advantages of knowing **x** it is necessary as well as sufficient to know the likelihood ratio. The point can be put formally thus:

THEOREM 1   If **y** is sufficient for **x**, then **y** is an extension of **r**.

PROOF. The theorem is virtually obvious in terms of the factorability criterion for sufficient statistics, for in the notation of (4.10)

$$(6) \qquad r_i(x) = \frac{R(y(x),\, i)}{\Sigma R(y(x),\, j)}$$

with probability one, exhibiting r_i as a function of y. ◆

COROLLARY 1   If **z** is sufficient for **x**, and if every **y** sufficient for **x** is an extension of **z**, then **z** is equivalent to **r**.

By ordinary analytic standards, the likelihood ratio seems to be a rather complicated statistic, at least in the case of n-fold partitions, where n is at all large; for, to one who takes seriously the idea that a multiple statistic should not also be regarded as a single statistic, the likelihood ratio seems at first sight to be n, or perhaps $(n - 1)$, statistics. Yet Theorem 1 and its corollary show that the likelihood ratio is, in a fundamental sense, the most compact sufficient statistic that a partition problem admits.

As an explicit example of a likelihood ratio, consider the twofold partition problem arising from Exercise 4.1 on confining attention to two different values of p, say p_1 and p_2. The likelihood ratio **r** is easily computed thus:

$$(7) \qquad P(x \mid B_i) = p_i^{y'(x)}(1 - p_i)^{n-y'(x)}$$

$$= (1 - p_i)^n \left(\frac{p_i}{1 - p_i}\right)^{y'(x)} = q_i^{\,n}\left(\frac{p_i}{q_i}\right)^{y},$$

so

$$(8) \qquad r_i(x) = \frac{q_i^{\,n}(p_i/q_i)^{y'(x)}}{\Sigma q_j^{\,n}(p_j/q_j)^{y'(x)}}.$$

Theorem 1 is thereby verified in the present instance; for (8) exhibits $\mathbf{r}$ explicitly as a contraction of $\mathbf{y}$, and $\mathbf{y}$ is easily exhibited as a contraction of $\mathbf{r}$ thus:

$$(9) \qquad y(x) = \frac{\log\left\{\dfrac{r_1(x)}{r_2(x)} \cdot \left(\dfrac{q_2}{q_1}\right)^n\right\}}{\log\dfrac{p_1 q_2}{p_2 q_1}}.$$

In this example, $\mathbf{y}$ is, in view of (8) and (9), equivalent to the likelihood ratio.

Exercises

1. Express $k(\beta(x))$ and $v(\mathbf{F}(\mathbf{x}))$ in terms of the likelihood ratio thus:

$$(10) \qquad \beta(i; r) =_{\mathrm{Df}} r_i \beta(i)/\sum_j r_j \beta(j),$$

$$(11) \qquad k(\beta(x)) = k(\beta(r(x))).$$

$$(12) \qquad v(F(x) \mid \beta) = \sum_r k(\beta(r)) \left[\sum_j P(r \mid B_j)\beta(j)\right].$$

2. This extended exercise develops the personalistic and behavioralistic theory of what, following the objectivistic and verbalistic traditions of statistics, is called the **testing of a simple dichotomy,** a type of decision problem that, though seldom very realistic, is a popular and instructive example with important implications for more realistic problems. Verbalistically such a problem is described as that of making the best guess on the basis of an observation as to whether it is B_1 or B_2 that obtains. Behavioralistically, this is generally interpreted as the problem of deciding, on the basis of observation, between two primary acts one of which is preferable to the other if B_1 obtains and vice versa if B_2 does. Here is one topic in which the assumption that i is confined to two values is rather more than simply a pedagogical simplification; a reader interested in relaxing the assumption will find pages 127–130 of [W3] stimulating.

Suppose that $\mathbf{F}$ contains only two acts $\mathbf{f}_1$ and $\mathbf{f}_2$ and is dominated by neither. Let $\phi_{ij} =_{\mathrm{Df}} E(\mathbf{f}_i \mid B_j)$.

(a) There is no loss of generality in supposing

$$(13) \qquad \delta_1 =_{\mathrm{Df}} \frac{\phi_{22} - \phi_{12}}{2} > 0, \qquad \delta_2 =_{\mathrm{Df}} \frac{\phi_{11} - \phi_{21}}{2} > 0,$$

which will henceforth be done. That is, it will be supposed that $\mathbf{f}_1$ is appropriate only to B_1 and vice versa.

(b) Show that

$$(14) \quad k(\beta) = \sum_j \phi_{1j}\beta(j) \qquad \text{for } \beta(1) \geq \delta_1/(\delta_1 + \delta_2) = \beta_0(1)$$

$$= \sum_j \phi_{2j}\beta(j) \qquad \text{for } \beta(2) \geq \delta_2/(\delta_1 + \delta_2) = \beta_0(2)$$

$$= \tfrac{1}{2}(\phi_{11} + \phi_{21})\beta(1) + \tfrac{1}{2}(\phi_{12} + \phi_{22})\beta(2) + | \, \delta_1\beta(2) - \delta_2\beta(1) \, |$$

$$= \sum_j \epsilon_j\beta(j) + | \, \delta_1\beta(2) - \delta_2\beta(1) \, |,$$

where β_0 and the ϵ_i's are defined by the context.

(c) $E(\mathbf{f}_i \,|\, \beta) = k(\beta)$, if and only if $\beta(i) \geq \beta_0(i)$. This condition obtains for both i's simultaneously, if and only if $\beta = \beta_0$.

(d) Show that

$$(15) \quad k(\beta(r)) = \left\{ \sum_j \epsilon_j r_j\beta(j) + | \, \delta_1 r_2\beta(2) - \delta_2 r_1\beta(1) \, | \right\} / \sum_j r_j\beta(j)$$

$$= \sum_j \phi_{ij}\beta(j; r) \qquad \text{for } r_i \geq r_i{}^*(\beta, \beta_0),$$

where

$$(16) \qquad\qquad r_i{}^*(\beta, \beta_0) =_{\mathrm{Df}} \frac{\beta_0(i)/\beta(i)}{\displaystyle\sum_j \beta_0(j)/\beta(j)} \, ;$$

and that

$$(17) \quad v(\mathbf{F}(\mathbf{x}) \,|\, \beta) = \sum_j \epsilon_j\beta(j) + \sum_r | \, \delta_1 P(r \,|\, B_2)\beta(2) - \delta_2 P(r \,|\, B_1)\beta(1) \, |$$

$$= \{\epsilon_1 + \delta_2[1 - 2P(r_1 < r_1{}^*(\beta, \beta_0) \,|\, B_1)$$

$$- P(r = r^*(\beta, \beta_0) \,|\, B_1)]\}\beta(1)$$

$$+ \{\epsilon_2 + \delta_1[1 - 2P(r_2 < r_2{}^*(\beta, \beta_0) \,|\, B_2)$$

$$- P(r = r^*(\beta, \beta_0) \,|\, B_2)]\}\beta(2).$$

(e) Any derived act $\mathbf{f}(\mathbf{x})$ determines a function $\mathbf{i}$ assigning an i to each x, $\mathbf{i}$ being implicitly defined thus: $\mathbf{f}(x) = \mathbf{f}_{i(x)}$. Conversely any $\mathbf{i}$ determines a derived act. Show that $E(\mathbf{f}(\mathbf{x}) \,|\, \beta) = v(\mathbf{F}(\mathbf{x}) \,|\, \beta)$, if and only if $r_{i(x)}(x) \geq r_{i(x)}{}^*(\beta, \beta_0)$ for every x. Such a function $i(x)$ is called a **likelihood-ratio test** associated with r^*. Show that at least one likelihood-ratio test is associated with every value of r^*, and that if $P(r = r^*) = 0$ (which is typically the case) there is only one.

(f) If $\mathbf{f}(\mathbf{x})$ is determined by a function of $\mathbf{i}$, the probability of deciding on the inappropriate value of $\mathbf{i}$ in case B_j obtains is generally called the probability of an **error of the j-th kind.** Analytically the probabilities of error of the first and second kind are, respectively,

$$(18) \qquad e_1 =_{Df} P(i(x) = 2 \mid B_1), \qquad e_2 =_{Df} P(i(x) = 1 \mid B_2).$$

If **i*** is a likelihood-ratio test associated with r^*, show that its errors of the first and second kind are subject to the bounds

$$(19) \qquad P(r_1 < r_1^* \mid B_1) \le e_1^* \le P(r_1 \le r_1^* \mid B_1)$$

$$(20) \qquad P(r_1 > r_1^* \mid B_2) \le e_2^* \le P(r_1 \ge r_1^* \mid B_2).$$

What about the typical case that $P(r = r^*) = 0$?

(g) Show that, if **i** is at least as good as **i*** in the sense that $e_i \le e_i^*$ for both i's, then **i** is a likelihood-ratio test and **i** is virtually **i*** in that $e_i = e_i^*$ for both i's. Hint: Consider an F and a β for which $r^*(\beta, \beta_0) = r^*$, showing that these exist, and note that, for this decision problem,

$$
\begin{aligned}
(21) \quad E(\mathbf{f}_{i*} \mid \beta) &= \{\epsilon_1 - \delta_2(1 - 2e_1^*)\}\beta(1) + \{\epsilon_2 - \delta_1(1 - 2e_2^*)\}\beta(2) \\
&= v(\mathbf{F}(\mathbf{x}) \mid \beta) \\
E(\mathbf{f}_i \mid \beta) &= \{\epsilon_1 - \delta_2(1 - 2e_1)\}\beta(1) + \{\epsilon_2 - \delta_1(1 - 2e_2)\}\beta(2) \\
&\ge v(\mathbf{F}(\mathbf{x}) \mid \beta),
\end{aligned}
$$

with equality if and only if **i** is a likelihood-ratio test.

This important conclusion about likelihood-ratio tests has been much emphasized, especially by the Neyman-Pearson school.

The concept of likelihood ratio, sometimes simply called likelihood, is now one of the most pervasive concepts of statistical theory. It seems to have been introduced in 1922 by R. A. Fisher (cf. index of [F3]), who emphasized it in connection with the important method of estimation named by him "the method of maximum likelihood." Its use in testing hypotheses was apparently first emphasized by J. Neyman and E. S. Pearson (see Vol. II, p. 303 of [K2]). In connection with likelihood ratios as necessary and sufficient statistics, mathematically advanced readers will be interested in Section 6 of [L6], [B2], and [M5]. One of the earliest contributions in this direction was made by C. A. B. Smith [S14].

6   Repeated observations

If $\mathbf{x}(n) = \{\mathbf{x}_1, \cdots, \mathbf{x}_n\}$, where, given B_i, the $\mathbf{x}_r$'s are independent identically distributed random variables, then $v(\mathbf{F}(\mathbf{x}(n)))$ is a non-decreasing function of n, for the $(n + 1)$-tuple is an extension of the n-tuple. If $k(\beta)$ is strictly convex—a condition that you now recognize as interesting—$v(\mathbf{F}(\mathbf{x}(n)))$ is easily seen to be strictly increasing in n, unless the individual $\mathbf{x}_r$'s are either utterly irrelevant or definitive.

It is to be expected, especially in the light of the approach to certainty discussed in § 3.6, that, as n becomes very large, $\mathbf{x}(n)$ will become practically definitive. Indeed, § 3.6 makes it possible to state and prove a formal theorem to that effect.

THEOREM 1

HYP.　　1. $\mathbf{x}(n) = \{\mathbf{x}_1, \cdots, \mathbf{x}_n\}$, where, given B_i, the $\mathbf{x}_r$'s are independent and identically distributed random variables.

　　2. The $\mathbf{x}_r$'s are not utterly irrelevant to B_i.

　　3. $v(\mathbf{F} \mid \beta) = k(\beta)$.

CONCL.　　$\displaystyle \lim_{n \to \infty} v(\mathbf{F}(\mathbf{x}(n)) \mid \beta) = l(\beta) =_{\mathrm{Df}} \beta(1)k(1, 0) + \beta(2)k(0, 1)$

uniformly in β.

PROOF.　Writing $\mathbf{x}$ as short for $\mathbf{x}(n)$,

(1) $$v(\mathbf{F}(\mathbf{x}) \mid \beta) = E[k(\beta(\mathbf{x}))].$$

For an arbitrary $\epsilon > 0$, let the closed interval I on which $\mathbf{k}$ is defined be partitioned into two subsets J and K, where J is the set of those β's such that

(2) $$k(\beta) \geq l(\beta) - \epsilon,$$

and K is the complement of J relative to I.

It follows from the continuity of the functions on each side of (2) that $\beta \, \varepsilon \, J$, if either component of β is sufficiently large.

The computation initiated in (1) can now be carried forward thus:

(3)
$$
\begin{aligned}
E[k(\beta(\mathbf{x}))] &= E[k(\beta(\mathbf{x})) \mid \beta(x(s)) \, \varepsilon \, J]P(\beta(x(s)) \, \varepsilon \, J) \\
&\quad + E[k(\beta(\mathbf{x})) \mid \beta(x(s)) \, \varepsilon \, K]P(\beta(x(s)) \, \varepsilon \, K) \\
&\geq E[l(\beta(\mathbf{x})) \mid \beta(x(s)) \, \varepsilon \, J]P(\beta(x(s)) \, \varepsilon \, J) \\
&\quad + \min_{\beta'} k(\beta') \cdot P(\beta(x(s)) \, \varepsilon \, K) - \epsilon \\
&= E[l(\beta(\mathbf{x}))] - \{E[l(\beta(\mathbf{x})) \mid \beta(x(s)) \, \varepsilon \, K] \\
&\quad - \min_{\beta'} k(\beta)\}P(\beta(x(s)) \, \varepsilon \, K) - \epsilon \\
&\geq l(\beta) - \max_{\beta'} | k(\beta') | \cdot P(\beta(x(s)) \, \varepsilon \, K) - \epsilon.
\end{aligned}
$$

Now, in view of the paragraph in which (3.6.15) occurs and the fact that, if either component of β is close to 1, $\beta \, \varepsilon \, J$; $P(\beta(x(s)) \, \varepsilon \, K)$ becomes arbitrarily small for sufficiently large n. ◆

7   Sequential probability ratio procedures

The present section digresses to discuss an interesting application of the ideas presented in this chapter to what is called sequential analysis. **Sequential analysis** refers in principle to the theory of observational programs in which the selection of what observations to make in later phases of the program depends on what has been observed in earlier phases. Such behavior is commonplace in everyday life; for example, you look for something until you find it, but not longer. Statistics itself has always used sequential procedures. For example, it is not rare to conduct a preliminary experiment to determine how a main experiment should be carried out. Thus, if one were required to estimate with a roughly preassigned precision the mean of a normal distribution of unknown mean and unknown variance, one might reasonably begin by taking ten or twenty observations, which would give some idea of the variance and would therefore determine about how many observations are necessary for achieving the requisite precision.

Commonplace though problems with sequential features are, A. Wald was the first to develop (1943) a systematic theory of a considerable body of problems of this sort. For early history see the Introduction of [W2] and the Foreword of Section I of [S17].

Some later ideas on sequential analysis, due mainly to Wald and Wolfowitz, are the subject of this section. It will not be practical to proceed with full rigor, primarily because random variables capable of assuming an infinite number of values are necessarily involved. Full details are given in [W3] and more compactly in [A7], but not in Wald's book on sequential analysis [W2].

Let $\mathbf{x} = \{\mathbf{x}(1), \cdots, \mathbf{x}(v), \cdots\}$, where the $\mathbf{x}(v)$'s are conditionally an infinite sequence of independent, relevant, identically distributed random variables. Rather informally, a sequential observational program with respect to $\mathbf{x}$ is a rule telling whether to observe $\mathbf{x}(1)$ or whether to make no observation at all; if the particular value $x(1)$ is observed, whether to observe $\mathbf{x}(2)$ or to discontinue observation; if the values $x(1)$ and $x(2)$ are observed whether to observe $\mathbf{x}(3)$ or to discontinue observation, etc.

More formally, let $\mathbf{N}$ be a function of the infinite sequence of values $x = \{x(1), \cdots, x(v), \cdots\}$ such that, if the sequence x' agrees with x in every component from the first through the $N(x)$th, then $N(x') = N(x)$. Such a function N determines a **sequential observational program**, which is a contraction of $\mathbf{x}$, call it $y(\mathbf{x}; \mathbf{N})$, defined thus:

(1) $$y(\mathbf{x}; \mathbf{N}) =_{\mathrm{Df}} \{\mathbf{x}(1), \cdots, \mathbf{x}(N(\mathbf{x}))\}.$$

It is to be understood that, if $N(x)$ is zero for some x, it is identically zero, and that $y(\mathbf{x}; 0)$ is a null observation.

It will be assumed that the random cost associated with a sequential observational program is proportional to the number of random variables observed, that is, $\mathbf{c} = N(\mathbf{x})\gamma$, $\gamma > 0$. No categorical defense of this assumption is suggested, but clearly there are interesting problems in which it is met at least approximately. The domain of applicability of the theory can actually be considerably extended by modifying the assumption to include a fixed overhead cost that applies except in case N is identically zero; this does not greatly complicate the analysis, as the interested reader will be able to see for himself. The theory would even remain virtually unchanged, if $\mathbf{c}$ were only assumed to be of the form

$$(2) \qquad \mathbf{c} = \mathbf{h} + \sum_{v=1}^{N(x)} \mathbf{c}(v), \qquad \text{if } N > 0,$$
$$= 0, \qquad \text{if } N = 0,$$

where $\mathbf{h}$, $\mathbf{c}(1)$, $\mathbf{c}(2)$, $\cdots$ are independent with finite expected values $E(\mathbf{h}) \geq 0$, $E(\mathbf{c}(r)) > 0$, and the $\mathbf{c}(v)$'s are identically distributed.

For any $\mathbf{F}$ there are some values of β for which it would be unwise to adopt any sequential observational program other than the null observation. Suppose, for example, that β is so close to an extreme value that $l(\beta) - k(\beta) < \gamma$; under this circumstance the most that could be gained by observing even $\mathbf{x}$ itself would be less than γ, but the cost of making so much as one observation is at least γ. Let the set of values of β for which it is not justified to make any but the null observation be denoted for a while by $J(\mathbf{F}; \gamma)$, or simply J, for short.

Now, if $\beta \, \varepsilon \, J$, the person's utility can, by the definition of J, be maximized by refraining from any observation but the null observation and accepting the utility $k(\beta)$; otherwise there will be some advantage to him in observing $\mathbf{x}(1)$. If the person does observe the particular value $x(1)$ of $\mathbf{x}(1)$, he finds himself with a posteriori probabilities $\beta(x(1))$ in place of the a priori β, he has paid (or at any rate entailed) a cost γ, and he must now decide whether to make any further observations. His new problem is simply the problem he would have faced at the outset had his a priori probabilities been $\beta(x(1))$ instead of β, except that all utilities are now reduced by γ. He justifiably accepts the utility $k(\beta(x(1))) - \gamma$, if $\beta(x(1)) \, \varepsilon \, J$; otherwise he will observe $\mathbf{x}(2)$. Continuing this line of argument step after step, it follows that optimal action consists in observing successive $\mathbf{x}(v)$'s until an a posteriori probability in J occurs, and then adopting a basic act consistent with the a posteriori probability.

In actual practice, it is far from easy to determine whether a particular value of β belongs to $J(\mathbf{F}; \gamma)$, because in principle the whole enormous variety of sequential observational programs has to be explored to determine whether any one of them has a derived value greater than $k(\beta)$. The practical advantage achieved in the preceding paragraph is that of greatly restricting the class of programs that merit consideration. Thus the problem of determining whether $\beta \, \varepsilon \, J(\mathbf{F}; \gamma)$ does not require a survey of all observational programs, but only of those defined in terms of some set J' according to the rule that $N(x)$ is the first integer for which $\beta(x(1), \cdots, x(n)) \, \varepsilon \, J'$.

If programs corresponding to all sets J' had to be examined, the process would still be mathematically impractical; indeed, in all but special cases, practical solutions have yet to be found. But, if any special conditions that J must necessarily satisfy are discovered, only sets J' satisfying those conditions need be examined. Some very general conditions are these: J contains the extreme points of I; J is topologically closed, that is, if a value β_0 is not in J, then the near neighbors of β_0 are also not in J. The first of these conditions requires no comment, and the second follows easily from the continuity as a function of β of

$$(3) \qquad E[k(\beta(y(\mathbf{x}; N))) - \gamma \mathbf{N} \mid \beta] - k(\beta).$$

These conditions alone do not go far toward narrowing to practical limits the variety of sets to be explored. Thus far in the development of the subject, really powerful conditions have been obtained only at the expense of considerable restrictions on the structure of $\mathbf{F}$ or, equivalently, of $\mathbf{k}$.

Suppose, then, that $\mathbf{F}$ is dominated by a finite number of acts or, what amounts to a little less, that the graph of $\mathbf{k}$ is polygonal, as it is for the $\mathbf{k}$ graphed in Figure 2.1. Technically, this restriction on $\mathbf{k}$ may be expressed by saying that the interval I is the union of a finite number of intervals of linearity of $\mathbf{k}$. Under the restriction, relatively much can be concluded about the structure of $J(\mathbf{F}; \gamma)$, for it is true in general, as will be shown in the next paragraph, that the intersection of J with any interval of linearity of $\mathbf{k}$ is a closed interval.

Suppose, indeed, that β_1 and β_2 belong to J and to a common interval of linearity of $\mathbf{k}$, but that β_0 on the interval between β_1 and β_2 does not belong to J. A contradiction follows according to the following computation, in which $\mathbf{h}$ is any act derived from a sequential observational program, cost included, that is advantageous at β_0.

$$(4) \qquad \sum_j E(\mathbf{h} \mid B_j)\beta_0(j) > k(\beta_0),$$

for $\mathbf{h}$ is supposed to be advantageous at β_0; and

(5)
$$\sum_i E(\mathbf{h} \mid B_i)\beta_m(i) \leq k(\beta_m), \qquad m = 1, 2,$$

for no derived act is supposed to be advantageous at β_m, since $\beta_m \; \varepsilon \; J$. Since β_0 is a weighted average, say $\Sigma\gamma_m\beta_m$, of the β_m's, and since $k(\beta)$ is linear in the interval between β_1 and β_2, it follows from (4) and (5) that

(6)
$$\sum_i E(\mathbf{h} \mid B_i)\beta_0(i) \leq k(\beta_0),$$

contradicting (4). The supposition that $\beta_0 \; \varepsilon \sim\!J$ has thus been reduced to absurdity.

The demonstration just given extends directly to n-fold problems. The general conclusion is that the intersection of J with any domain of linearity of $\mathbf{k}$ is convex, so that, if $\mathbf{k}$ is polyhedral, J is the union of a finite number of closed convex sets, each lying wholly in a domain of linearity of $\mathbf{k}$. The practical implications of the conclusion are enormously greater for twofold than for higher-fold problems, because twofold problems lead to one-dimensional bounded, closed, convex sets, which present no great variety, all of them being closed bounded intervals. But threefold problems, for example, lead to closed bounded two-dimensional convex sets, a restriction that leaves great room for variety.

If $\mathbf{k}$ is polygonal, the variety of sets J' to be surveyed is enormously reduced, for J' must be the union of a known number of intervals, each of which is confined to a known interval. Suppose that this number is m; the class of sequential observational programs to be surveyed can be characterized by the two end points of each of the m intervals, except that the possibility that some of the intervals are vacuous must be borne in mind. Since the extremes of I are necessarily in J, and therefore necessarily appear as end points of intervals in J, the exploration has been reduced to a $2(m - 1)$ parameter family of possibilities.

The possibility that $m = 1$, which almost means that $\mathbf{F}$ is dominated by a single element of itself, is trivial; for then all β's are in J, and observation is never called for. This can be seen in many ways. In particular, it follows as an illustration of the machinery that has just been developed, thus: The end points, or extremes, of I are both in J, as always, and, since $m = 1$, they are both in the same interval of linearity of J; therefore the interval between them, namely every value of β, lies in J.

The possibility that $m = 2$—in ordinary statistical usage, the sequential testing of a simple dichotomy—is of particular importance.

It occurs typically when **F** is dominated by two acts, neither of which dominates the other, as in Exercise 5.2. One of the two acts is appropriate to one "hypothesis" B_1, and the other is appropriate to B_2. In case $m = 2$, it is easily seen, by methods that have now been indicated more than once, that each of the two closed intervals that constitute J has as one end point one of the extremes of I. Neither of the two intervals can be vacuous, nor can either consist only of a single point. It is relatively easy to find, at least approximately, the two values of β that determine $J(\mathbf{F}; \gamma)$, and the theory of this situation has correspondingly been brought to a relatively high degree of perfection; for details, see [S17], [W2], [W3], and [A7].

Following (or at least paraphrasing) Wald [W2], a sequential observational program characterized by making successive observations until the a posteriori probabilities fall into some set J, followed by adopting a basic act appropriate to the a posteriori probability, is called a **sequential probability ratio procedure.** The reason for this nomenclature is that to observe until the a posteriori probabilities fall into J is to observe until the numbers

$$(7) \qquad \beta(i \mid x(1), \cdots, x(v)) = \frac{\beta(i)P(x(1), \cdots, x(v) \mid B_i)}{\sum_j \beta(j)P(x(1), \cdots, x(v) \mid B_j)}$$

lie in a certain set, or, what amounts to the same thing, satisfy certain conditions. But, the particular value of β having been assigned, this is tantamount to requiring the ratios of probabilities

$$(8) \qquad \frac{P(x(1), \cdots, x(N) \mid B_i)}{P(x(1), \cdots, x(N) \mid B_j)}$$

to satisfy certain conditions.

Since (7) and (8) are ways of expressing the likelihood ratio, the observational program together with the act derived from it might also be referred to as a sequential likelihood-ratio procedure. Indeed, but for the precedent established by Wald, that would seem the better name.

As an actual example of a sequential probability ratio procedure, suppose that the distribution of $\mathbf{x}(v)$ given B_i attaches the probabilities p_i and $q_i = 1 - p_i$ to the values 1 and 0, respectively. The expression (8) can in any case be written in the factored form

$$(9) \qquad \prod_{v=1}^{N} \left\{ \frac{P(x(v) \mid B_i)}{P(x(v) \mid B_j)} \right\},$$

and in the present example this takes the special form

(10)
$$\left(\frac{p_1}{p_2}\right)^{y(N)}\left(\frac{q_1}{q_2}\right)^{N-y(N)} = \left(\frac{q_1}{q_2}\right)^{N}\left(\frac{p_1 q_2}{p_2 q_1}\right)^{y(N)},$$

where

(11)
$$y(N) = \sum_{v=1}^{N} x(v).$$

It is noteworthy, in connection with sufficient statistics, that the condition that the a posteriori probability be in J is in this case expressible, according to (10), as a condition on $y(N)$ and N. Specializing the example further, suppose that J is of the sort appropriate to testing a simple dichotomy. The condition that the a posteriori probability be in $\sim J$ is then expressed by each of the following equivalent pairs of inequalities, where $\alpha(1)$ and $\alpha(2)$ are positive numbers such that $\alpha(1) + \alpha(2) < 1$.

(12)
$$\beta(1 \mid x(1), \cdots, x(N)) < 1 - \alpha(1),$$
$$\beta(2 \mid x(1), \cdots, x(N)) < 1 - \alpha(2).$$

(13)
$$\frac{\beta(1)Q}{\beta(1)Q + \beta(2)} < 1 - \alpha(1),$$
$$\frac{\beta(2)}{\beta(1)Q + \beta(2)} < 1 - \alpha(2),$$

where Q for the moment denotes the likelihood ratio (10).

(14)
$$Q < \frac{\beta(2)(1 - \alpha(1))}{\beta(1)\alpha(1)} = Q^*,$$
$$Q > \frac{\beta(2)\alpha(2)}{\beta(1)(1 - \alpha(2))} = Q_*,$$

where Q^*, Q_* are defined by the context. Since, according to (13), the structure of $\sim J$ is superficially determined by three parameters, say by $\beta(1)$, $\alpha(1)$, and $\alpha(2)$, it is worthy of some note that the corresponding condition is ultimately expressed in terms of only two special parameters, Q^* and Q_*; this is only natural, considering that $\sim J$ is an open interval determined by its two end points. The act that would be appropriate to B_1 is called for by values of $Q \geq Q^*$, and the one appropriate to B_2 is called for by values of $Q \leq Q_*$.

Thus far, the particular form (10) of the likelihood ratio has not really been exploited in the calculation, so (14) applies to the testing of simple dichotomies generally. Taking account of (10), (14) can by elementary manipulation be put in the following form.

(15)
$$y(N) < \{\log Q^* + N \log (q_2/q_1)\}/\log (p_1q_2/p_2q_1),$$
$$y(N) > \{\log Q_* + N \log (q_2/q_1)\}/\log (p_1q_2/p_2q_1),$$

where, for definiteness, it is supposed that $p_1 > p_2$. Thus, the region in the (N, y) plane determined by $\sim J$, the region in which further observations are called for, is a band bounded by two parallel lines of positive slope.

8 Standard form, and absolute comparison between observations

If $\mathbf{x}$ and $\mathbf{y}$ are such that, for every $\mathbf{F}$ and β, $v(\mathbf{F}(\mathbf{x}) \mid \beta) \geq v(\mathbf{F}(\mathbf{y}) \mid \beta)$; then $\mathbf{x}$ imitates, so to speak, an extension of $\mathbf{y}$, and it may appropriately be said that $\mathbf{x}$ is a *virtual extension* of $\mathbf{y}$. Correspondingly, if $\mathbf{x}$ is a virtual extension of $\mathbf{y}$, and $\mathbf{y}$ is a virtual extension of $\mathbf{x}$, it may be said that $\mathbf{x}$ and $\mathbf{y}$ are *virtually equivalent*.

No matter what a priori probabilities a person may have, or what basic acts are available to him, he will have no preference between a pair of virtually equivalent observations, so virtually equivalent observations are indeed equivalent for many practical purposes. Where combinations of observations are under consideration, however, the relation of virtual equivalence does not resemble true equivalence. For example, if $\mathbf{x}$ and $\mathbf{y}$ are equivalent, then each is equivalent to the multiple observation $\{\mathbf{x}, \mathbf{y}\}$, but if $\mathbf{x}$ and $\mathbf{y}$ are only virtually equivalent, they may well be independent, in which case neither will typically be equivalent to $\{\mathbf{x}, \mathbf{y}\}$.

This section explores the notions of virtual extension and virtual equivalence. In particular, an interesting standard representative of the class of observations virtually equivalent to a given observation $\mathbf{x}$ is defined and discussed. This material is scarcely referred to later in the book, and it may without much loss be skipped or glossed over. It will be couched frankly in the language of n-fold as opposed to twofold partitions, but readers with the rest of the chapter behind them will easily be able to concentrate on the twofold situation, if they find it more understandable.

Most of the ideas to be presented in this section were originated by H. F. Bohnenblust, L. S. Shapley, and S. Sherman in a private memorandum dated August 1949, which I was privileged to see at that time.

This work was extended and brought to the attention of the public by David Blackwell in [B16].

It is obvious that, if **y** is a sufficient statistic for **x**, then **x** and **y** are virtually equivalent. In particular the likelihood ratio **r** derived from **x** is virtually equivalent to **x**. Moreover, the reader may anticipate, and it will be formally shown in the course of this section, that if and only if observations are virtually equivalent do their likelihood ratios have the same distribution for every value of β, or, what comes to the same thing, given each B_i, $i = 1, \cdots, n$. Thus the n conditional distributions of the likelihood ratio given each B_i could be taken to characterize the observations virtually equivalent to a given one, say **x**. Actually, as will be shown, the class of observations virtually equivalent to **x** can be represented by the distribution of the likelihood ratio for any single non-extreme value of β. For definiteness, the particular value $\beta^* = \{1/n, \cdots, 1/n\}$ will be used, but the interested reader will find it a simple exercise to extend all the considerations based on β^* to any other non-extreme β, as would be necessary in any extension of the theory to infinite partitions.

Let $m(r)$ be the probability that the likelihood ratio in the standard form (5.5) attains the particular value r when $\beta = \beta^*$. With self-evident abbreviations,

$$(1) \qquad m(r) = P(r \mid \beta^*)$$

$$= \sum_j P(r \mid B_j)(1/n)$$

$$= \frac{1}{n} \sum_j \sum_{r(x)=r} P(x \mid B_j).$$

The second line of (1) exhibits $m(r)$ expressed in terms of the n distributions $P(r \mid B_i)$. It is rather more interesting to see that those n distributions can themselves all be expressed in terms of the single distribution m, as follows from the definition (5.5) of **r** and the third line of (1) thus:

$$(2) \qquad P(r \mid B_i) = \sum_{r(x)=r} P(x \mid B_i)$$

$$= \sum_{r(x)=r} r_i(x) \sum_j P(x \mid B_j)$$

$$= nr_i m(r).$$

Similarly,

$$(3) \qquad P(r \mid \beta) = n \left\{ \sum_j r_j \beta(j) \right\} m(r).$$

Regarded as a probability measure on the set of all n-tuples of numbers r, m has the following three important properties.

$$P(r_i \geq 0 \mid m) = 1;$$

(4)
$$P\left(\sum_j r_j = 1 \mid m\right) = 1;$$

$$E(\mathbf{r}_i \mid m) = n^{-1}.$$

Of these, the first two are obvious from the definition of $\mathbf{r}$, and the third follows by calculation from (2) thus:

(5)
$$1 = \sum_r P(r \mid B_i) = n \sum_r r_i m(r)$$

$$= nE(\mathbf{r}_i \mid m).$$

Conversely, suppose that m is any mathematical probability defined on the set of n-tuples r of numbers, subject to the conditions (4), then, as can easily be verified, n mathematical probabilities are formally defined by the equation $P(r \mid B_i) = nr_i m(r)$. Mathematically, r distributed thus can be regarded as an observation. The following calculation demonstrates the expected conclusion that the likelihood ratio of this observation is the observation itself and that its distribution given β^* is m.

$$\frac{P(r \mid B_i)}{\sum_j P(r \mid B_j)} = \frac{nr_i m(r)}{n \sum_j r_j m(r)} = r_i.$$

(6)
$$P(r \mid \beta^*) = \sum_j nr_j m(r)(1/n) = m(r).$$

It is interesting and fruitful to compute $v(\mathbf{F}(\mathbf{x}) \mid \beta)$ in terms of m.

(7)
$$v(\mathbf{F}(\mathbf{x}) \mid \beta) = E(k(\beta(\mathbf{x})) \mid \beta)$$

$$= E[k(\{\mathbf{r}_i \beta(i)/\sum_j \mathbf{r}_j \beta(j)\}) \mid \beta]$$

$$= nE\left[k(\{\mathbf{r}_i \beta(i)/\sum_j \mathbf{r}_j \beta(j)\}) \sum_j \mathbf{r}_j \beta(j) \mid m\right].$$

Temporarily adopt the convention that, if α is any n-tuple of positive numbers and $\mathbf{h}$ any function of r (not necessarily convex), $T(\alpha)\mathbf{h}$ is a function of r defined thus:

(8)
$$T(\alpha)h(r) =_{\text{Df}} h(\{r_i \alpha(i)/\sum_j r_j \alpha(j)\}) \Sigma r_j \alpha(j).$$

Then (7) takes the abbreviated form

(9)
$$E(k(\beta(\mathbf{x})) \mid \beta) = nE(T(\beta)k(\mathbf{r}) \mid m).$$

To see the implications of (9), it is necessary to know something about what the operation $T(\beta)$ does to the function $\mathbf{k}$, in particular to know that $T(\beta)\mathbf{k}$ is convex in r. The derivation of these necessary facts is straightforward and is left to the reader as a sequence of exercises.

Exercises

1a. $T(\alpha)T(\beta)\mathbf{h} = T(\{\alpha(1)\beta(1), \cdots, \alpha(n)\beta(n)\})\mathbf{h} = T(\beta)T(\alpha)\mathbf{h}$.

1b. $\mathbf{h} = T(\{\alpha(1)^{-1}, \cdots, \alpha(n)^{-1}\})T(\alpha)\mathbf{h}$.

2. $T(\beta^*)\mathbf{h} = \dfrac{1}{n}\mathbf{h}$.

3. If $h(r) \geq g(r)$ for r between r' and r''; then $T(\alpha)h(r) \geq T(\alpha)g(r)$ for r between $r_i'\alpha(i)/\sum_j r_j'\alpha(j)$ and $r_i''\alpha(i)/\sum_j r_j''\alpha(j)$.

4. If $\mathbf{h}$ is linear, then so is $T(\alpha)\mathbf{h}$.

5. If $\mathbf{h}$ is convex (strictly convex), then so is $T(\alpha)\mathbf{h}$.

Exercise 5 is obvious in the light of Exercises 3 and 4, but some may prefer the demonstration suggested by the following calculation, where $\lambda + \mu = 1$; $\lambda, \mu \geq 0$; and obvious abbreviations are used.

$$(10) \quad T(\alpha)h(\lambda r + \mu r')$$

$$= h\left(\frac{\lambda\alpha\cdot r}{\alpha\cdot(\lambda r + \mu r')}\frac{r}{\alpha\cdot r}\alpha + \frac{\mu\alpha\cdot r'}{\alpha\cdot(\lambda r + \mu r')}\frac{r'}{\alpha\cdot r'}\alpha\right)\alpha\cdot(\lambda r + \mu r')$$

$$\leq \lambda h\left(\frac{r}{\alpha\cdot r}\alpha\right)\alpha\cdot r + \mu h\left(\frac{r'}{\alpha\cdot r'}\alpha\right)\alpha\cdot r'$$

$$= \lambda T(\alpha)h(r) + \mu T(\alpha)h(r').$$

It is amusing to establish once more that observation generally pays, this time by means of (10), (4), and Exercises 5 and 2.

$$(11) \quad nE(T(\beta)k(\mathbf{r}) \mid m) \geq nT(\beta)k(E(\mathbf{r} \mid m))$$

$$= nT(\beta)k(\beta^*)$$

$$= k(\beta).$$

If $\mathbf{x}$ and $\mathbf{x}'$ are observations and m and m' are the corresponding distributions, it is now easy to say in terms of m and m' when $\mathbf{x}$ is utterly irrelevant, when it is definitive, and when $\mathbf{x}$ is virtually an extension of $\mathbf{x}'$.

More exercises

6. The observation $\mathbf{x}$ is utterly irrelevant if and only if $P(r = \beta^* \mid m) = 1$.

7. The observation $\mathbf{x}$ is definitive; if and only if $P(r_i = 1 \mid m) = 1/n$, or, equivalently, if and only if $P(r_i = 0 \mid m) = (n - 1)/n$.

8a. The observation **x** is a virtual extension of **x′**, if and only if, for every convex function **h** defined for r,

$$(12) \qquad E(h(\mathbf{r}) \mid m) \geq E(h(\mathbf{r}) \mid m').$$

8b. The two observations are virtually equivalent, if and only if, for every convex function[+] **h**,

$$(13) \qquad E(h(\mathbf{r}) \mid m) = E(h(\mathbf{r}) \mid m').$$

The conclusion reached in Exercise 8b can be much improved. Indeed, it will be shown that the two observations are virtually equivalent, if and only if m and m' are the same probability measures. This will be achieved if, for example, it is shown that m and m' have the same moments, for it is well known that two different countably additive probability measures confined to a bounded set of n-tuples of numbers cannot have the same moments.[†] The moments in question are expected values of monomials of the form

$$(14) \qquad g(r) = r_1{}^{\epsilon_1} r_2{}^{\epsilon_2} \cdots r_n{}^{\epsilon_n},$$

where the ϵ_i's are non-negative integers. In general, **g** will not be convex, so it cannot be concluded immediately that **g** has the same expected value with respect to m and m'. If, however, a highly convex function is added to **g**, then the sum will be convex and its expected value will be the same with respect to m and m'. Since, by hypothesis, this is also true of the convex term of the sum, it must also be true of the not necessarily convex term. Specifically, let

$$(15) \qquad h(r) = g(r) + \lambda \sum_j r_j{}^2,$$

where λ is a positive number to be determined later. To test **h** for convexity, let s be for the moment an arbitrary n-tuple of numbers and σ a real variable, and compute the second derivate of $h(r + \sigma s)$ with respect to σ at $\sigma = 0$.

$$(16) \qquad \frac{d^2 h(r + \sigma s)}{d\sigma^2}\bigg|_{\sigma=0} = \sum_{i,j} \frac{\partial^2 g(r)}{\partial r_i\,\partial r_j} s_i s_j + \lambda \sum_j s_j{}^2.$$

Considering that each r_i is between 0 and 1, the absolute values of the derivatives of g that appear in (16) have a common upper bound, say

† See, for example, Corollary 1.1, p. 11, of [S13].

Under our usual simplifying assumption that **x** is confined to a finite number of values, m is certainly countably additive. Actually, the whole theory can be developed mutatis mutandis assuming only that the distribution of **x** is countably additive on some suitable Borel field.

+ Morse and Sacksteder (1966) show, in effect, that the test can be confined to the very special convex functions max $\rho_i r_i$, where the ρ_i are arbitrary positive numbers.

μ; so, if $\lambda \geq \mu n^2$, **h** is convex in the region where each r_i lies between 0 and 1 and is a fortiori convex in the intersection of that region with the hyperplane $\Sigma r_j = 1$.

Now that it has been established that m and m' represent virtually equivalent observations, if and only if m and m' are identical, it is apparent that m—or, more exactly, the set of conditional distributions $P(r \mid B_i) = nr_im(r)$—is a unique standard form for all observations virtually equivalent to **x**.

If **x** virtually extends **y**, it is to be expected that, no matter what reasonable definition of "informative" may be suggested, **x** will be at least as informative as **y**. In particular, it is to be expected that the information of B_i with respect to B_j (as defined in § 3.6) will be at least as large for **x** as for **y**, which the following calculation verifies, supposing for simplicity that, for both observations, infinite information is impossible. The point in question depends on the convexity of the function **h** defined by

$$(17) \qquad\qquad h(r) = r_i(\log r_i - \log r_j),$$

because

$$(18) \qquad\qquad I_{i,j} = E(\log r_i - \log r_j \mid B_i)$$
$$= nE[r_i(\log r_i - \log r_j) \mid m].$$

The required convexity can be demonstrated much as it was in $(15)^+$ for a different function also momentarily called **h**:

$$(19) \quad \frac{d^2}{d\sigma^2} h(r + \sigma s)\bigg|_{\sigma=0} = \frac{\partial^2 h(r)}{\partial r_i{}^2} s_i{}^2 + 2\frac{\partial^2 h(r)}{\partial r_i\,\partial r_j} s_i s_j + \frac{\partial^2 h(r)}{\partial r_j{}^2} s_j{}^2$$

$$= \frac{s_i{}^2}{r_i} - \frac{2s_i s_j}{r_j} + \frac{r_i s_j{}^2}{r_j{}^2}$$

$$= \frac{1}{r_i r_j{}^2}(r_j s_i - r_i s_j)^2 \geq 0.$$

It would be interesting to know whether every virtual extension is realized by an actual extension, that is, whether whenever **x** is a virtual extension of **y** there exist random variables **x'** and **y'** such that **x** and **x'** are virtually equivalent, **y** and **y'** are virtually equivalent, and **x'** extends **y'**. To the best of my knowledge that conclusion has thus far been established only in the case of twofold problems, the demonstration for that case being given by Blackwell in [B16].

$^+$ Actually, this calculation depends only on the convexity of $(\log r_i - \log r_j)$ in r_j/r_i.

CHAPTER 8

Statistics Proper

1 Introduction

I think any professional statistician, whether or not he found himself in sympathy with the preceding chapters, would feel that, even allowing for the abstractness expected in a book on foundations, those chapters do not really discuss his profession. He would not, I hope, find the same shortcoming in this and the succeeding chapters, for they are concerned with what seems to me to be statistics proper. The purpose of the present short chapter is to explain this transition and to serve as a general introduction to its successors.

2 What is statistics proper?

So far as I can see, the feature peculiar to modern statistical activity is its effort to combat two inadequacies of the theory of decision, as I have thus far discussed it. In the first place, there are the vagueness difficulties associated with what in § 4.2 were called "unsure probabilities." Second, there are the special problems that arise from more than one person's participating in a decision.

From the personalistic point of view, **statistics proper** can perhaps be defined as the art of dealing with vagueness and with interpersonal difference in decision situations. Whether this very tentative definition is justified, later sections and chapters will permit the statistical reader to judge. At any rate, vagueness and interpersonal difference are the concepts that, directly or indirectly, dominate the rest of this book.

I will not try to discuss vagueness in this chapter, but something may profitably be said here about interpersonal differences.

3 Multipersonal problems

As I have already frequently said, it seems to me that multipersonal considerations constitute much of the essence of what is ordinarily called statistics, and that it is largely through such considerations that the achievements of the British-American School can be interpreted in

terms of personal probability. This is a view that can best be defended
by illustration, and the requisite illustrations will be scattered throughout later chapters; but some support is lent to it by those critics of
personal probability who say that personal probability is inadequate
because it applies only to individual people, whereas the methods of
science are, more or less by definition, those methods that are acceptable to all rational people.

The sort of multipersonal problems I mean to call attention to are
those arising out of differences of taste and judgment, as opposed to
those, so familiar in economics, arising out of conflicting interests. As a
matter of fact, the latter type of multipersonal situation can, if one
chooses, be regarded as among the former; it may, for example, be
said that you and I have different tastes for the process of taking a dollar from me and giving it to you.

Though modern statisticians do not at all deny the existence of different tastes in different people, only occasionally do they take that
difference explicitly into account. In particular, the theory of utility
has scarcely ever entered explicitly into the works of statisticians. Our
intellectual ancestors who believed in the principles of mathematical
expectation were less tolerant than modern statisticians in so far as
they denied rationality in those whose tastes departed from that principle, and some of their bigotry is occasionally met with today.

In dealing with multipersonal situations, it is clearly valuable to
recognize those in which the people involved may all reasonably be
expected to have the same *tastes*, that is, utilities, with respect to the
alternatives involved in the situation. Explicit attempts to discover
general circumstances under which people's tastes will be identical are
rare. The most important and fruitful attempt of this sort is represented by D. Bernoulli's idea that utility functions will typically be
approximately linear within sufficiently confined ranges of income.
Consciously or unconsciously, that principle is repeatedly appealed to
throughout statistics; it was, for example, brought out in § 6.5 that the
very idea of an observation depends for its practical value on Bernoulli's
principle of approximate linearity.

Relatively inexplicit exploitations of similarity of taste are sometimes
made in statistics. The idea is often expressed, for example, that the
penalty for making an estimate discrepant from the number to be estimated will, for everyone concerned, be proportional (within a reasonable range) to the square of the discrepancy; an argument for this principle as a rule of thumb appropriate to many contexts will be given in
§ 15.5. Again, there are situations in which it is agreed that the penalty will depend only on the discrepancy and not on the true value of

the number to be estimated. Of course, there are problems in which both rules are invoked simultaneously, the penalty being supposed to be proportional to the square of the discrepancy and independent of the value to be estimated.

Turn now to **differences in judgment,** that is, to differences in the personal probability, for different people, of the same event. Though modern objectivistic statisticians may recognize the existence of differences of judgment, they argue in theoretical discussions that statistics must be pursued without reference to the existence of those differences, indeed without reference to judgment at all, in order that conclusions shall have scientific, or general, validity. To put the same idea in personalistic terms, I would say that statistics is largely devoted to exploiting similarities in the judgments of certain classes of people and in seeking devices, notably relevant observation, that tend to minimize their differences.

The tendency of observation to bring about agreement has been illustrated in § 3.6. Some of the other general circumstances in which different people may be expected to agree, or at least nearly agree, in some of their judgments have also been mentioned. For example, it may well happen that different people are faced with partition problems that are the same in that the same variable is to be observed by each person, but differ in that each person has his own a priori probabilities β and his own set of available acts F. If, however, the conditional distribution of x given B_i is the same for each person, then the people will, for example, agree as to whether a contraction y of x is sufficient, which is often of great practical value. Again, there are circumstances under which each of these same people will agree that certain derived acts are nearly optimal.

4 The minimax theory

In recent years there has been developed a theory of decision, here with due precedent to be called the minimax theory, that embraces so much of current statistical theory that the remaining chapters can largely be built around it. The minimax theory was originated and much developed by A. Wald, whose work on it is almost completely summarized in his book [W3]. Wald's minimax theory, of course, derives from, and reflects the body of statistical theory that had been developed by others, particularly the ideas associated with the names of J. Neyman and E. S. Pearson. It seems likely that, in the development of the minimax theory, Wald owed much to von Neumann's treatment of what von Neumann calls zero-sum two-person games, which though conceptually remote from statistics, is mathematically all but identical with study of the minimax rule, the characteristic feature of the minimax theory.

Wald in his publications, and even in conversation, held himself aloof from extramathematical questions of the foundations of statistics; and therefore many of the opinions expressed in later chapters on such points in connection with the minimax theory were neither supported nor opposed by him. It may fairly be said, however, that he was an objectivist and that his work was strongly motivated by objectivistic ideas.

My policy here of holding difficulties of mathematical technique to a minimum by making stringent simplifying assumptions will be adhered to in connection with the minimax theory. A large part of Wald's book [W3] is concerned with overcoming the difficulties in technique that are here avoided by simplifying assumptions, but that must be faced in many practical problems. Despite Wald's able effort, important problems of analytic technique still remain in connection with the minimax theory. It should also be appreciated that the individual mathematical problems raised by applications of the minimax theory are often very awkward, even when stringent simplifying assumptions are complied with; consequently much work on specific applications of the theory is still in progress.

CHAPTER 9

Introduction to
the Minimax Theory

1 Introduction

This chapter explains what the minimax theory is, almost without reference to the theory of personal probability. This course seems best, because the theory was originated from an objectivistic point of view and as the solution of an objectivistic problem. Moreover, a philosophically more neutral presentation seems to result, if the ideas of personal probability are here kept out of the foreground.

The minimax theory begins with some of the ideas with which the theory of personal probability, as developed in this book, also begins. In particular, the notions of person, world, states of the world, events, consequences, acts, and decisions presented in §§ 2.2–5 apply as well to the minimax theory—from which they were in fact derived—as to the theory of personal probability.

The point at which the two theories depart from each other is § 2.6, which postulates that the person's preferences establish a simple order among *all* acts. That assumption is necessarily rejected by objectivists, for it, together with the sure-thing principle (which they presumably accept), implies the existence of personal probability. For objectivists, of course, conditional probability does not apply to all ordered pairs of events. More specifically, it seems to be a tacit assumption of objectivistic statistics that the world envisaged in any one problem is partitioned into events with respect to each of which the conditional probabilities of all events (ignoring the mathematical technicality of measurability considerations) are defined, but that conditional probability with respect to sets other than unions of elements of the partition are not defined. That, incidentally, is why partition problems dominate objectivistic statistics. The partition in question is in general infinite, but, for mathematical simplicity, it will here be assumed to be a finite partition B_i.

The objectivistic position is not in principle opposed to the concept of utility. In particular, the minimax theory is predicated on the idea that the consequences of those acts with which it deals are measured numerically by a quantity the expected value of which the person wishes to have as large as possible, whenever (from the objectivistic point of view) the concept of expected value applies. It will therefore be doing the minimax theory little or no injustice to postulate here, as elsewhere, that the consequences of acts are measured in utility.

These preliminaries disposed of, the general **objectivistic decision problem** is to decide on an act f in some given **F**, by criteria depending only on the conditional expectations $E(f \mid B_i)$, and therefore without reference to the "meaningless" $P(B_i)$.

Taking any personalistic or necessary point of view literally, it is nonsensical to pose an objectivistic decision problem, that is, to ask which f of **F** is best for the person, without reference to the $P(B_i)$. On the other hand, many, if not all, holders of objectivistic views, like Wald, find themselves logically compelled by two widely held tenets to consider such problems meaningful. First, for reasons I have alluded to in Chapter 2 and will soon expand upon, many theoretical statisticians today agree, at least tacitly, that the object, or at any rate one object, of statistics is to recommend wise action in the face of uncertainty—a point of view that Wald was particularly active in bringing to the fore. Second, statisticians of the British-American School, of which Wald is to be considered a member, are objectivists and are therefore committed to the view that the probabilities $P(B_i)$ are meaningless, or, at any rate, that they cannot be legitimately used in solutions of statistical problems.

So far as I know, Wald is the only one who has proposed any solution to the general objectivistic decision problem, barring minor variations. His proposal, which is here called the minimax theory, is rather complicated to state. In view of its complexity and the importance of this theory for the rest of this book, and for statistical theory generally, I hope the reader will have particular patience with the present chapter.

2　The behavioralistic outlook

Prior to Wald's formulation of what is here called the objectivistic decision problem, the problems of statistics were almost always thought of as problems of deciding what to say rather than what to do, though there had already been some interest in replacing the verbalistic by the behavioralistic outlook. The first emphasis of the behavioralistic outlook in statistics was apparently made by J. Neyman in 1938 in [N3], where he coined the term "inductive behavior" in opposition to "inductive inference." In the verbalistic outlook, which still dominates most everyday statistical thought, the basic acts are supposed to be assertions; and schemes based on observation are sought that seldom lead to false, or at any rate grossly inaccurate, assertions.

The verbalistic outlook in statistics seems to have its origin in the verbalistic outlook in probability criticized in § 2.1, which in turn is traceable to the ancient tradition in epistomology that deductive and inductive inference are closely analogous processes.

I, and I believe others sympathetic with Wald's work, would analyze the verbalistic outlook in statistics thus: Whatever an assertion may be, it is an act; and deciding what to assert is an instance of deciding how to act. Therefore decision problems formulated in terms of acts are no less general than those formulated in terms of assertions.

If, on the other hand, a sufficiently broad interpretation is put on the notion of assertion, perhaps every decision to adopt an act can be regarded as an assertion to the effect that that act is the best available, in which case the difference between the verbalistic and the behavioralistic outlooks is only terminological; but I do think that, even under such an interpretation, the behavioralistic outlook with its tendency to emphasize consequences offers the better terminology.

Fallacious attempts to analyze away the difference between the verbalistic and behavioralistic viewpoints are also sometimes put forward, especially in informal discussion. For example, it is sometimes said that one should act as though his best estimate of a quantity were in fact the quantity itself. But on that basis few of us would buy life insurance for next year, for we do not typically estimate the year of our death to be so close. Other examples are discussed by Carnap in Section 50 of [Cl].

If assertions are, indeed, to be interpreted as a special class of acts of particular importance to statistics, I have no clear idea what that class may be; but it would presumably exclude certain acts, such as the design of an experiment, that surely are of importance to statistics. Actually the verbalistic outlook has led to much confusion in the foundations of statistics, because the notion of assertion has been used in several different, but always ill-defined, senses, and because emphasis on assertion distracts from the indispensable concept of consequences. I conclude that the behavioralistic outlook is clearer, fuller, and better unified than the verbalistic; and that such value as any verbalistic concept may have it owes to the possibility of one or more behavioralistic interpretations.

This analysis is really too brief and must be supplemented by certain remarks. To begin with, the reader may wonder whether the verbalistic outlook has adherents who defend it against the behavioralistic, and if so what their arguments may be. Actually, the statistical public seems to greet the behavioralistic outlook as a relatively new idea—how old it may actually be is beside the point here—which as such must be regarded with some skepticism. To the best of my knowledge, however, only one objection against the behavioralistic outlook has been presented. It must be discussed next.

It has been seen as an objection to the behavioralistic outlook that the consequences of some assertions, particularly those of pure science, are extremely subtle and difficult to appraise. As a function of the true but unknown velocity of light, what, for example, will be the consequences of asserting that the velocity of light is between 2.99×10^{10} and 3.01×10^{10} centimeters per second? But, if some acts do have subtle consequences, that difficulty cannot properly be met by denying that they are acts or by ignoring their consequences. Certain practical solutions of the difficulty are known. For example, considerations of symmetry or continuity may, as is illustrated in Chapters 14 and 15, make a wise decision possible even in some cases where the explicit consequences of the available acts are beyond human reckoning. Again, analysis sketched in the next two paragraphs tends to show that assertions with extremely subtle consequences play a smaller role in science and other affairs than might at first be thought.

No worker would actually publish—indeed no journal would accept —as research the hypothetical assertion about the velocity of light mentioned in the paragraph above. The consequences might be subtle, if he did; but they would not be very important, for no one would take him seriously. An actual worker would do as much as was practical to say what observations relevant to the velocity of light he, and perhaps others, had performed and what had been observed. To be sure, his statement of the observations would typically be much condensed; he would resort to sufficient statistics or other devices to put his reader rapidly in position to *act* as though the reader himself had made the observations. Assertions about the velocity of light, and countless others of that sort, are of course published in textbooks and handbooks. These assertions do indeed have complicated consequences, so judgment is called for in the compilation of such books; but the seriousness of the consequences of their assertions is limited because of the possibility of referring to original research publications, a possibility serious textbooks and handbooks facilitate by the inclusion of bibliographies.

On the other hand, it is obvious that many problems described according to the verbalistic outlook as calling for decisions between assertions really call only for decisions between much more down-to-earth acts, such as whether to issue single- or double-edged razors to an army, how much postage to put on a parcel, or whether to have a watch re-adjusted.

It is time now to turn back to objectivistic decision problems.

3 Mixed acts

Speaking with pedantic strictness, it might be said that Wald does not propose a solution for the general objectivistic decision problem, because, before undertaking a solution, he insists that $\mathbf{F}$ be subject to a certain condition. On the other hand, he argues that the condition is typically met in practice; he might fairly have insisted that it is the very heart of much actual statistical practice. Before discussing the issue in detail, let me give a small but typical illustration of it.

Suppose that in a rental library I am confronted with the choice between two detective stories, each of which looks more horrifying than the other. At first sight it would seem that only two acts are open to me, namely, to rent one book or the other, but Wald points out that there are other possibilities, not ordinarily thought of as such. In particular, I can eliminate one of the books by flipping a coin. More accurately and more generally, I can let my choice depend on the outcome of a random variable that is utterly irrelevant to the fundamental partition—in this example, a random variable the outcome of which is independent of the relative merits of the two books. The random variable may as well be confined at the outset to two values corresponding to the rental of one or the other of the books, and random variables assigning the same probabilities to the books are equivalent for the purpose at hand. In practice, especially serious statistical practice, such random variables are, taking reasonable precautions, readily provided by coins, cards, dice, tables of random numbers, and other devices.

In terms of the general objectivistic decision problem, Wald's point can (except for mathematical technicalities) be formulated thus: If $\mathbf{f}_r$ represents a finite number of elements of $\mathbf{F}$, and $\phi(r)$ is a corresponding set of non-negative numbers such that $\Sigma\phi(r) = 1$, then the person can make the **mixed act**

$$(1) \qquad\qquad \mathbf{f} = \sum_r \phi(r)\mathbf{f}_r$$

available to himself by observing at no appreciable cost a random variable taking the values r with corresponding probabilities $\phi(r)$ irrespective of which B_i obtains, so $\mathbf{F}$ may be assumed to include $\mathbf{f}$. Technically, the sum in (1) should, for full generality, be replaced by an integral with respect to a probability measure. But such integrals become superfluous under the simplifying asssumption, which is herewith made, that there are in **F** a finite set of acts $\mathbf{f}_r$, to be called **primary acts**, with respect to which every act in **F** can be represented in the form (1). In the rental-library example, the two acts corresponding to the two books can be regarded as primary.

Since mixed acts are also available from the personalistic point of view, it may well be asked whether it is advantageous to consider them in connection with that point of view, and, if not, how they can be of advantage from one point of view but not the other. The answer to the first part of the question is easy. Indeed, if **f** is defined by (1) then it is personalistically impossible that **f** should be definitely preferred to every $\mathbf{f}_r$, that is, that

$$(2) \qquad E(\mathbf{f}) = \sum_r \phi(r)E(\mathbf{f}_r) > \max_r E(\mathbf{f}_r),$$

for a weighted mean cannot be greater than all its terms. Technical explanation of the efficacy of mixed acts from the objectivistic point of view can best be presented after the whole statement of the minimax rule, but those at all familiar with modern statistical practice will derive some insight from the remark that the usual preference of statisticians for random samples represents a preference for certain mixed acts.

4 Income and loss

It is sometimes suggestive, and in conformity with some statistical (though not quite with economic) usage, to refer to $E(\mathbf{f} \mid B_i)$ as the **income** of **f** when B_i obtains, and, correspondingly, to use the notation $I(\mathbf{f}; i)$. An important concept associated with the income is that which I shall refer to as the **loss** (symbolized by $L(\mathbf{f}; i)$) incurred by the act **f** when B_i obtains. By that I mean the difference between the income the person could attain if he were able to act with the certain knowledge that B_i obtained and that which he will attain if he decides on **f** when B_i does in fact obtain. Formally,

$$(1) \qquad L(\mathbf{f}; i) =_{\mathrm{Df}} \max_{\mathbf{f}'} I(\mathbf{f}'; i) - I(\mathbf{f}; i).$$

If the person decides on **f** when B_i obtains, $L(\mathbf{f}; i)$ measures in terms of income the error he has made. If he were himself informed of B_i after **f** had been chosen, which is not typically the case, $L(\mathbf{f}; i)$ would, so to speak, measure his cause for regret. On that account, some have proposed to call loss "regret," but that term seems to me charged with emotion and liable to lead to such misinterpretation as that the loss necessarily becomes known to the person. On the other hand, the term "loss" has been used by Wald in the sense of negative income, but in contexts where loss as defined here is, of the two senses, the only defensible one, as will be explained in § 8. I hope the sense proposed here will not cause serious confusion.

Exercises

1. For each i, there is at least one primary act f_r such that

$$(2) \qquad\qquad I(f_r; i) = \max_f I(f; i).$$

Such a primary act may fairly be called *correct* for i.

2. $L(f; i) = \Sigma\phi(r)L(f_r; i) \geq 0$, equality holding if and only if f is a mixture of acts correct for i.

3. $L(f; i) = \max_{r'} I(f_{r'}; i) - I(f; i)$.

4. $L(f; i) = -I(f; i)$, if and only if

$$(3) \qquad\qquad \max_r I(f_r; i) = 0.$$

5   The minimax rule, and the principle of admissibility

The most characteristic feature of the minimax theory is a certain rule of behavior, or recommendation to the person. This rule, to be called the **minimax rule,** can now be formulated thus: Decide on an act f', such that

$$(1) \qquad\qquad \max_i L(f'; i) = \min_f \max_i L(f; i),$$

where f and f' are, of course, confined to $\mathbf{F}$.

In words, the minimax rule recommends the choice of such an act that the greatest loss that can possibly accrue to it shall be as small as possible. An f satisfying the recommendation of the minimax rule will be called a **minimax act,** and the greatest loss that can accrue to a minimax act will be called the **minimax value** of the (objectivistic) decision problem and written L^*. Under the simplifying assumptions that have been made, it is not technically difficult to show that at least one minimax act exists. The statement of the rule can be reasonably extended to mathematically more general situations, but a digression about this possibility is not appropriate here. The name of the rule is presumably derived from the abbreviation "min max" in (1) or from the Latin phrase "minimum maximorum" thus abbreviated.

It may well happen that **F** contains more than one act that is minimax for the problem, in which case the minimax rule recommends, not a particular act, but only that the choice be narrowed to the set of minimax acts. Some other criterion must then be invoked to narrow the choice further. In particular, it can be shown that at least one of the minimax acts is admissible, in the sense of § 6.4. As Wald indicates, it would, therefore, be an inexcusable violation of the sure-thing principle not to narrow the choice to admissible acts. This application of the sure-thing principle will be called the **principle of admissibility.** The minimax rule and the principle of admissibility constitute the subject matter of, and thereby define, the **minimax theory.**

6   Illustrations of the minimax rule

It would be hard to imagine an objectivistic decision problem simpler than that of whether to make an even-money (or more accurately, even-utility) bet in favor of a certain event or to refrain from betting. That problem, therefore, provides a convenient first example of the minimax rule and the concepts associated with it. Supposing, as one may without loss of generality, that the bet is for one utile, the objectivistic decision problem is completely described by Table 1, which gives the in-

TABLE 1.   THE INCOME OF AN EVEN-MONEY BET, $I(\mathbf{f}_r; i)$

Act	Event	
	B_1	B_2
Bet, $\mathbf{f}_1$	1	-1
Don't bet, $\mathbf{f}_2$	0	0

come of each of the two primary acts for each of the two elements of the partition corresponding to the event in question and its complement.

In view of Exercises 4.2 and 4.3 the corresponding loss function is described by Table 2. Therefore,

$$(1) \qquad \max_i L(\mathbf{f}; i) = \max_i \Sigma\phi(r)L(\mathbf{f}_r; i)$$

$$= \max_i \phi(i) \geq \tfrac{1}{2},$$

equality obtaining if and only if $\phi(1) = \phi(2) = \frac{1}{2}$. Therefore, $L^* = \frac{1}{2}$, and the only minimax act is $\mathbf{f} = \frac{1}{2}\mathbf{f}_1 + \frac{1}{2}\mathbf{f}_2$.

TABLE 2. THE LOSS OF AN EVEN-MONEY BET, $L(\mathbf{f}_r; i)$

Act	Event	
	B_1	B_2
$\mathbf{f}_1$	0	1
$\mathbf{f}_2$	1	0

In this problem, therefore, the minimax rule recommends that the person decide, in effect, by flipping a fair coin. If the odds in the bet had not been even, the minimax rule would have recommended the use of a coin with a certain bias; this more general example will be worked out in detail in § 12.4. It is noteworthy in connection with the present problem—for it happens in many others—that, for the minimax act $\mathbf{f}$, $L(\mathbf{f}; i) = L^*$ for every value of i.

The following more elaborate example, illustrating the mechanism of observation, is paraphrased from a slightly incorrect example in [S2]. Of three numbered coins, two are pennies and one is a dime, or else one is a penny and two are dimes. This gives rise to a sixfold partition B_i, because any of the three coins may be the singular one, and in two ways. The available primary acts are described in two stages thus: First, the person may select one of the coins by number for observation, or he may refrain from so doing; second, he must guess at the denomination of the singular coin. His income in utiles is defined by the following conditions:

1. If the singular coin is a penny, he must pay a tax of 10; if it is a dime, he receives a bonus of 20.

2. If he chooses to observe a coin, he must pay an inspection fee of 1, regardless of the particular coin selected for observation.

3. If his guess is incorrect he pays a penalty of 8.

It is easy to see that the first of the three terms in the person's income is irrelevant to his loss, since his decision does not affect the magnitude of that term. His loss is therefore the sum of two terms. The first of these is 1 or 0 depending on whether he decides to make an observation; the second is 0 or 8, depending on whether his guess is correct.

If the person chooses not to pay the inspection fee, it is clear from the preceding example that, no matter what he does, his loss may be as high as 4, and that it is certain to be that small if and only if he governs his guess (essentially) by the flip of a fair coin.

Suppose next that the person decides to make an observation. If he selects any particular coin for observation, he is as badly off as he was before the observation, and he has in addition incurred the inspection fee. Thus, even if the person knows that the first coin is a penny, there is nothing he can do to be sure that his total loss will not be more than 5, and, as before, he can guarantee that small a loss only by governing his guess with the flip of a fair coin.

I think every practicing statistician would say that, if an observation is to be made at all, one of the three coins should be selected at random (i.e., the probability 1/3 should be attached to observing each of them) and after the observation the person should guess that the singular coin is opposite in denomination to the one observed. It will be shown in the next paragraph that this common-sense act is minimax.

In the first place, the loss $L(\mathbf{f}_0; i)$ for the act $\mathbf{f}_0$ in question is, for each i, equal to $1 + \frac{1}{3} \times 8 = 3\frac{2}{3}$, which is less than 4; for the inspection fee is 1 and the probability of making a wrong guess, which would result in the loss of 8, is 1/3. To show that $\mathbf{f}_0$ is minimax, it will be enough to show that every act can result in a loss of at least $3\frac{2}{3}$. One possibility for doing this (which in § 12.3 will be shown to be a natural one to try) is to show that, for a certain set of weights, the weighted average of $L(\mathbf{f}; i)$ with respect to i is at least $3\frac{2}{3}$ for all $\mathbf{f}$. In fact, it is sufficient, in view of Exercise 4.2, to establish such an inequality for the primary acts. In the present example, it happens that the weights can be chosen to be equal. What is to be shown, then, is that the following inequality obtains for every primary $\mathbf{f}$.

(1) $$L(\mathbf{f}) =_{\text{Df}} \frac{1}{6} \sum_i L(\mathbf{f}; i) \geq 3\frac{2}{3}.$$

Now, if the primary act $\mathbf{f}$ does not involve observation, $L(\mathbf{f}) = 4$; because three of the six terms to be averaged are then 8, and the other three are 0. Suppose next, for definiteness, that $\mathbf{f}$ involves the observation of the first coin; there are then three possibilities to consider. First, the guess is made without regard for the denomination observed, in which case the observation is, so to speak, thrown away, making $L(\mathbf{f}) = 5$. Second, the denomination guessed may be the same as the denomination observed, in which case the guess will be wrong for four of the six values of i, making $L(\mathbf{f}) = 6\frac{1}{3}$. Finally, the denomination guessed may be the opposite of the one observed, in which case the guess will be wrong for two of the six values of i, making $L(\mathbf{f}) = 3\frac{2}{3}$. This argument shows that $L^* \geq 3\frac{2}{3}$; and, since $L(\mathbf{f}_0; i) = 3\frac{2}{3}$ for every i, $\mathbf{f}_0$ is a minimax act and $L^* = 3\frac{2}{3}$. It would not be difficult to show that $\mathbf{f}_0$ is the only minimax act for this problem.

7   Objectivistic motivation of the minimax rule

The minimax rule recommends an act for the person to choose; more strictly, it recommends a sharp narrowing of his choice. But how can this particular recommendation be motivated? To the best of my knowledge no objectivistic motivation of the minimax rule has ever been published. In particular, Wald in his works always frankly put the rule forward without any motivation, saying simply that it might appeal to some. Though my heart is no longer in the objectivistic point of view, I will in the next few paragraphs suggest a relatively objectivistic motivation of the rule.

I evolved this far from satisfactory argument at a time when I took the objectivistic view for granted. Now, as a personalist, it still seems interesting to me in that it shows, or at least suggests, how statistical devices combat vagueness, a topic I find very difficult to discuss directly. On a different level, the argument may shed light on the personalistic view by suggesting how personalistic ideas entered the mind of at least one objectivist.

A categorical defense of the minimax rule seems definitely out of the question. Suppose, for example, that the person is offered an even-money bet for five dollars—or, to be ultra-rigorous, for five utiles—that internal combustion engines in American automobiles will be obsolete by 1970. If there is any event to which an objectivist would refuse to attach probability, that corresponding to the obsolescence in question is one. As the example centering around Tables 6.1–2 makes clear, the minimax rule recommends that the bet be taken or rejected according as a fair coin falls heads or tails. Yet, I think I may say without presumption that you would regard the bet against obsolescence as a very sound investment, agreeing that provision for adequate interest and compensation for changes in the value of money is implicit in measurement of income in utiles.

On the other hand, there are practical circumstances in which one might well be willing to accept the rule—even one who, like myself, holds a personalistic view of probability. It is hard to state the circumstances precisely, indeed they seem vague almost of necessity. But, roughly, the rule tends to seem acceptable when L^* is quite small compared with the values of $L(\mathbf{f}; i)$ for some acts $\mathbf{f}$ that merit serious consideration and some values of i that do not in common sense seem nearly incredible. Suppose, for example, that I were faced with such a decision problem, in which it may be assumed for simplicity that there is only one minimax act **f**, and consider how I might defend the choice of that act to someone who proposed another to me. He might, for example, tell me that he knows from long experience, or by a tip from his broker, that some act **g** is preferable to **f**. "Well," I might say, "I have all the respect in the world for you and your sources of information, but you can see for yourself—for it is objectively so—that the most I can lose if I adopt **f** is L^*." He will not be able to say the same for **g**, and in many actual situations the greatest possible loss under **g** may be many times as great as L^* and of such a magnitude as to make a serious difference to me should it occur, which may well end the argument so far as I am concerned.

It is of interest, however, to imagine that my challenger presses me more closely, reminding me that I am a believer in personal probability, and that in fact I myself attach an expected loss L to **g** that is several times smaller than L^*. Even then, depending on the circumstances, I might answer frankly that in practice the theory of personal probability is supposed to be an idealization of one's own standards of behavior; that the idealization is often imperfect in such a way that an aura of vagueness is attached to many judgments of personal probability; that indeed in the present situation I do not feel I know my own mind well enough to act definitely on the idea that the expected loss for **g** really is L; but that I do, of course, feel perfectly confident that **f** cannot result in a loss greater than L^*, a prospect that in the case at hand does not distress me much.

It seems to me that any motivation of the minimax principle, objectivistic or personalistic, depends on the idea that decision problems with relatively small values of L^* often occur in practice. The mechanism responsible for this is the possibility of observation. The cost of a particular observation typically does not depend at all on the uses to which it is to be put, so when large issues are at stake an act incorporating a relatively cheap observation may sometimes have a relatively small maximum loss. In particular, the income, so to speak, from an important scientific observation may accrue copiously to all mankind generation after generation.

8 Loss as opposed to negative income in the minimax rule

As a variant to the minimax rule as I have stated (or perhaps I should say interpreted) it, one might consider the possibility of letting the negative of income play the role of the loss in (5.1). Indeed, strictly speaking, Wald himself always proposed the minimax rule in that form. I believe he never made written allusion to the rule formulated in terms of loss (as "loss" is defined here); orally he took the position that loss and the form of the minimax rule based on it were inventions of mine, toward which he was tentatively sympathetic. There is virtually no mathematical difference between the two rules, and it was characteristic of Wald's approach to the foundations of statistics to be reluctant to commit himself with respect to any other differences.

Though the minimax rule founded on the negative of income seems altogether untenable, as will soon be explained, and though no one but myself seems to question that I originated the variant of the theory based on loss, little or no originality is attributable to me in this respect. Wald more than foreshadowed the idea, for, though he based his minimax rule on the negative of income, he made it clear in publications, including [W3], that he regarded as typical problems in which the income has, for every i, the property specified in Exercise 4.4. Therefore, in the situations Wald regarded as typical, the distinction between the two forms of the rule vanishes, so, until hearing his explicit disavowal, I considered the idea of loss as opposed to negative income his.

To see that the minimax rule founded on the negative of income is utterly untenable for statistics, consider, for example, a twofold partition problem with two primary acts in which the income is as in Table 1.

TABLE 1. $I(f_r; i)$

Act	Event	
	B_1	B_2
f_1	-1	-1
f_2	-10	1

Now, if the person were interested in minimizing the maximum of the negative income, he would have no recourse but to decide on f_1, in which case (but in no other) he could be sure that the negative income would be at most 1, whichever B_i obtained. This may not in itself seem objectionable, but suppose now that the person has available free of cost an observation, however relevant to B_i. Then, no matter what derived act he chooses, if B_1 obtains, his negative income will be at least 1 utile; and, to be sure that it is not more, he again has no recourse but to decide on f_1. In short, for the problem at hand, the person's behavior would not be influenced by any observation, however relevant. This seems to me absurd on the face of it, but perhaps the absurdity can be brought out by a less abstract situation paralleling the example just given. A person has a ladder, and, just as he is about to use it, it occurs to him that the ladder may possibly be dangerously defective. He envisages two basic primary acts: f_1, to throw the ladder away and buy a new one, which will cost 1 utile in either event; and f_2, to use the ladder, which will, if the ladder is defective, result in his injury to the extent of 10 utiles, and will, if the ladder is sound, accomplish his object, which is worth 1 utile. Now, if the person acts on the principle of minimizing the maximum of negative income, he will throw the ladder away, no matter what tests tend to show that it is sound.

CHAPTER 10

A Personalistic Reinterpretation of the Minimax Theory

1 Introduction

In this chapter a reinterpretation of the minimax theory, based on the theory of personal probability and the idea that statistical problems are typically multipersonal, is tentatively put forward. The reinterpretation is based on a model or scheme that captures, I believe, much of the essence of actual statistical situations, but it may be possible to effect that end with other equally simple and even more realistic models; for the one to be presented here leaves much to be desired. In structure, this chapter is kept roughly parallel with Chapter 9, to enable the reader to examine as closely as he may wish the parallelism between the objectivistic interpretation given there and the personalistic one given here. In particular, the liberty is taken of giving old symbols new meanings in order to bring out the parallelism between the two interpretations.

2 A model of group decision

Consider a group of people, indexed by numbers i. These people are supposed to have the same utility function, at least for the consequences to be considered in the present context, but their personal probabilities are not necessarily the same. The group of people is placed in a situation in which it must, acting in concert, choose an act f from a finite set of available acts F, the consequences of the acts being measured in terms of the common utility of the members of the group.

The situation just described will be called a **group decision problem.** It is epitomized by a jury. The members of the jury, in legal theory, are supposed to have common value judgments in connection with the legal matters at hand; for these are incorporated in the law as stated in the instructions of the court. But it is part of the very concept of a jury that its members may be of different opinions; that their judgments as to questions of fact may differ; that, to put it technically, they may have different systems of personal probability. Still other situations resembling the group decision problem are widespread in science and industry, though the group decision problem does by no means represent the only sort of social interaction tending to make the theory of personal probability, confined to a single person, inadequate. Whenever a hospital or a factory modifies its procedures, whenever a doctrine is adopted with little reservation by virtually all the workers in a science, or whenever a panel of experts drafts a report, something like group decision is taking place.

Since the members of the group in a group decision problem, though required to act in concert, typically differ from one another in their probability judgments, it is too much to expect that any rule can be formulated that will be acceptable to, or in any sound sense proper for, all groups under all circumstances. On the other hand, there may be one or more rules of thumb that will lead the group to an acceptable compromise in many practical circumstances. Two such suggestions, the group minimax rule and the group principle of admissibility, will be made and explored in the next section.

3 The group minimax rule, and the group principle of admissibility

In the first place, the possibility of using mixed acts is to be pointed out. If, for example, you and I, walking together, disagree about which branch of a fork in the road leads home, we can, and in fact may, decide which to try by flipping a coin.

In general, mixed acts are available in a group decision problem for reasons analogous to their availability in objectivistic decision problems, for, though the members of a group may generally differ in the probabilities they personally assign to some events, there is in practice an abundance of events associated with coins, cards, random numbers, and the like that make it possible for the group to mix the primary acts in any proportion, all members of the group being in agreement about what the proportions are. The example of the fork in the road illustrates how the use of mixed acts can effect such a compromise as to make decision possible in what might otherwise be an impasse. As in the account of the objectivistic decision problems, it will therefore be taken for granted from now on that F contains all mixtures of its elements, and once more, for mathematical simplicity, it will be assumed that there are a finite number of primary acts f_r in F, of which all others are mixtures.

The ith person in the group attaches a certain expected utility, or (**personal**) **income**, to the act f; call it $I(f; i)$. In the judgment of the

ith person, adoption of the act f would represent a **(personal) loss,**

$$(1) \qquad\qquad L(f; i) = \max_{f'} I(f'; i) - I(f; i).$$

(possibly zero) as compared with the income or expected utility that in his opinion would result from an act he considers most promising.

The group minimax rule is the suggestion that an act be adopted such that the largest loss faced by any member of the group will be as small as possible. To put it formally, the suggestion is that an f' be adopted such that

$$(2) \qquad\qquad \max_i L(f'; i) = L^* =_{\mathrm{Df}} \min_f \max_i L(f; i).$$

The parallelism between the group minimax rule and the minimax rule stated in § 9.5 is great. In particular, (2) is identical in appearance with (9.5.1). This is really only a pun, though a fruitful one, because L, i, and even f have altogether different meanings in the two contexts.

As indicated at the outset, it cannot be expected that the group minimax rule will, or reasonably should, be accepted by every group faced with every problem. But, much as in the corresponding objectivistic decision problems, it may happen that, if L^* is small, in a rather vague sense, the group will accept the group minimax rule. Indeed, if L^* is small, the group minimax rule requires no member of the group to face a large loss, so no member will feel that the suggestion is a serious mistake. In any event, no member of the group can suggest an alternative that will not make some member's loss as great as L^*, for there is none. Moreover, in many problems the group minimax rule will lead to the same loss L^* for every member of the group (as is explained in § 12.3), a circumstance which, when it occurs, may add to the acceptability of the suggestion by making it seem fair.

Of course it is possible that, as in the objectivistic interpretation, more than one act fulfilling the minimax principle exists. Here, a paraphrase of the principle of admissibility will further narrow the choice, for if

$$(3) \qquad\qquad L(g; i) \leq L(f; i)$$

for every i, with inequality obtaining for some i, the group cannot seriously consider f.

4   Critique of the group minimax rule

Some of the criticisms that have been, or may be, raised against the minimax rule can as well be discussed in connection with one interpretation as with the other, and Chapter 13 will be devoted to such criticisms. But some that bear specifically on the multipersonal interpretation in this chapter should be discussed here.

In the first place, the group minimax rule is flagrantly undemocratic. In particular, the influence of an opinion, under the group minimax rule, is altogether independent of how many people in the group hold that opinion. In general, it is difficult to give a formal analysis of the concept of democratic decision, a point discussed at length by Arrow [A5], Hildreth [H4a], and others. Perhaps, considering that the people in the group are postulated to have a common utility function, a satisfactory analysis of democratic decisions could be given in the case of a group decision problem by some such procedure as minimizing the average with respect to i of $L(\mathbf{f}; i)$. But, in many situations in which I envisage application of the group minimax principle, the group will in fact be a rather nebulous body of people, for example the group of all specialists in some field. The principle would in such a case be administered by a single member of the group somewhat in the following fashion. In planning an investigation, the results of which he intends to publish, he will endeavor to take account of all opinions, so far as he can know or guess them, that are considered at all reasonable in his field of investigation. And when he publishes his results he will say, in effect, "Whatever reasonable opinions have heretofore been held by members of this specialty, in the light of my investigation and the minimax rule, it is now proper for the members of the specialty, in so far as they are called upon to act in concert, to agree to such and such an action." To put it a little differently, in such an application the group is rather fictitious, and the individual investigator is admitting as reasonable a rather large class of opinions, but excluding many that he is sure his confreres will agree are utterly absurd. He will, for example, feel quite free to exclude those opinions that almost all educated people regard as superstitious.

The group minimax rule is also objectionable in some contexts, because, if one were to try to apply it in a real situation, the members of the group might well lie about their true probability judgments, in order to influence the decision generated by the minimax rule in the direction each considers correct. This objection is, however, scarcely serious in the fictitious sort of application suggested above.

It is appropriate, in terminating this section, to discuss a certain distinction, neglect of which can, as was pointed out to me orally by Bruno de Finetti, lead to serious misunderstanding of the group minimax rule. Voluminous observation typically tends to make any one person almost certain of the truth, and also, when a group of people is involved, it typically tends to make L^* small. These two tendencies, though related, are separate phenomena, as an illustration will bring out.

Suppose that Peter and Paul are required to bet 1 utile in concert either that the majority of a large electorate has voted for, or that it has voted against, a certain issue; but that before betting they are to be allowed to examine a random sample of 1,001 ballots.

If specific opinions about the division of the electorate are assigned to Peter and Paul, the situation can be regarded as a group decision problem. To start with an interesting extreme possibility, suppose that it is Peter's unequivocal opinion that 55% of the electorate is for and 45% is against the issue and Paul's that the division is 45% for and 55% against; that is, Peter, for example, is supposed to act as though he *knows* that the division is 55%–45%.

If, finally, it is understood that the group decision problem consists in the two people, Peter and Paul, deciding, before the sample is actually observed, how their bet is to be determined by the composition of the sample; then the unique minimax act is to bet that the electorate majority is whatever the sample majority happens to be. Granting this easily established solution of the minimax problem, it is obvious that the two people both face the minimax loss L^*. Peter, to be specific, regards L^* as the probability that through random fluctuation the sample will accidentally fail to corroborate his "knowledge" that the majority is for the issue. Numerically, L^* is about 0.0008.

Peter and Paul, recognizing that the possibility of observing the sample reduces the minimax loss to about 0.0008 as compared with the 0.5 that it would be if no sample were available, may well find the minimax act a satisfactory compromise; at any rate, it is hard to see in this situation how they could arrive at any other.

Though the incorporation of the sample into the problem has greatly reduced L^*, observation of the sample does not affect the opinion of either person in the slightest, for unequivocal opinions such as they hold are not subject to modification in the light of evidence. At least one of the two people is immovably wrong, and the observation of no sample, however large, can bring them both close to the truth. This brings out a contrast between the reduction of L^* and the approach to certainty of the truth, both of which typically occur with the accumulation of evidence.

The same contrast is expressed by remarking that, though the two people may readily adopt the minimax act, each feeling that at the expense of a small risk he is diverting the obstinacy of his colleague to their common good; after the observation of the sample, one or the other of them is bound to feel that the prize has been lost by a sad and improbable accident.

The wary will ask, "Who will feel how, when the actual majority is disclosed and settlement made? What if Peter's unequivocal opinion turns out to be false?" Such questions suggest that paradox lurks in an example in which different people unequivocally hold mutually inconsistent opinions, so there is some interest in considering a modification of the example, free of that objectionable feature.

Suppose then that Peter and Paul, though strongly opinionated about the division of the electorate, are not absolutely unequivocal in their opinions. To be quite definite, suppose that Peter attaches probability $1-10^{-10}$ to the division 55%–45% and probability 10^{-10} to the division 45%–55%, and that Paul attaches the same probabilities but in the opposite order to the two divisions. Here, as in the example of the unequivocal opinions, the unique minimax act is to let the bet be chosen in accordance with the sample majority; L^* is a trifle lower than before. Observation of the sample does now generally affect the opinions of the two people, but, though it radically reduces the minimax loss, it does not typically bring the two people into close agreement. If, for example, the division is in fact 45%–55%, Paul's strong a priori belief that that is the actual division is almost sure to be strengthened by the sample, and Peter's equally strong but false belief is almost sure to be weakened. Still, the probability is only about 1/2 that Peter will be led by the sample to attach an a posteriori probability even as great as 0.05 to the actual division. Thus, speaking loosely but practically, the approach to certainty of the truth is here not typically nearly so far advanced by observation as is the reduction of the minimax loss.+

It may not be superfluous to point out that the preceding paragraph alludes not only to the two different personal probability systems of Peter and of Paul, but also to certain conditional probabilities that you and I have accepted hypothetically in setting up the example.

Whichever division does actually obtain, it is rather probable that, once the sample is observed, either Peter or Paul will wish he could break his contract. This seems to me to reflect a serious objection to the group minimax principle, especially in those applications in which the members of the group are not literally consulted, for people cannot be expected to abide by disappointing contracts they might have made but didn't.

For other approaches to the group decision problem see de Finetti [D6], [D7a], de Finetti (1954), Staël von Holstein (1970, p. 65 and ff.), and Winkler (1968).

+ As de Finetti has remarked, the separation between the two phenomena is more clearly brought out if Peter and Paul decide which bet to make on the basis of a tennis match between themselves. For, if each thinks himself much the superior player, L^* will be depressed, though the opinions of Peter and Paul about the election remain completely unaffected by the outcome of the match.

CHAPTER 11

The Parallelism between the Minimax Theory and the Theory of Two-Person Games

1 Introduction

John von Neumann, in 1928 [V3], developed a theory of games in which two people play each other for money.† This theory is mathematically so closely akin to that of the minimax rule and has had such influence on its development that it would be artificial to give an exposition of the minimax rule without saying something of the theory of what von Neumann calls zero-sum two-person games, though the account given here must necessarily be highly compressed. The most convenient references in English to the theory of zero-sum two-person games, should the reader be interested in a fuller account, are [B18], [M3], and Chapters II and III of [V4]; though those who read German may find it best to start with the expository sections of the paper [V3] in which von Neumann first discussed the subject.

The sort of systematic punning by which the formal parallelism between the objectivistic and personalistic minimax theories was emphasized in Chapter 10 will be used once more, to bring out the formal parallelism between those theories and that of zero-sum two-person games. Logic will be still further sacrificed to clarity and convenience by calling the two people who play the game "you" and "I."

2 Standard games

A certain sort of game, here called a *standard game*, is defined thus: You secretly choose a number r from a finite set of possibilities, and I secretly choose a number i, also from a finite set of possibilities. The numbers r and i having been chosen, you pay me the sum of money (possibly negative) $L(r; i)$, where $\mathbf{L}$ is an arbitrary function of r and i, known to both of us. It is assumed that, for the sums involved, each of us finds money proportional to utility.

† In this completely independent development he was to some extent anticipated by Emil Borel. Consult [F9], [F10], and [B21] for details and further references.

At first sight, standard games look very dull, though it is immediately recognized that some such games are played. A tiny but typical example is the game of "Button, button, who's got the button?"; "Stone, paper, scissors" is almost as familiar an example; and others could be mentioned. But, and this seems remarkable at first, any game, except possibly those dependent on physical skill, can be viewed as a standard game. The great generality of standard games is demonstrated in detail in Chapter II of [V4], but informal discussion of a single example will render the idea intuitively clear. Suppose then that you and I are to play a game of poker (of a specified variety). At first sight poker does not seem to be a standard game, because it involves several random events, and several decisions on the part of each of us, some to be made in the light of others. But, it can be argued, there are only a finite number of different situations that can arise in the course of a game of poker. You could, therefore, in principle write into a notebook exactly which choice you would make in each of the possible situations with which you might be faced in playing poker with me. The number of possible ways of compiling such notebooks, or policies of play, is finite; so, except for limitations of time and patience, you will be at no disadvantage in playing one game with me, if you simply chose once and for all that one of the many possible policies of play that seems best to you. Similarly, from my point of view, the game consists, in principle, in choosing one policy of play. Once you have chosen one of the policies possible for you, say the rth, and I have chosen one of the policies possible for me, say the ith, the amount you will have to pay me at the termination of the game is a random variable. Since it is agreed that the payments are effectively in utiles for both of us, your payment to me is effectively the expected value of this random variable, which may be called $L(r; i)$ and which is in principle known to both of us as a function of r and i. The elaborate game of two-person poker is thus exhibited, at some expense to realism, as a standard game.

Regarding the choice of an r by you or an i by me as a primary act, both of us are at liberty to use mixed acts. Indeed, explicit attention apparently was first called to the possibility of using mixed acts by Borel (see [B21]), in just this context.

Let **f** and **g** represent mixed acts assigning probabilities $\phi(r)$ and $\gamma(i)$ to the values r and i, respectively. The standard game is now replaced by a somewhat different game in which you choose an **f**; I choose a **g**; and you pay me the amount $L(\mathbf{f}; \mathbf{g})$, where

$$(1) \qquad L(\mathbf{f}; \mathbf{g}) =_{\mathrm{Df}} \sum_{r,\,i} L(r; i)\phi(r)\gamma(i).$$

3 Minimax play

Von Neumann adduces an argument, the statement of which will be briefly postponed, that, if you have respect for my intelligence, you will see to it that the most I can possibly take from you shall be as small as possible, that is, you will choose an $\mathbf{f}'$ for which

$$(1) \qquad \max_{\mathbf{g}} L(\mathbf{f}'; \mathbf{g}) = L^* =_{\mathrm{Df}} \min_{\mathbf{f}} \max_{\mathbf{g}} L(\mathbf{f}; \mathbf{g}).$$

Symmetrically, according to his argument, I should choose a $\mathbf{g}'$ such that

$$(2) \qquad \min_{\mathbf{f}} L(\mathbf{f}; \mathbf{g}') = L_* =_{\mathrm{Df}} \max_{\mathbf{g}} \min_{\mathbf{f}} L(\mathbf{f}; \mathbf{g}).$$

Since, making the recommended choice, you are sure that you will not pay me more than L^*, and I am correspondingly sure that you will not pay me less than L_*; it follows that $L_* \leq L^*$. This inequality would, of course, have obtained even if mixed acts were not permitted. It is a remarkable mathematical fact (not to be proved in this book) that, permitting mixed acts, equality always obtains; so the special symbol L_* is superfluous here.

The argument for the recommended choices rests on the equality of L^* and L_*. You realize that I can take at least L^* from you and that, if you are not careful, I may take more. On the other hand, I realize that you can prevent my taking more than L^* from you and that, if I am not careful, I may get less. This suggests to many that a pair of intelligent players, each respecting the intelligence of the other, will each adopt one of the recommended acts.

4 Parallelism and contrast with the minimax theories

Some formal parallelism between the minimax theories of decision and the theory of zero-sum two-person games is evident, but the parallelism is much more complete than may appear at first sight. The mixtures $\mathbf{g}$ are without counterpart in the two minimax theories of decision, and the appearance of $\mathbf{g}$ in (3.1) at the place where i appears in (9.5.1) may seem to mar the parallelism between these two equations. But, letting

$$(1) \qquad L(\mathbf{f}; i) =_{\mathrm{Df}} \sum_r L(r; i)\phi(r),$$

in the game theory (in close parallelism with the decision theories),

$$(2) \qquad L(\mathbf{f}; \mathbf{g}) = \sum_i L(\mathbf{f}; i)\gamma(i) \leq \max_i L(\mathbf{f}; i),$$

and

$$(3) \qquad \max_{g} L(\mathbf{f}; \mathbf{g}) = \max_{i} L(\mathbf{f}; i).$$

Therefore (3.1) is equivalent to

$$(4) \qquad \max_{i} L(\mathbf{f}'; i) = \min_{\mathbf{f}} \max_{i} L(\mathbf{f}; i) = L^{*}.$$

Thus from the point of view of the minimax theories of decision the **g**'s represent no material innovation and are at worst useless baggage. Actually, though of little if any relevance in the interpretation of the minimax theories, the **g**'s constitute a useful mathematical device. Their usefulness has in fact been illustrated in working out the second example in § 9.6 and will be systematically demonstrated in the next chapter, along with the usefulness of the apparently irrelevant "maximin" problem posed by (3.2) and of the fact that $L_{*} = L^{*}$.

Some remarks on the possibility of interpreting the **g**'s in the minimax theories are postponed to the end of this section.

In the game theory, **L** may be any function whatsoever of its arguments r and i, but, in the decision theories, **L** is subject to the condition that, for every i,

$$(5) \qquad \min_{r} L(r; i) = 0,$$

where $L(r; i)$ is of course to be interpreted as $L(\mathbf{f}_r; i)$. Here is the only mathematical difference between the game theory and the decision theories, the former being mathematically slightly more general than the latter.

Though the mathematical differences are negligible, the intellectual difference between the situations leading to the game theory on the one hand and to the decision theories on the other is great. Serious misunderstandings of the (objectivistic) minimax theory have often resulted from identifying it with the game theory. Among other things, loss is then confounded with negative income, and the misconception that the (objectivistic) minimax rule is ultrapessimistic is created. I have even heard it stated on this account that the minimax rule amounts to the assumption that nature is malevolently opposed to the interests of the deciding person.

Though mathematical convenience seems to be the basic reason for introducing the **g**'s in the minimax theories, it is tempting to ask whether the **g**'s have also some natural interpretation in those theories. At the moment, I do not see a convincing interpretation in either theory, but completeness demands an account of an interpretation suggested by

Wald for his version of the objectivistic theory, especially since this interpretation influenced some of Wald's most widely used terminology.

The objectivistic problem of deciding on an act in ignorance of which partition element B_i obtains, the $P(B_i)$ being regarded as meaningless, suggests a new problem that may perhaps also be called objectivistic. The new problem arises on postulating that $P(B_i)$ is meaningful but utterly unknown, that is, $P(B_i) = \gamma(i)$, where the $\gamma(i)$'s are the components of a $\mathbf{g}$ here interpreted as the a priori distribution unknown to the deciding person.

Since for Wald "loss" was synonymous with "negative expected income," he naturally calculated the loss of the new problem thus:

$$(6) \qquad L(\mathbf{f}; \mathbf{g}) = -E(\mathbf{f} \mid \mathbf{g})$$
$$= \sum_i -E(\mathbf{f} \mid B_i)P(B_i)$$
$$= \sum_i L(\mathbf{f}; i)\gamma(i),$$

arriving thus at the very function suggested by the game theory. In Wald's version of the theory, the new problem therefore amounts to the formal introduction of the $\mathbf{g}$'s in connection with the old one, which neatly fulfills the reasonable expectation that there should be no material difference between regarding $P(B_i)$ as meaningless and regarding it as meaningful but utterly unknown.

The suggested interpretation of a $\mathbf{g}$ as an unknown—or, to mirror Wald more faithfully, fictitious—a priori distribution does not work, however, if the loss function of the new problem is defined by (9.4.1), for the new function $\tilde{L}(\mathbf{f}; \mathbf{g})$ is not then generally the same as the function $L(\mathbf{f}; \mathbf{g})$ suggested by the game theory; thus

$$(7) \qquad \tilde{L}(\mathbf{f}; \mathbf{g}) = \max_{\mathbf{f}'} E(\mathbf{f}' - \mathbf{f} \mid \mathbf{g})$$
$$= \max_{\mathbf{f}'} \sum_i E(\mathbf{f}' - \mathbf{f} \mid B_i)\gamma(i)$$
$$= \max_{\mathbf{f}'} \sum_i \{L(\mathbf{f}; i) - L(\mathbf{f}'; i)\}\gamma(i)$$
$$= L(\mathbf{f}; \mathbf{g}) - \min_{\mathbf{f}'} L(\mathbf{f}'; \mathbf{g})$$
$$\leq L(\mathbf{f}; \mathbf{g}),$$

equality holding for a typical $\mathbf{g}$ (i.e., a $\mathbf{g}$ such that $\gamma(i) > 0$ for every i) only in the altogether trivial situation that $\mathbf{F}$ is dominated by one of its elements.

Does this mean that, contrary to expectation, there is a material difference between the new problem with loss $\tilde{L}$ and the old one? The following exercises show that it does not.

Exercises

1. $\max\limits_{g} \tilde{L}(f; g) = \max\limits_{i} L(f; i)$.

2. $\min\limits_{f} \max\limits_{g} \tilde{L}(f; g) = L^*$.

3. $\max\limits_{g} \tilde{L}(f; g) = L^*$, if and only if $\max\limits_{i} L(f; i) = L^*$.

CHAPTER 12

The Mathematics of Minimax Problems

1 Introduction

Since the two different minimax decision theories and the theory of zero-sum two-person games have a common mathematical core, it will be worth while to digress for a chapter even at the expense of some repetition, to discuss this common core mathematically, that is, virtually without reference to its various possible interpretations. The discussion will have to be drastically confined relative to the large body of relevant literature, but the reader who wishes to pursue the subject much further will find [B18], [V4], [W3], and [M3] to be key references.

2 Abstract games

To begin with a very general situation, which will later be specialized to the one of main interest, let $\mathbf{f}$ and $\mathbf{g}$ denote generic elements of any two abstract sets, and let $L(\mathbf{f}; \mathbf{g})$ be the value of an essentially arbitrary real-valued function. It will, however, be assumed for simplicity that for every $\mathbf{f}'$ and $\mathbf{g}'$ the quantities

(1)
$$\max_{\mathbf{g}} L(\mathbf{f}'; \mathbf{g}), \qquad \min_{\mathbf{f}} L(\mathbf{f}; \mathbf{g}')$$
$$L^* =_{\mathrm{Df}} \min_{\mathbf{f}} \max_{\mathbf{g}} L(\mathbf{f}; \mathbf{g}), \qquad L_* =_{\mathrm{Df}} \max_{\mathbf{g}} \min_{\mathbf{f}} L(\mathbf{f}; \mathbf{g})$$

exist. To say that a maximum, for example, exists is not only to say that the function in question is bounded from above, but also that the maximum value is actually attained for at least one value of the argument. For want of a more neutral term, call the function $L(\mathbf{f}; \mathbf{g})$ an **abstract game.**

An $\mathbf{f}'$ is called **minimax,** if and only if

(2)
$$\max_{\mathbf{g}} L(\mathbf{f}'; \mathbf{g}) = L^*;$$

and a $\mathbf{g}'$ is called **maximin,** if and only if

(3)
$$\min_{\mathbf{f}} L(\mathbf{f}; \mathbf{g}') = L_*.$$

The existence of minimax and maximin values of the variables is implicit in (1). It is an easy exercise to show that $\mathbf{f}'$ is minimax, if and only if

(4) $$L(\mathbf{f}'; \mathbf{g}) \leq L^*$$

for every $\mathbf{g}$.

The corresponding characterization of maximin $\mathbf{g}''$'s as those such that

(5) $$L(\mathbf{f}; \mathbf{g}') \geq L_*$$

for every $\mathbf{f}$ could similarly be shown. But the symmetry of the situation is such that it would be superfluous to derive this characterization of a maximin explicitly. Indeed, every theorem, or general conclusion, about $L(\mathbf{f}; \mathbf{g})$ obviously has a **dual**, which arises on applying the theorem to the new abstract game $\tilde{L}(\mathbf{g}; \mathbf{f})$ with $\tilde{L}(\mathbf{g}; \mathbf{f}) = -L(\mathbf{f}; \mathbf{g})$. This is typical of what is known in mathematics as a **duality principle**. Henceforth the duals of demonstrated conclusions, even when not explicitly stated, will be as freely used as the demonstrated conclusions themselves. Some conclusions are of course self dual. Incidentally, another example of a duality principle was used in § 5.4, and a very important one was pointed out in connection with Boolean algebra in § 2.4.

An argument showing that $L_* \leq L^*$ was given in connection with the theory of games. More formally, if $\mathbf{f}'$ and $\mathbf{g}'$ are, respectively, minimax and maximin, then from (4) and (5)

(6) $$L^* \geq L(\mathbf{f}'; \mathbf{g}') \geq L_*.$$

It is possible, indeed typical, that $L_* < L^*$. Suppose, for example, that $\mathbf{f}$ and $\mathbf{g}$ are variables that take only two values and that $L(\mathbf{f}; \mathbf{g})$ is described by Table 1. Here, as the reader should verify, both $\mathbf{f}$'s

TABLE 1. $L(\mathbf{f}; \mathbf{g})$

		g	
		1	2
	1	0	1
f			
	2	1	0

and both $\mathbf{g}$'s are minimax and maximin, respectively, and $L^* = 1$, $L_* = 0$.

The following theorem is frequently applicable to the identification of minimax and maximin values of $\mathbf{f}$ and $\mathbf{g}$, and of L^* and L_*.

THEOREM 1   If $\mathbf{f}'$, $\mathbf{g}'$, and the number C are such that $L(\mathbf{f}'; \mathbf{g}) \leq C$ $\leq L(\mathbf{f}; \mathbf{g}')$ for every $\mathbf{f}$ and $\mathbf{g}$; then $L^* = L_* = C = L(\mathbf{f}'; \mathbf{g}')$, $\mathbf{f}'$ is minimax, and $\mathbf{g}'$ is maximin.

PROOF.   First, $C \geq L^*$, because

$$(7) \qquad C \geq \max_{\mathbf{g}} L(\mathbf{f}'; \mathbf{g}) \geq \min_{\mathbf{f}} \max_{\mathbf{g}} L(\mathbf{f}; \mathbf{g}) = L^*;$$

and, dually, $C \leq L_*$. But $L_* \leq L^*$; so $C \leq L_* \leq L^* \leq C$, that is, $L^* = L_* = C$. Now (4) and (5) apply. ◆

COROLLARY 1   If $\mathbf{f}'$ and $\mathbf{g}'$ are such that $L(\mathbf{f}'; \mathbf{g}) \leq L(\mathbf{f}; \mathbf{g}')$ for every $\mathbf{f}$ and $\mathbf{g}$; then $\mathbf{f}'$ and $\mathbf{g}'$ are, respectively, minimax and maximin, and $L^* = L_* = L(\mathbf{f}'; \mathbf{g}')$.

3 Bilinear games

If one stumbles somehow onto a pair $\mathbf{f}'$, $\mathbf{g}'$ satisfying the hypothesis of Corollary 2.1, then he has discovered a minimax, a maximin, and the values (in this case equal to each other) of L^* and L_*. But that possibility of discovery does not exist unless $L^* = L_*$, which at the level of generality of the last section is unusual. Almost all real interest, however, centers on a very special class of abstract games, here to be called bilinear games, for which it is demonstrable that L^* is invariably equal to L_*.

The definition of bilinear games involves several steps. First, consider an abstract game, $L(r; i)$, based on a pair of variables, r and i. The two variables are here assumed for simplicity to have only a finite number of possible values, an assumption that can, and for statistics must, be considerably relaxed. Next, let $\mathbf{f}$ and $\mathbf{g}$ be non-negative functions of r and i, respectively, arbitrary except for the constraint that

$$(1) \qquad \sum_r f(r) = \sum_i g(i) = 1,$$

in short, probability measures on the r's and i's, respectively. Finally, the **bilinear game** $L(\mathbf{f}; \mathbf{g})$ is defined thus.

$$(2) \qquad L(\mathbf{f}; \mathbf{g}) = {}_{\mathrm{Df}} \sum_{r, i} L(r; i) f(r) g(i).$$

It is important to recognize that the duality principle continues to hold, that is, if $L(\mathbf{f}; \mathbf{g})$ is a bilinear game, then $\tilde{L}(\mathbf{g}; \mathbf{f}) = -L(\mathbf{f}; \mathbf{g})$ is also one.

In terms of the auxiliary functions

(3)
$$L(\mathbf{f}; i) =_{\mathrm{Df}} \sum_r L(r; i) f(r),$$

$$L(r; \mathbf{g}) =_{\mathrm{Df}} \sum_i L(r; i) g(i),$$

the following equalities and inequalities can easily be verified by the reader.

(4)
$$\max_{\mathbf{g}} L(\mathbf{f}; \mathbf{g}) = \max_i L(\mathbf{f}; i),$$

$$\min_{\mathbf{f}} L(\mathbf{f}; \mathbf{g}) = \min_r L(r; \mathbf{g}).$$

(5) $$\min_r \max_i L(r; i) \geq \min_{\mathbf{f}} \max_i L(\mathbf{f}; i) = L^* \geq L_*$$

$$= \max_{\mathbf{g}} \min_r L(r; \mathbf{g}) \geq \max_i \min_r L(r; i).$$

But more can be said in connection with (5), for it has been shown by von Neumann [V3] that for the special class of functions now under discussion L^* is actually equal to L_*. This important equality cannot conveniently be proved here, but the interested reader can refer to the relatively simple proof given by von Neumann and Morgenstern in Section 17.6 of [V4] (reading first, if necessary, the introduction to the mathematics of convex sets that constitutes Chapter 16 of that book) or to the version of it presented in [B18].

In the light of the equality of L^* and L_*, (5) becomes

(6) $$\min_r \max_i L(r; i) \geq \min_{\mathbf{f}} \max_i L(\mathbf{f}; i) = L^*$$

$$= \max_{\mathbf{g}} \min_r L(r; \mathbf{g}) \geq \max_i \min_r L(r; i).$$

In view of (4) and (6), Theorem 2.1 can be much improved upon for bilinear games:

THEOREM 1   For bilinear games, the following three conditions on $\mathbf{f}'$, $\mathbf{g}'$, and C are equivalent:
  1. $\mathbf{f}'$ minimax, $\mathbf{g}'$ maximin, and $L^* = C$.
  2. $L(\mathbf{f}'; \mathbf{g}) \leq C \leq L(\mathbf{f}; \mathbf{g}')$ for every $\mathbf{f}$ and $\mathbf{g}$.
  3. $L(\mathbf{f}'; i) \leq C \leq L(r; \mathbf{g}')$ for every i and r.

PROOF.   Condition 2 implies 1, by Theorem 2.1; 1 implies 3 by (6); and 3 implies 2 by (4). ◆

COROLLARY 1   A necessary and sufficient condition that **f** be minimax is that, for some **g**, $L(\mathbf{f}; i) \leq L(r; \mathbf{g})$ for every r and i. Under that condition $L^* = L(\mathbf{f}; \mathbf{g})$, and **g** is maximin.

Corollary 1 seems an especially appropriate expression of Theorem 1 in connection with the minimax decision theories, where the **g**'s are, after all, not really of interest in themselves. Theorem 1, and equivalently Corollary 1, are of great practical value. To be sure, there are algorithms, or rules (given by Shapley and Snow in [S12]), by which L^* and all minimax values of **f** can in principle be computed, but these algorithms are so awkward to apply that in practice one generally guesses one or more minimax **f**'s, and also a maximin **g**, on the basis of some clues, verifying the guess and evaluating L^* by Corollary 1. To finish the job, one then finds, if one can, an argument to show that the minimax **f**'s thus discovered are all there are. This rather imperfect procedure is especially important, since it can relatively easily be extended to many situations in which r and i are not confined to finite ranges, as does not seem to be true of the algorithms.

As was mentioned in § 10.3 and as the examples that have been given illustrate, if **f** is minimax, then $L(\mathbf{f}; i)$ is in practice often actually equal to L^* for all, or at least many, values of i. Insight into that phenomenon is given by the following theorem.

THEOREM 2   If i is such that there exists a maximin **g** for which $g(i) > 0$, then $L(\mathbf{f}; i) = L^*$ for every minimax **f**.

PROOF. $L(\mathbf{f}; i) \leq L^*$, because **f** is minimax. Therefore $L(\mathbf{f}; \mathbf{g})$, being a weighted average of the $L(\mathbf{f}; i)$'s, is at most L^*; and it is actually less, if any term with positive weight is not equal to L^*. But $L(\mathbf{f}; \mathbf{g}) \geq L^*$, because **g** is maximin. ◆

It can happen, and in statistical practice it often does happen, that every i satisfies the hypothesis of Theorem 2, in which case $L(\mathbf{f}; i) = L^*$ for every i and every minimax **f**.

Theorem 2 often provides a basis for guessing a minimax **f**, a maximin **g**, and the value of L^*, which can then be checked by application of Corollary 1. To take a simple example, suppose that there are n values of r, and n of i. There may be some reason to conjecture that each i is used by some maximin **g**, that is, that each i satisfies the hypothesis of Theorem 2. If the conjecture is in fact true, then $f(r)$ and L^* satisfy the system of equations

(7)
$$\sum_r 1f(r) + 0L^* = 1$$
$$\sum_r L(r; i)f(r) - 1L^* = 0.$$

Typically, (7) as a system of $n + 1$ linear equations in $n + 1$ variables will have exactly one solution $(f(r), L^*)$. This solution, if the conjecture is valid, will actually consist of the components of a minimax $\mathbf{f}$ (in this case the only one) and the value of L^*. But the conjecture is not yet confirmed. In particular, if *any* $f(r)$ in the solution of (7) is negative, it is contradicted; if not, the investigation can proceed. The candidates for maximin values of $\mathbf{g}$ are now, by the dual of Theorem 2, among the solutions of the system.

$$\sum_i 1g(i) + 0L^* = 1$$

(8)

$$\sum_i L(r; i)g(i) - 1L^* = 0,$$

where r is confined to the values for which $f(r) > 0$. To consider only the simplest and most typical case, suppose $f(r) > 0$ for every r. Regarding L^* as known, (8) consists of $n + 1$ equations for n variables, which at first sight might be expected generally to have no solution. To put the matter differently, if one forgets for the moment that L^* has been determined by (7), it might seem possible that (8) could lead to a different value, say $L^{*\prime}$. But, using the latter part of (8) and then the first part of (7), it is seen that

(9) $$\sum_{r, i} L(r; i)f(r)g(i) = \sum_r f(r)L^{*\prime} = L^{*\prime},$$

and dually the double sum equals L^*; so discrepancy between L^* and $L^{*\prime}$ is not among the real snags in the tentative program—irrespective of the number of r's participating in (8). Finally, if (8) leads to even one set of positive $g(i)$'s, it follows from Corollary 1 that the $\mathbf{f}$ and L^* derived from (7) are the unique minimax and the true value of L^*, respectively.

The converse of Theorem 2 has been proved by Bohnenblust, Karlin, and Shapley in [B19], though their proof cannot be reproduced here. As is pointed out by these authors, the converse does not extend at all readily to situations involving infinite ranges of r and i. Theorem 2 and its converse can be summarized thus:

THEOREM 3   There exists a maximin $\mathbf{g}$ for which $g(i) > 0$, if and only if $L(\mathbf{f}; i) = L^*$ for every minimax $\mathbf{f}$.

4 An example of a bilinear game

It is now convenient to discuss a certain example, or rather a class of examples, of bilinear games, namely those in which i takes only two values, say 1 and 2. Two preliminary remarks will help to orient the discussion. First, bilinear games in which i takes only one value are devoid of interest, for the minimax problem in that case is simply a problem of finding an ordinary minimum. Second, the discussion of bilinear games in which i takes only two values includes, in effect, because of the duality principle, the discussion of those in which r takes only two values.

If i takes only the two values 1 and 2, the values $\mathbf{g} = \{g(1),\ g(2)\}$ can be represented graphically by points on an interval, as illustrated at the foot of Figure 1. For every r, $L(r;\mathbf{g})$ is linear as a function of

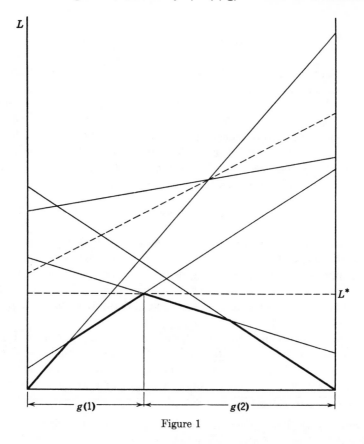

Figure 1

$\mathbf{g}$, as is $L(\mathbf{f};\mathbf{g})$ for every $\mathbf{f}$. It is, of course, just because the $L(\mathbf{f};\mathbf{g})$ of a bilinear game is linear in this sense and its dual that I use the term "bilinear." In Figure 1 the five slanting solid lines represent the five linear functions $L(r;\mathbf{g})$ of a bilinear game in which r (for illustration) takes five values and i takes two. The dashed lines represent two values of $\mathbf{f}$, each of which has for simplicity been so chosen as to use, or mix, only two values of r.

As may be verified by inspection, the particular bilinear game represented by Figure 1 has the special property that min $L(r; i) = 0$ for each i, which is the distinguishing property of those bilinear games that arise in connection with the minimax decision theories described in Chapters 9 and 10.

Figure 1 bears a more than accidental resemblance to Figure 7.2.1. In particular, the concave function

$$(1) \qquad\qquad \min_{r} L(r; \mathbf{g})$$

marked by heavy line segments in Figure 1 is closely analogous to the convex function so marked in Figure 7.2.1. The particular $\mathbf{g}$ emphasized by Figure 1 is that for which the function (1) attains its maximum value, which according to (3.6) is L^*. This $\mathbf{g}$ is therefore the unique maximin. It has been shown quite generally in [B19] that bilinear games with more than one minimax or maximin are, in a sense, unusual; Figure 1 makes it graphically clear that the special bilinear games now under consideration do usually have a unique maximin, because there is more than one maximin only in case (1) happens to have a horizontal segment.

What are the minimax $\mathbf{f}$'s for the bilinear game represented by Figure 1? According to the dual of Theorem 3.2, an r cannot be used in the formation of a minimax $\mathbf{f}$ unless $L(r; \mathbf{g}) = L^*$ for the (in this case unique) maximin $\mathbf{g}$. That consideration eliminates all but two of the r's from consideration, and it is graphically clear that this will usually be the case for bilinear games in which i takes only two values. Theorem 3.2 itself, applied to the particular game under discussion, shows that the graph of $L(\mathbf{f}; \mathbf{g})$ as a function of $\mathbf{g}$ must be horizontal for any minimax $\mathbf{f}$. The two preceding conditions together eliminate all values of $\mathbf{f}$ except the one corresponding to the horizontal dashed line in Figure 1; and that $\mathbf{f}$ is indeed minimax, because $L(\mathbf{f}; i) = L^*$ for both values of i.

To specialize still further, suppose that r as well as i takes only two values. Such a game can, of course, be represented graphically in the spirit of Figure 1. Several qualitatively different situations can occur, which might, for example, be classified by the relation of the two linear functions $L(r, \mathbf{g})$ to each other. The reader should graph and consider many or all of these possibilities for himself. The only one treated here will be that in which the two functions cross each other at an interior $\mathbf{g}$, with one function sloping up and the other down. It is graphically clear that there will then be a unique minimax and a unique maximin, as will now be shown analytically.

The condition postulated can be expressed without loss of generality thus:

(2)
$$L(1; 2) > L(1; 1), \qquad L(2; 1) > L(2; 2),$$
$$L(2; 1) > L(1; 1), \qquad L(1; 2) > L(2; 2).$$

Or, more mnemonically,

(3)
$$L(1; 2), L(2; 1) > L(1; 1), L(2; 2).$$

It is conjectured, in this case on graphical grounds, that the program outlined in connection with (3.7–8) applies, and the reader can indeed verify that that program leads to the conclusion

(4)
$$L^* = \{L(1; 2)L(2; 1) - L(1; 1)L(2; 2)\}/\Delta,$$

where

(5)
$$\Delta = L(1; 2) + L(2; 1) - L(1; 1) - L(2; 2);$$

and that the unique minimax $\mathbf{f}$ and maximin $\mathbf{g}$ are

(6)
$$\begin{cases} f(1) = [L(2; 1) - L(2; 2)]/\Delta \\ f(2) = [L(1; 2) - L(1; 1)]/\Delta, \end{cases}$$

(7)
$$\begin{cases} g(1) = [L(1; 2) - L(2; 2)]/\Delta \\ g(2) = [L(2; 1) - L(1; 1)]/\Delta. \end{cases}$$

If the game arises from an application of the minimax decision theory, (3) almost always applies. More precisely, in this case, except possibly for the order of numbering,

(8) $L(1; 1) = L(2; 2) = 0$ and $L(1; 2), L(2; 1) \geq 0;$

so, if only the inequalities in (8) are both strict, (3) applies. Then (4–7) specialize to

(9)
$$L^* = L(1; 2)L(2; 1)/\Delta,$$

where

(10)
$$\Delta = L(1; 2) + L(2; 1);$$

(11) $f(1) = L(2; 1)/\Delta, \qquad f(2) = L(1; 2)/\Delta,$

(12) $g(1) = L(1; 2)/\Delta, \qquad g(2) = L(2; 1)/\Delta.$

5 Bilinear games exhibiting symmetry

Mathematically the solution of a bilinear game is often simplified by considerations of symmetry. For statistical applications, the implications of symmetry for bilinear games are of fundamental importance in so far as they represent a counterpart in the minimax theory of the disreputable but irrepressible principle of insufficient reason. This section discusses these implications in an elementary, but formal, way. It can be skimmed over or skipped outright without much detriment to the understanding of later sections.

Any discussion of symmetry involves, at least implicitly, the branch of mathematics known as the theory of groups. Though what is to be said here about games exhibiting symmetry is intended to be clear without prior knowledge of the theory of groups, it may be mentioned that introductions to that subject are to be found in many places, for example in [B14].

It can, and in practice often does, happen that a bilinear game has some *symmetry*.† This means that there are permutations, here symbolized by T, T', etc., of the values of r among themselves and the values of i among themselves such that

$$(1) \qquad\qquad L(Tr; Ti) = L(r; i)$$

for every r and i, where, of course, Tr and Ti are the values into which T carries r and i respectively. Permutations satisfying (1) are said to *leave the game invariant*, or to belong to the *group (of symmetries) of the game*. The permutation U that leaves every r and every i fixed must be counted among the permutations in the group of the game, but the game has no symmetry (worthy of the name) unless there are other permutations besides U in its group.

An example of a game with high symmetry is the game implicit in the second example of § 9.6, for, to any permutation whatsoever of the six i's in that game among themselves, there is a corresponding permutation of the r's such that the two permutations taken together leave the game invariant. It was, of course, the exploitation of symmetry that made the treatment of that example relatively simple.

Returning to bilinear games in general, if T and T' are in the group of the game, then the product TT' defined by the condition that

$$(2) \qquad\qquad (TT')r =_{\mathrm{Df}} T(T'r), \qquad (TT')i =_{\mathrm{Df}} T(T'i)$$

is obviously also a permutation in the group of the game. This multi-

† This concept must not be confused with that of "symmetrical games," which are symmetrical in the sense that the equation $L(r; i) = -L(i; r)$ is meaningful and true for every r and i.

plication of permutations somewhat resembles the ordinary multiplication of numbers. In particular, $(TT')T''$ is evidently the same as $T(T'T'')$, though it is not necessarily true that $TT' = T'T$.

Relative to this multiplication the permutation U plays the role of the unit, or number 1, in arithmetic, for it is obvious that $TU = UT = T$ for any permutation T.

For every permutation T, there is evidently a permutation T^{-1}, and one only, that undoes T, that is, one such that $T^{-1}T = U$. It is easy to see also that $TT^{-1} = U$ and that, if T is in the group of the game, T^{-1} is too. The notation T^{-1} is of course motivated by the consideration that, relative to the multiplication of permutations, T^{-1} plays the role of the reciprocal of T.

It will be adopted as a definition that $T\mathbf{f}$ and $T\mathbf{g}$ are the functions such that $T\mathbf{f}(r) = f(T^{-1}r)$ and $Tg(i) = g(T^{-1}i)$ for every permutation of T and for every r and i. The intervention of T^{-1} in this definition may at first seem arbitrary, but it is motivated by the following considerations. First, if $\mathbf{f}$ is, for example, the function such that $f(r_0) = 1$ and $f(r) = 0$ for $r \neq r_0$, then $T\mathbf{f}$ should be such that $T\mathbf{f}(Tr_0) = 1$ and $Tf(r) = 0$ for $r \neq Tr_0$. Second, $S(T\mathbf{f})$ should be $(ST)\mathbf{f}$ rather than $(TS)\mathbf{f}$. The definition having been adopted, $L(T\mathbf{f}; T\mathbf{g})$ can be calculated thus:

$$(3) \qquad L(T\mathbf{f}; T\mathbf{g}) = \sum_{r,i} L(r; i)f(T^{-1}r)g(T^{-1}i)$$

$$= \sum_{r,i} L(Tr; Ti)f(T^{-1}Tr)g(T^{-1}Ti)$$

$$= \sum_{r,i} L(Tr; Ti)f(r)g(i),$$

where the basic fact is exploited that, if r, i runs once through all pairs of values, then Tr, Ti also does so. It follows from (1) and (3) that, if T is in the group of the game, then

$$(4) \qquad L(T\mathbf{f}; T\mathbf{g}) = L(\mathbf{f}; \mathbf{g}).$$

An $\mathbf{f}$ ($\mathbf{g}$) is called *invariant under the group of the game*, if and only if $T\mathbf{f} = \mathbf{f}$ ($T\mathbf{g} = \mathbf{g}$) for every T in the group. There is a natural way to construct from any $\mathbf{f}$ an $\bar{\mathbf{f}}$ invariant under the group, and dually for $\mathbf{g}$. Namely, let

$$\bar{\mathbf{f}} =_{\text{Df}} \frac{1}{n} \sum_{T} T\mathbf{f},$$

$$(5)$$

$$\bar{\mathbf{g}} =_{\text{Df}} \frac{1}{n} \sum_{T} T\mathbf{g},$$

where (here and throughout this section) n is the number of elements in the group and the summation is over all elements of the group. The definition (5) accomplishes its objective, because

$$(6) \qquad \sum_r \bar{f}(r) = \frac{1}{n} \sum_T \sum_r f(T^{-1}r)$$

$$= \frac{1}{n} \sum_T 1 = \frac{n}{n} = 1,$$

and

$$(7) \qquad T'\bar{f}(r) = \bar{f}(T'^{-1}r)$$

$$= \frac{1}{n} \sum_T f(T^{-1}T'^{-1}r)$$

$$= \frac{1}{n} \sum_T T'Tf(r) = \bar{f}(r)$$

for every r and for every T' in the group. In (7) use is made of the easily established facts that $T^{-1}T'^{-1} = (T'T)^{-1}$ and that as T runs once through the group so does $T'T$. The justification of $\bar{g}$ is, of course, dual to that of $\bar{f}$. It is noteworthy that $\mathbf{f} = \bar{\mathbf{f}}$, if and only if $\mathbf{f}$ is invariant under the group of the game.

Suppose R (I) is a set of the r's (i's). Then, by definition, $r \, \varepsilon \, TR$ ($i \, \varepsilon \, TI$), if and only if $T^{-1}r \, \varepsilon \, R$ ($T^{-1}i \, \varepsilon \, I$); and the set R (I) is *invariant under the group of the game*, if and only if $TR = R$ ($TI = I$) for every T in the group.

Exercises

1a. If R is invariant, so is $\sim R$.

1b. If R and R' are invariant, so are $R \cap R'$ and $R \cup R'$.

1c. The vacuous set and the set of all r's are invariant.

2. For every R, let $\bar{R} = _{\text{Df}} \bigcup_T TR$, where T is of course confined to the group; and, for every r, define the *trajectory of r* as $[\overline{r}]$, where $[r]$ is, as is customary, the set whose only element is r.

(a) $\bar{R}$ is the smallest invariant set containing R.

(b) $\bar{R}$ is the intersection of all invariant sets containing R.

(c) $\bar{R} = \bigcup_{r \, \varepsilon \, R} [\overline{r}]$.

(d) $[\overline{r}]$ is the smallest invariant set of which r is an element.

3a. If R is invariant, and $R \cap [\overline{r}] \neq 0$, then $R \supset [\overline{r}]$.

3b. If R is invariant, and $r \, \varepsilon \, R$, then $R \supset [\overline{r}]$.

3c. If $[\overline{r}] \cap [\overline{r'}] \neq 0$, then $[\overline{r}] = [\overline{r'}]$.

4a. The following conditions are equivalent:

α. R is invariant.

β. $R = \bar{R}$.

γ. For every $r \, \varepsilon \, R$, $[\bar{r}] \subset R$.

δ. R is partitioned into sets each of which is a trajectory.

4b. The following conditions are equivalent:

α. $\mathbf{f}$ is invariant.

β. The set of r's for which $\mathbf{f}$ takes any given value is invariant.

γ. $\mathbf{f}$ is constant on every trajectory.

5a. If $T'r = r$, then $(TT'T^{-1})Tr = Tr$.

5b. If $\{r\}$ denotes the number of elements of the group that leave r fixed, then $\{r\} = \{Tr\}$.

5c. If $\| \, r \, \|$ denotes the number of elements in $[\bar{r}]$, then $n = \{r\} \| \, r \, \|$.

5d. Both $\{r\}$ and $\| \, r \, \|$ are divisors of n.

5e. The value of $\bar{\mathbf{f}}$ everywhere on the trajectory of r is

$$(8) \qquad \frac{1}{\| \, r \, \|} \sum_{r \, \varepsilon \, [\bar{r}]} f(r).$$

6. Note the dual of each of the preceding exercises.

In the establishment of all these preliminaries, the theory of bilinear games has been almost lost sight of, but it is now possible to say much about the significance of invariant functions and sets for bilinear games. I begin with a theorem valued for some of its corollaries rather than for any charm of its own.

THEOREM 1   If $L(\mathbf{f}'; T\mathbf{g}) \leq L(\mathbf{f}''; T\mathbf{g})$ for every T, then $L(\bar{\mathbf{f}}'; \mathbf{g}) \leq L(\mathbf{f}''; \bar{\mathbf{g}})$. If in addition $L(\mathbf{f}'; \mathbf{g}) < L(\mathbf{f}''; \mathbf{g})$, then $L(\bar{\mathbf{f}}'; \mathbf{g}) < L(\mathbf{f}''; \bar{\mathbf{g}})$.

PROOF.

$$(9) \qquad L(T^{-1}\mathbf{f}'; \mathbf{g}) = L(\mathbf{f}'; T\mathbf{g}) \leq L(\mathbf{f}''; T\mathbf{g}).$$

Therefore

$$(10) \qquad L(\bar{\mathbf{f}}'; \mathbf{g}) = \frac{1}{n} \sum_{T} L(T^{-1}\mathbf{f}'; \mathbf{g})$$

$$\leq \frac{1}{n} \sum_{T} L(\mathbf{f}''; T\mathbf{g}) = L(\mathbf{f}''; \bar{\mathbf{g}}).$$

If $L(\mathbf{f}'; \mathbf{g}) < L(\mathbf{f}''; \mathbf{g})$, then (9) is strict for $T = U$, and therefore (10) is also strict. ◆

COROLLARY 1   If $L(\mathbf{f}'; T\mathbf{g}) = L(\mathbf{f}''; T\mathbf{g})$ for every T, then $L(\bar{\mathbf{f}}'; \mathbf{g}) = L(\mathbf{f}''; \bar{\mathbf{g}})$.

COROLLARY 2   If $L(\mathbf{f}'; \mathbf{g}) = L(\mathbf{f}''; \mathbf{g})$ for every $\mathbf{g}$, then $L(\bar{\mathbf{f}}'; \mathbf{g}) = L(\bar{\mathbf{f}}''; \bar{\mathbf{g}})$ for every $\mathbf{g}$.

COROLLARY 3 $L(\bar{\mathbf{f}}; \mathbf{g}) = L(\mathbf{f}; \bar{\mathbf{g}}) = L(\bar{\mathbf{f}}; \bar{\mathbf{g}})$ for every $\mathbf{f}$ and $\mathbf{g}$.

COROLLARY 4   If $\mathbf{f}$ is invariant under the group of the game, $L(\mathbf{f}; \mathbf{g}) = L(\mathbf{f}; \bar{\mathbf{g}})$ for every $\mathbf{g}$.

Paraphrasing some of the nomenclature of § 6.4, if $L(\mathbf{f}'; \mathbf{g}) \leq L(\mathbf{f}''; \mathbf{g})$ for every $\mathbf{g}$, say that $\mathbf{f}'$ *dominates* $\mathbf{f}''$; if $\mathbf{f}'$ dominates $\mathbf{f}''$, but $\mathbf{f}''$ does not dominate $\mathbf{f}'$, say that $\mathbf{f}'$ *strictly dominates* $\mathbf{f}''$; if $\mathbf{f}'$ dominates $\mathbf{f}''$, and $\mathbf{f}''$ dominates $\mathbf{f}'$, say that $\mathbf{f}'$ and $\mathbf{f}''$ are *equivalent;* if $\mathbf{f}'$ is not strictly dominated by any $\mathbf{f}$, say that $\mathbf{f}'$ is *admissible.*

COROLLARY 5   If $\mathbf{f}'$ dominates, strictly dominates, or is equivalent to $\mathbf{f}''$, then $\bar{\mathbf{f}}'$ dominates, strictly dominates, or is equivalent to $\bar{\mathbf{f}}''$, respectively.

COROLLARY 6   If $L(\mathbf{f}; T\mathbf{g}) \leq L(\bar{\mathbf{f}}; T\mathbf{g})$ for every T, then $L(\mathbf{f}; \mathbf{g}) = L(\bar{\mathbf{f}}; \mathbf{g})$.

COROLLARY 7   If $L(\mathbf{f}; i) \leq L(\bar{\mathbf{f}}; i)$ for every $i \in I$, where I is invariant under the group of the game, then $L(\mathbf{f}; i) = L(\bar{\mathbf{f}}; i)$ for $i \in I$.

COROLLARY 8   It is impossible that $\mathbf{f}$ strictly dominates $\bar{\mathbf{f}}$.

THEOREM 2 $\max_\mathbf{g} L(\bar{\mathbf{f}}; \mathbf{g}) \leq \max_\mathbf{g} L(\mathbf{f}; \mathbf{g})$, equality holding, if and only if the right-hand maximum is attained for a $\mathbf{g}$ invariant under the group of the game.

PROOF.

$$(11) \qquad \max_\mathbf{g} L(\bar{\mathbf{f}}; \mathbf{g}) = \max_\mathbf{g} L(\mathbf{f}; \bar{\mathbf{g}})$$

$$\leq \max_\mathbf{g} L(\mathbf{f}; \mathbf{g}).$$

The inequality in (11) follows from the fact that every $\bar{\mathbf{g}}$ is a $\mathbf{g}$; equality holds, if and only if the final maximum is attained for some $\bar{\mathbf{g}}$, that is, for some invariant $\mathbf{g}$. ◆

COROLLARY 9   If $\mathbf{f}$ is minimax, so is $\bar{\mathbf{f}}$.

COROLLARY 10   There exists a minimax $\mathbf{f}$ invariant under the group of the game.

If a game has more than one minimax $\mathbf{f}$, it is tempting to suppose that in statistical, if not in all, applications of the theory an invariant, or symmetrical, minimax **f** would recommend itself at least as highly as any other minimax **f**. This supposition, being vague, cannot be really proved, but certain facts tend to support it. In particular, the following theorem is a reassuring improvement of Corollary 10.

THEOREM 3   There is at least one admissible, invariant, minimax **f**.

PROOF. It is a direct consequence of a theorem (Theorem 2.22, p. 54, of [W3]) of Wald's, too technical for statement or proof here, that at least one invariant minimax **f** is strictly dominated by no invariant **f'**. If that **f** were strictly dominated by any **f''** (invariant or not), it would also, according to Corollary 5, be dominated by $\bar{\mathbf{f}}''$, which is impossible. Therefore **f** is admissible. ◆

If the bilinear game has high symmetry or, more explicitly, if the number of trajectories into which the r's or the i's, or both, are partitioned is small; the search for invariant minimax **f**'s and invariant maximin **g**'s is relatively simple. An invariant minimax is characterized as an invariant **f'** such that

$$(12) \qquad \max_{\mathbf{g}} L(\mathbf{f}'; \mathbf{g}) = \min_{\mathbf{f}} \max_{\mathbf{g}} L(\mathbf{f}; \mathbf{g}) = L^*.$$

But, since at least one invariant minimax exists, the criterion (12) is not changed if the minimization on its right side is confined to invariant **f**'s; with **f** so confined, the criterion remains unchanged, if both maximizations are confined to invariant **g**'s (as Corollary 3 shows). Thus the search for invariant minimax **f**'s and invariant maximin **g**'s amounts to the solution of an abstract game that arises from the original bilinear game by ruling out certain values of **f** and **g**, namely the un-invariant ones.

This new and smaller abstract game can be exhibited as a bilinear game thus: Let it be understood for the moment that r' ranges over such a set of the r's that there is exactly one r' in every trajectory $[r]$; dually for i'. For invariant f and g,

$$(13) \qquad L(\mathbf{f}; \mathbf{g}) = \sum_r \sum_i L(r; i) f(r) g(i)$$

$$= \sum_{r'} \sum_{i'} \sum_{r \,\varepsilon\, [r']} \sum_{i \,\varepsilon\, [i']} L(r; i) f(r) g(i)$$

$$= \sum_{r'} \sum_{i'} f(r') g(i') \sum_{r \,\varepsilon\, [r']} \sum_{i \,\varepsilon\, [i']} L(r; i)$$

$$= \sum_{r'} \sum_{i'} L'(r'; i') f'(r') g'(i'),$$

where

(14) $$L'(r'; i') =_{\mathrm{Df}} \frac{1}{\| r' \| \| i' \|} \sum_{r \,\epsilon\, [r']} \sum_{i \,\epsilon\, [i']} L(r; i),$$

and

(15) $$f'(r') =_{\mathrm{Df}} \| r' \| f(r'); \, g'(i') =_{\mathrm{Df}} \| i' \| g(i').$$

Finally, it is easily verified that, except for the conditions $f'(r') \geq 0$, $g'(i') \geq 0$, and $\Sigma f'(r') = \Sigma g'(i') = 1$, the coefficients $f'(r')$ and $g'(i')$ are arbitrary. The new game is therefore to all intents and purposes a bi-linear game with only as many r''s and i''s as there are r-trajectories and i-trajectories, respectively, in the original game. The new game, incidentally, may well have symmetry of its own.

If there is only one r- or one i-trajectory, the new game is so simple it scarcely deserves to be called a game. This occurs, for example, in the second example of § 9.6, where there is only one i-trajectory. In that situation there is only one invariant $\mathbf{g}$, and it is equal at every i to the reciprocal of the total number of i's (which is here the value of $\| i \|$ for every i). That $\mathbf{g}$ must therefore be an admissible maximin. The value of L^* is therefore given by

(16) $$L^* = \min_r \frac{1}{\| i \|} \sum_i L(r; i).$$

The invariant minimax $\mathbf{f}$'s are those and only those invariant $\mathbf{f}$'s such that $f(r) = 0$ for every r that fails to minimize the sum in (16). More-over, here the minimax $\mathbf{f}$'s (invariant or not) are all equivalent, as can be argued thus: Any invariant minimax $\mathbf{f}$ is such that

(17) $$L(\mathbf{f}; \mathbf{g}) = L(\mathbf{f}; \bar{\mathbf{g}}) = L^*$$

for every $\mathbf{g}$. If any minimax $\mathbf{f}$ whatsoever failed to satisfy (17), it would strictly dominate $\bar{f}$; but according to Corollary 8 that is impossible. Therefore in the very special situation at hand all minimax $\mathbf{f}$'s satisfy (17) and are accordingly equivalent.

It is, of course, important to extend consideration of symmetry to bilinear games with infinite sets of r's and i's, and infinite groups of symmetries, but the task has not yet proved straightforward. Two key references bearing on it are [L4] and [B17].

CHAPTER 13

Objections to the Minimax Rules

1 Introduction

I have already expressed and supported my opinion that neither the objectivistic nor the personalistic minimax rule can be categorically defended (§ 9.7 and § 10.3). On the other hand, certain objections have been leveled against the objectivistic rule (that being the well-known one) that seem to me to call for reinterpretation, if not outright refutation.

2 A confusion between loss and negative income

Some objections valid against the minimax rule based on negative income are irrelevant to that based on loss. The notions that the minimax rule is ultrapessimistic and that it can lead to the ignoring of even extensive evidence have already been discussed as examples of such objections.

Another example I would put in the same category has been suggested by Hodges and Lehmann [H5]. In this example a person who has observed n independent tosses of a coin for which the probability of heads has an unknown value p is required to predict the outcome of the $(n + 1)$th toss. Hodges and Lehmann here interpret prediction in the following somewhat sophisticated, but reasonable, sense. The person is, in the light of his observation, required to choose a number ρ between 0 and 1 and to pay a fine of $(1 - \rho)^2$ or ρ^2 according as the $(n + 1)$th toss is in fact heads or tails. Thus the (expected) income attached to the primary act ρ and event p is

(1)
$$I(\rho; p) = -p(1 - \rho)^2 - (1 - p)\rho^2$$
$$= -(\rho - p)^2 - p(1 - p).$$

As Hodges and Lehmann show, the only derived act (mixed or pure) that yields the minimax of the negative income is to set $\rho = \frac{1}{2}$ irrespective of the observation. But it is, in common sense, absurd thus to ignore the observation of the first n tosses. In view of this absurdity, almost everyone would agree that applying the minimax rule directly to the negative of (1) is a foolish act for the person to employ. The absurdity of minimizing the maximum of negative income in this example is of course no valid argument against minimizing the maximum loss. It is easy to see that the loss corresponding to (1) is

$$(2) \qquad L(\rho; p) = (\rho - p)^2.$$

As Hodges and Lehmann happen to show in the same paper [H5] (though in a different context), and as will be discussed in some detail in § 4, the unique minimax derived act does use the observations to advantage, resulting in a loss of

$$(3) \qquad \frac{1}{4(1 + n^{1/2})^2}$$

irrespective of p. The absurd act of setting $\rho = \frac{1}{2}$ irrespective of the observation results in the loss $(p - \frac{1}{2})^2$, which in any ordinary context would be inferior to (3), especially for large n.

Incidentally, the minimax derived from (2), though not nearly so bad as setting ρ identically equal to $\frac{1}{2}$, is itself open to a serious objection, which will be explained in § 4.

3 Utility and the minimax rule

Some objections to the objectivistic, and mutatis mutandis to the group, minimax rule are in effect objections to the concept of utility, which underlies the minimax rules. Criticisms of the concept of utility have already been discussed in Chapter 5, particularly in § 5.6, but certain aspects of the discussion need to be continued here.

It is often said, and I think with justice, that, even granting the validity of the utility concept in principle, a person can seldom write down his income function $I(r; i)$ with much accuracy. This idea is put forward sometimes with one interpretation and sometimes with another. Of these, only the first is strictly an objection to the utility concept.

That one is a dilemma raised by the phenomenon of vagueness. Vagueness may so blur a person's utility judgments that he cannot accurately write down his income function. I suppose that no one will seriously deny this; I would be particularly embarrassed to do so, for it is almost a recapitulation of the very argument that leads me, though in principle a personalist, to see some sense in the objectivistic decision problem. On the other horn, if all meaning is denied to utility (or some extension of that notion) no unification of statistics seems possible.

Three special circumstances are known to me under which escape from the dilemma is possible. First, there are problems in which some straightforward commodity, such as money, lives, man hours, hospital bed days, or submarines sighted, is obviously so nearly proportional to utility as to be substitutable for it. Second, there are problems in which exact or approximate minimax decisions can be calculated on the basis of only relatively little, and easily available, information about the income function, such as symmetry, monotoneity, or smoothness. The possibility of cheap extensive observation, which (when it occurs) makes the minimax principle attractive, also tends to make many decision problems fall into both of the two types in which the difficulty of vagueness is alleviated. For example, in a monetary decision problem with cheap observation available, it often happens that the weak law of large numbers, and the like, can be invoked to justify regarding cash income as proportional to utility income.

Third, there are many important problems, not necessarily lacking in richness of structure, in which there are exactly two consequences, typified by overall success or failure in a venture. In such a problem, as I have heard J. von Neumann stress, the utility can, without loss of generality, be set equal to 0 on the less desired and equal to 1 on the more desired of the two consequences.

The second sense in which it may, though not quite properly, be said to be impossible to write down the income function is typified by this example. A manufacturer of small short-lived objects, say paper napkins, is faced with the problem of deciding on a program of sampling to control the quality of his product. He complains that, though for this problem his utility is adequately measured by money, he cannot write down his income function because he does not know how the public will react to various levels of quality—that, in particular, the minimax rule does not tell him at all how much he ought to spend on the sampling program, though it may say how any given amount can best be employed. The manufacturer has a real difficulty, though he expresses it inaccurately. He forgets that the lack of knowledge that gives rise to the decision problem involves not only the state of his product, but also the state of the public; taking the state of the public into account, there is no real difficulty in writing down the income function. But, if it is not practical for the manufacturer to make observations bearing on the state of the public as well as those bearing on the state of the product, the minimax rule is not a practical solution to his problem; for, rigorously applied, it would remove him from the paper-napkin business. I believe that in practice the personalistic method often is, and must be, used to deal with the unknown state of the public, while objectivistic methods, particularly the minimax principle, are now increasingly often used to deal with the state of the product—a sort of dualism having some parallel in almost all serious applications of statistics. This is not to deny that relatively objectivistic methods of market research can sometimes be used, nor that there are personalistic elements aside from those concerning the state of the public in much of even the most advanced quality control practice.

4   Almost sub-minimax acts

Another sort of objection to the objectivistic minimax rule is illustrated by the following example attributed to Herman Rubin and published by Hodges and Lehmann [H5].   An integer-valued random variable x subject to the **binomial distribution**

$$(1) \qquad P(x \mid p) = \binom{n}{x} p^x (1-p)^{n-x}$$

is observed by a person who knows n but not p.   His decision problem is to decide on a function $\hat{\mathbf{p}}$ of x subject to the loss function:

$$(2) \qquad L(\hat{\mathbf{p}}; p) = E((\hat{\mathbf{p}} - p)^2 \mid p)$$

$$= \sum_x (\hat{p}(x) - p)^2 \binom{n}{x} p^x (1-p)^{n-x}.$$

In other terms, he must estimate p on the basis of an observation of x and subject to a loss equal to the square of his error.   The traditional estimate of p is defined by $\hat{p}_0(x) = x/n$.   This estimate has many virtues; it is the maximum-likelihood estimate, the only unbiased estimate, and (as is shown in [G1]) the only minimax estimate for a somewhat different problem from that posed by (2).   But for (2) the unique minimax is (as is shown in [H5]) defined by

$$(3) \qquad \hat{p}_1(x) = \hat{p}_0(x) + \frac{(\frac{1}{2} - \hat{p}_0(x))}{1 + n^{1/2}}.$$

As it is straightforward to verify for every p,

$$(4) \qquad L(\hat{\mathbf{p}}_0; p) = \frac{p(1-p)}{n};$$

and

$$(5) \qquad L(\hat{\mathbf{p}}_1; p) = \frac{1}{4(1 + n^{1/2})^2},$$

which constant is, therefore, L^*. The ratio of the first of these functions to the second is

(6) $$4p(1 - p) \left(1 + \frac{1}{n^{1/2}}\right)^2,$$

the maximum of which occurs at $p = 1/2$ and is

(7) $$\left(1 + \frac{1}{n^{1/2}}\right)^2.$$

Thus, for large n, the maximum loss of $\hat{\mathbf{p}}_0$ is larger than L^* by only a slight fraction. Moreover, the loss of $\hat{\mathbf{p}}_0$ is less than L^* except when p lies in the interval where

(8) $$4p(1 - p) \geq (1 + n^{-1/2})^{-2},$$

that is, where

(9) $$\left| p - \tfrac{1}{2} \right| \leq \tfrac{1}{2}\{1 - (1 + n^{-1/2})^{-2}\}^{1/2} \simeq (4n)^{-1/4}.$$

To take a numerical example, consider $n = 10^5$ (which the practical will note is rather big for a sample). The advantage of $\hat{\mathbf{p}}_1$ over $\hat{\mathbf{p}}_0$ at $p = 1/2$ is then only 0.64%, and, once p departs by as much as 0.04 from $1/2$ in either direction, the advantage is with $\hat{\mathbf{p}}_0$. It amounts, for example, to 3.5%, 15.5%, ∞% in favor of $\hat{\mathbf{p}}_0$, when p is 0.6, 0.8, 1.0, respectively.

Many agree that in such an example good judgment will, under ordinary circumstances, prefer $\hat{\mathbf{p}}_0$ to the recommendation of the minimax rule, $\hat{\mathbf{p}}_1$. To my mind, this example constitutes a valid objection against the minimax rule, in the sense that it demonstrates once more that, whatever value that rule may have, it is at best a rule of thumb.

The example is a good illustration of the role of personal probability in ordinary statistical thinking, for the source of the dissatisfaction a person would ordinarily feel for $\hat{\mathbf{p}}_1$ as opposed to $\hat{\mathbf{p}}_0$ stems from the fact that he would not ordinarily attach enough personal probability to the immediate neighborhood of $p = 1/2$ to justify preference for $\hat{\mathbf{p}}_1$. It follows from the numbers given above, for example, that, if the person attaches a probability of less than 0.84 to the interval [0.4, 0.6], he will prefer $\hat{\mathbf{p}}_0$ to $\hat{\mathbf{p}}_1$; the same conclusion can be derived from the supposition that the standard deviation of the personal distribution of p is at least 0.04. Of course, situations can be imagined in which the personal probabilities would be so concentrated about $1/2$ as to justify preference for $\hat{\mathbf{p}}_1$; the point of the example is only that there are situations in which that would clearly not be the case.

Interesting material and important references bearing on the phenomenon illustrated by the decision problem under discussion are given

by Wolfowitz in [W17]. It seems to be suggested there that the difficulty can be met by postulating some small amount ϵ by which the
person does not mind having his income decreased. Taken literally,
this postulate implies on repeated application that all incomes are
equivalent for the person, but Wolfowitz makes it clear that he does
not mean to propose the postulate in a sense that allows repeated applications. The idea is reminiscent of those theories of probability
that permit the neglect of an occasional improbable event (mentioned
in the last paragraph of § 4.4) and seems to me open to an objection
similar to the one raised in connection with them. In particular, the
choice of the ϵ would be not only personal, but ill defined as well.

5 The minimax rule does not generate a simple ordering

Finally, an objection made by Chernoff [C7] to the objectivistic minimax theory must be discussed. This will entail statement and illustration of the phenomenon on which the objection is based, and statement and analysis of the objection itself.

The phenomenon pertains to the relation between two objectivistic
decision problems, to be called for the moment the narrow and the
wide problems. The narrow problem is determined by certain primary
acts $\mathbf{f}_r$; and the wide one is determined by those primary acts and one
more, say $\mathbf{f}_0$. In other words, the wide problem presents the person
with one more choice than the narrow. Calling the two income func-
tions $I(\mathbf{f}; i)$ and $I_0(\mathbf{f}; i)$, it is to be understood, of course, that $I(\mathbf{f}; i)$
$= I_0(\mathbf{f}; i)$ for any $\mathbf{f}$ that does not use, that is, give positive weight to,
$\mathbf{f}_0$. The corresponding equation does not necessarily obtain for the
loss functions; indeed it clearly does so, if and only if the maximum of
$I_0(\mathbf{f}; i)$ in $\mathbf{f}$ can be attained for each i without using $\mathbf{f}_0$. Even in case
no minimax of the wide game uses $\mathbf{f}_0$, it is therefore to be expected that
the minimax $\mathbf{f}$'s of the wide game will be different from those of the
narrow game. In fact, it can happen that no minimax of the wide game
uses either $\mathbf{f}_0$ or any $\mathbf{f}_r$ used by a minimax of the narrow game; this is
the phenomenon to be discussed in this section.

To see how the phenomenon can occur, suppose that Figure 12.4.1
represents the loss function of the narrow problem; and consider what
the corresponding figure is for the wide problem, supposing that $\mathbf{f}_0$ is
such that

$$\Delta = {}_{\mathrm{Df}} I(\mathbf{f}_0; 2) - \max_r I(\mathbf{f}_r; 2) > 0,$$

(1)

$$\Sigma = {}_{\mathrm{Df}} \max I(\mathbf{f}_r; 1) - I(\mathbf{f}_0; 1) > 0.$$

It is clear that Δ and Σ can attain any positive values, irrespective of the structure of the narrow problem. The figure for the wide problem is constructed thus: The graph corresponding to each $\mathbf{f}_r$ is left fixed at its right end and raised by the amount Δ at its left, and $\mathbf{f}_0$ is represented by a line sloping up with slope Σ from the lower left-hand corner. It is easy to see that the raising of the left ends of the graphs of the $\mathbf{f}_r$'s can make any $\mathbf{f}_r$ with a positive slope horizontal. If, further, such an $\mathbf{f}_r$ minimizes $L(\mathbf{f}; \mathbf{g})$ for some $\mathbf{g}$, it can be made a minimax by choosing Σ sufficiently large. Thus, speaking specifically of Figure 12.4.1, the $\mathbf{f}_r$ corresponding to the left segment of the heavy concave graph, which is not used in the minimax of the narrow problem, can become the unique minimax. Figure 12.4.1 is a little special in that the heavy concave graph has only one vertex to the left of the maximin of the narrow problem. If there were more than one, the phenomenon could also be exhibited by making the second vertex to the left the unique maximin, which would occur for all Δ's and Σ's in a certain range. Thus the phenomenon occurs not only for isolated values of Δ and Σ but typically for whole domains of values.

Suppose, to take a striking case, that one $\mathbf{f}_r$, say $\mathbf{f}_{r'}$, is the unique minimax for the narrow problem and a different one, $\mathbf{f}_{r''}$, is the unique minimax for the wide problem. It is absurd, as Chernoff says in effect, to recommend $\mathbf{f}_{r'}$ as the best act among the $\mathbf{f}_r$'s when only the $\mathbf{f}_r$'s are available and then to recommend $\mathbf{f}_{r''}$ as the best for an even wider class of possibilities. Fancy saying to the butcher, "Seeing that you have geese, I'll take a duck instead of a chicken or a ham."

It is absurd, then, to contend that the objectivistic minimax rule selects the *best* available act. But that is not so devastating to the rule as might at first appear, for it is not contended by anyone known to me that the rule does select the best. On the contrary, the rule is invoked only as a sometimes practical rule of thumb in contexts where the concept of "best" is impractical—impractical for the objectivist, where it amounts to the concept of personal probability, in which he does not believe at all; and for the personalist, where the difficulty of vagueness becomes overwhelming. To have a consistent concept of "best," that is, to have a mode of decision that does not exhibit the phenomenon, amounts, as Chernoff himself points out, to the establishment of a simple ordering of preference among acts. In so far as that can be done consistently with the sure-thing principle, personal probability is practically defined thereby. If the sure-thing principle is violated, the ordering is absurd as an expression of preference. For example, the rule of minimizing the maximum of the negative of income does not exhibit the phenomenon. It amounts to considering $\mathbf{f} \leq \mathbf{f'}$, if and only if

$$(2) \qquad\qquad \max_{i} I(\mathbf{f}; i) \leq \max_{i} I(\mathbf{f'}; i).$$

This establishes a simple ordering, but one that violates the sure-thing principle by violating P2.

The phenomenon has a particularly natural interpretation for the group minimax rule. It would not be strange, for example, if a banquet committee about to agree to buy chicken should, on being informed that goose is also available, finally compromise on duck.

CHAPTER 14

The Minimax Theory
Applied to Observations

1 Introduction

In this chapter the concept of observation is re-explored from the point of view of the minimax rule. In principle, objectivistic and group minimax problems should here be treated on an equal footing. But, since mathematically the two theories are identical, it seems wisest to focus on one, interjecting occasional digressions about the other. I have chosen to focus on the objectivistic problems. That choice, being in accordance with other literature on the minimax rule, will facilitate the reader's further study of the subject, and it also renders more obvious the intimate connection between the minimax rules and the theory of partition problems presented in Chapter 7. The present chapter can indeed be regarded largely as a paraphrase of Chapter 7, so there will unavoidably be many references to the notations and conclusions of that chapter.

2 Recapitulation of partition problems

Paralleling the treatment of observation in Chapters 6 and 7, an *objectivistic observational problem* will be roughly defined to consist of an objectivistic problem, regarded as basic; an observation; and a second objectivistic problem, derived from the basic one and the observation.

More explicitly, the *basic problem* may be any objectivistic problem. It will be characterized by the values of $E(\mathbf{f} \mid B_i)$, where $\mathbf{f}$ ranges over a set of acts $\mathbf{F}$ subject to the conditions laid down in § 9.3, and B_i is a partition.

The observation is a random variable $\mathbf{x}$ (confined, as usual in this book, to a finite set of values), subject to the conditional distributions $P(x \mid B_i)$, and so articulated with $\mathbf{F}$ that $E(\mathbf{f} \mid B_i, x) = E(\mathbf{f} \mid B_i)$ for every x such that $P(x \mid B_i) > 0$. The last condition is (7.2.7); as mentioned in connection with that equation, the condition will in particular be met, if every **f** is constant on every B_i, a specialization costing but little in real generality.

The *derived problem* (paralleling § 6.2) consists of $\mathbf{F}(\mathbf{x})$, the set of all functions assigning elements **f** of the basic acts **F** to values x of the observation **x**. The values of $E(\mathbf{f}(\mathbf{x}) \mid B_i)$ for $\mathbf{f}(\mathbf{x}) \, \varepsilon \, \mathbf{F}(\mathbf{x})$ are computable from the $E(\mathbf{f} \mid B_i)$ and the $P(x \mid B_i)$ thus:

$$
\begin{aligned}
(1) \qquad E(\mathbf{f}(\mathbf{x}) \mid B_i) &= E(E(\mathbf{f}(\mathbf{x}) \mid B_i, \mathbf{x})) \\
&= \sum_x E(\mathbf{f}(x) \mid B_i, x) P(x \mid B_i) \\
&= \sum_x E(\mathbf{f}(x) \mid B_i) P(x \mid B_i)
\end{aligned}
$$

It will now be shown that the set of derived acts $\mathbf{F}(\mathbf{x})$ satisfies the technical conditions imposed on the set of basic acts **F**, so that the derived problem is also an objectivistic decision problem. In fact, if every $\mathbf{f} \, \varepsilon \, \mathbf{F}$ is expressible in the form $\Sigma f(r) \mathbf{f}_r$ (with the usual condition on the $f(r)$'s), primary acts for $\mathbf{F}(\mathbf{x})$ analogous to the $\mathbf{f}_r$'s can be defined by attaching to every function $\mathbf{r} = r(\mathbf{x})$ an element $\mathbf{f}(\mathbf{x}; \mathbf{r})$ of $\mathbf{F}(\mathbf{x})$, where

$$
(2) \qquad \mathbf{f}(x; \mathbf{r}) =_{\mathrm{Df}} \mathbf{f}_{r(x)}.
$$

There are only a finite number of $\mathbf{f}(\mathbf{x}; \mathbf{r})$'s, and all elements of $\mathbf{F}(\mathbf{x})$ are expressible as weighted averages of them; the first assertion is obvious, and the second poses the problem of finding, for any system of probability measures $\phi(r; x)$ on the r's, at least one probability measure on the set of functions **r** with respect to which $P(r(x) = r) = \phi(r; x)$ for every r and x. The problem typically has many solutions; the simplest is to let the $r(x)$'s, regarded for each x as functions of **r**, be independent random variables on the set of **r**'s considered as a probability space, that is, to set

$$
P(\mathbf{r}) = \prod_x \phi(r(x); x).
$$

Formally, this particular solution leads to the identity

$$
\begin{aligned}
(3) \qquad \mathbf{f}(x) &= \sum_r \phi(r; x) \mathbf{f}_r \\
&= \sum_\mathbf{r} \left\{ \prod_{x'} \phi(r(x'); x') \right\} \mathbf{f}_{r(x)}.
\end{aligned}
$$

The identity and the fact that the coefficients in braces are non-negative and add up to 1, are easy to check analytically, if it is recognized that summation with respect to **r** means multiple summation with respect to $r(1)$, $r(2)$, $\cdots$ (the x's being for definiteness supposed to take integral values). Equation (3) shows incidentally that it is immaterial whether it is before or after the observation that mixed acts are introduced.

Turn momentarily to the idea of observation in group decision problems. Here the $E(\mathbf{f}; B_i)$'s are replaced by $I(\mathbf{f}; i)$'s, the expected income of $\mathbf{f}$ in the opinion of the ith person. There is no partition B_i, except in a special, though theoretically important, case, namely that of the ith person holding unequivocally that B_i obtains.

The $P(x \mid B_i)$'s are here replaced by $P(x; i)$'s, the personal distribution of x for the ith person. It is postulated that, for each person, the conditional expectation of $\mathbf{f}$ is unaffected by knowledge of x.

The derived acts are formally the same as for an objectivistic decision problem, and the income function of the derived group decision problem is

$$(4) \qquad I(\mathbf{f}(\mathbf{x}); i) = \sum_x I(\mathbf{f}(x); i)P(x; i).$$

Returning to objectivistic problems, (9.4.1) defines the loss function of the basic objectivistic problem and, mutatis mutandis, that of the derived problem also, thus:

$$(5) \qquad L(\mathbf{f}(\mathbf{x}); i) = \max_{\mathbf{f}'(\mathbf{x})} E(\mathbf{f}'(\mathbf{x}) \mid B_i) - E(\mathbf{f}(\mathbf{x}) \mid B_i).$$

The right side of (5) admits some simplification, for, if the person knew which B_i obtained, observation would be valueless to him. Accordingly,

$$(6) \qquad L(\mathbf{f}(x); i) = \max_{\mathbf{f}'} E(\mathbf{f}' \mid B_i) - E(\mathbf{f}(\mathbf{x}) \mid B_i).$$

Analytically, the simplification is justified thus:

$$(7) \qquad \max_{\mathbf{f}} E(\mathbf{f} \mid B_i) \leq \max_{\mathbf{f}(\mathbf{x})} E(\mathbf{f}(\mathbf{x}) \mid B_i)$$

$$= \max_{\mathbf{f}(\mathbf{x})} \sum_x E(\mathbf{f}(x) \mid B_i)P(x \mid B_i)$$

$$\leq \max_{\mathbf{f}} E(\mathbf{f} \mid B_i).$$

In discussing application of the minimax rule to the basic and derived loss functions, it is doubly advantageous to introduce mixtures of the i's, for thereby the theory of bilinear games presented in Chapter 12 and that of partition problems (with some reinterpretation) can both be brought to bear. Letting β denote a generic system of weights $\beta(i)$, $\beta(i) \geq 0$ and $\Sigma\beta(j) = 1$, and using the notation of Chapter 7, the bilinear games associated with the primary and derived problems are, respectively,

(8) $$L(\mathbf{f}; \beta) = l(\beta) - E(\mathbf{f} \mid \beta),$$

(9) $$L(\mathbf{f}(\mathbf{x}); \beta) = l(\beta) - E(\mathbf{f}(\mathbf{x}) \mid \beta)$$
$$= l(\beta) - \sum_j \sum_x E(\mathbf{f}(x) \mid B_j) P(x \mid B_j) \beta(j)$$
$$= l(\beta) - \sum_x E(\mathbf{f}(x) \mid \beta, x) P(x \mid \beta).$$

If necessary, (9) can be interpreted and verified by comparison with (7.3.7) and (7.2.8), in that order.

In Chapter 7, $\beta(i)$ was generally required not only to be non-negative, but also strictly positive; on examination, this slight difference from the present context will be found innocuous. Again, in Chapter 7, the statement and derivation of conclusions were, for simplicity, nominally confined to twofold partition problems. Here the extension of those conclusions to n-fold problems will be freely used, though some readers may prefer here, as there, to focus on twofold problems.

Letting L^* denote the minimax (and maximin) value of the basic, and $L^*(\mathbf{x})$ that of the derived problem, it is obvious, since $\mathbf{F}(\mathbf{x}) \supset \mathbf{F}$, that $L^*(\mathbf{x}) \leq L^*$; but there is some interest in viewing this inequality as a consequence of (7.3.4):

(10) $$L^*(\mathbf{x}) = \max_\beta \min_{\mathbf{f}(\mathbf{x})} L(\mathbf{f}(\mathbf{x}); \beta)$$
$$= \max_\beta [l(\beta) - v(\mathbf{F}(\mathbf{x}) \mid \beta)]$$
$$\leq \max_\beta [l(\beta) - v(\mathbf{F} \mid \beta)]$$
$$= \max_\beta \min_\mathbf{f} L(\mathbf{f}; \beta) = L^*.$$

It is clear that the maximin β's for the basic and derived problems are the β's that maximize the concave functions

(11) $$h(\beta) =_{\mathrm{Df}} l(\beta) - v(\mathbf{F} \mid \beta) = l(\beta) - k(\beta)$$

and

(12) $$h(\beta; \mathbf{x}) =_{\mathrm{Df}} l(\beta) - v(\mathbf{F}(\mathbf{x}) \mid \beta) = l(\beta) - E(k(\beta(\mathbf{x})) \mid \beta),$$

respectively. The search for minimax $\mathbf{f}(\mathbf{x})$'s, for example, is greatly narrowed by the consideration that, if $\mathbf{f}(\mathbf{x})$ is minimax, $E(\mathbf{f}(\mathbf{x}) \mid \beta) = v(\mathbf{F}(\mathbf{x}) \mid \beta)$ for some β, indeed for every maximin β. According to § 7.3, equality obtains in (10), if and only if there is a maximin β_0 of the basic problem such that

(13) $$\beta_0(x) = {}_{Df} \left\{ \frac{P(x \mid B_i)\beta_0(i)}{\sum_j P(x \mid B_j)\beta_0(j)} \right\}$$

is also a maximin of the basic problem for every x such that

$$\Sigma P(x \mid B_j)\beta_0(j) > 0.$$

The most typical possibility, and the only one to be explored here, is that the basic problem has a unique maximin β_0 with $\beta_0(j) > 0$ for all j. Under this assumption, $L^*(\mathbf{x}) = L^*$, if and only if $\mathbf{x}$ is utterly irrelevant, as is easily shown.

In the same spirit, as can easily be shown, $L^*(\mathbf{x}) = 0$, if $\mathbf{x}$ is definitive, but not typically otherwise; and, if $\mathbf{x}$ extends $\mathbf{y}$, then $L^* (\mathbf{x}) \leq L^* (\mathbf{y})$ with equality if, and typically only if, $\mathbf{y}$ is sufficient for $\mathbf{x}$.

3 Sufficient statistics

Digressing from the minimax rule for a moment, something more fundamental can be said about a sufficient statistic $\mathbf{y}$ of $\mathbf{x}$. Namely, for every $\mathbf{f}(\mathbf{x}) \, \varepsilon \, \mathbf{F}(\mathbf{x})$, there exists an $\mathbf{f}(\mathbf{y}) \, \varepsilon \, \mathbf{F}(\mathbf{y})$ such that $I(\mathbf{f}(\mathbf{y}); \, i) = I(\mathbf{f}(\mathbf{x}); \, i)$ for every i. Indeed $\mathbf{f}(y) = \sum_x \mathbf{f}(x)P(x \mid y)$ defines such an act. Without appeal to so weak a step as the minimax rule, this remark demonstrates that even an objectivist loses nothing by exchanging knowledge of an observation for knowledge of a sufficient statistic of it. The remark might as well have been expressed in § 7.4, except that there it would have involved some circumlocution, mixed acts not yet having been introduced.

4 Simple dichotomy, an example

Much of what has been said thus far is well illustrated by the minimax counterpart of Exercise 7.5.2. The reader is accordingly asked to review that exercise and continue it thus:

Exercises

1. For the problem in question:

   (a) $$h(\beta) = \delta_2\beta(1) + \delta_1\beta(2) - \mid \delta_1\beta(2) - \delta_2\beta(1) \mid.$$

   (b) $h(\beta; \mathbf{x})$

   $$= \delta_2\beta(1) + \delta_1\beta(2) - \sum_r \mid \delta_1 r_2\beta(2) - \delta_2 r_1\beta(1) \mid \left\{ \sum_j P(r \mid B_j) \right\}$$

   $$= \delta_2[2P(r_1 < r_1{}^*(\beta, \beta_0) \mid B_1) + P(r = r^*(\beta, \beta_0) \mid B_1)]\beta(1)$$
   $$+ \delta_1[2P(r_2 < r_2{}^*(\beta, \beta_0) \mid B_2) + P(r = r^*(\beta, \beta_0) \mid B_2)]\beta(2).$$

2a. A β is maximin, if and only if $r^*(\beta, \beta_0)$ is such that

(1) $\delta_2 P(r_1 < r_1^*(\beta, \beta_0) \mid B_1) \leq \delta_1 P(r_2 \leq r_1^*(\beta, \beta_0) \mid B_2)$

and

(2) $\delta_2 P(r_1 \leq r_1^*(\beta, \beta_0) \mid B_1) \geq \delta_1 P(r_2 < r_1^*(\beta, \beta_0) \mid B_2)$.

2b. There is typically only one maximin, but there may be a closed interval of them.

3. Though the acts of $\mathbf{F}$ and $\mathbf{F}(\mathbf{x})$ as defined by Exercise 7.5.2 do not provide for mixed acts, it will suffice to consider mixtures of the $\mathbf{f}(\mathbf{x})$'s. Each of these will be determined by an $\mathbf{i}$, and nothing will be lost by requiring $\mathbf{i}$ to be of the form $i(r(x))$.

4a. Any minimax will be equivalent to a mixture of $\mathbf{f}(\mathbf{x})$'s each corresponding to a likelihood-ratio test associated with $r^*(\beta, \beta_0)$ for every maximin β.

4b. In view of Exercise 3, the only likelihood-ratio tests that need be considered for a minimax β are:

$i(r) = 1$, if and only if $r_1 \leq r_1^*(\beta, \beta_0)$.

$i(r) = 1$, if and only if $r_1 < r_1^*(\beta, \beta_0)$.

These are not necessarily different tests.

5a. If the maximin β is unique, the minimax act is unique (except possibly for equivalent acts) and is a mixture of exactly two $\mathbf{f}(\mathbf{x})$'s corresponding to the two likelihood-ratio tests defined in Exercise 4b.

This conclusion calls for some comment, for, in ordinary statistical practice, one or the other of the extreme likelihood-ratio tests is used, never a mixture. This practice is not in serious conflict with the minimax rule, because the maximum loss associated with either extreme is typically only slightly greater than $L^*(\mathbf{x})$. Moreover, vagueness about the exact magnitude of δ_1 and δ_2 would usually frustrate any attempt to calculate the coefficients of the mixture. Incidentally, mixture is not called for at all when $\mathbf{r}$ is continuously distributed, for $h(\beta, \mathbf{x})$ is then smooth rather than polygonal; that is, if $P(r = r' \mid B_i) = 0$ for every r' and both i's, then $h(\beta; \mathbf{x})$ has a continuous first derivative in β. To show this and to show that the derivative is $\delta_2 P(r_1 \leq r_1^* \mid B_1) - \delta_1 P(r_2 \leq r_2^* \mid B_2)$ may be taken as an exercise only slightly beyond the usual mathematical level of this book.

5b. If there is more than one maximin β, then any one that is not extreme has only one likelihood-ratio test associated with it, and the same one for all. The $\mathbf{f}(\mathbf{x})$ corresponding to that test is essentially the only minimax.

5   The approach to certainty [+]

In concluding the paraphrase of § 7.1–6 that has thus far been the subject of the present chapter, it should be mentioned that the approach to certainty studied in § 7.6 obviously implies that the corresponding $L^*(\mathbf{x}(n))$ approaches zero with increasing n.

6   Cost of observation

A cost $\mathbf{c}$ associated with an objectivistic observational problem diminishes the income by $E(\mathbf{c} \mid B_i)$ for each i, regardless of $\mathbf{f}$; that is, allowing for the cost, $I(\mathbf{f}; i) = E(\mathbf{f} - \mathbf{c} \mid B_i)$. But the cost, being unavoidable, does not affect the loss function, so the minimax problem associated with the observation is independent of the cost. The costs do intervene, however, in an essential way in the problem of deciding which to choose of several available observations, say $\mathbf{x}_a$ at cost $\mathbf{c}_a$; it is important to bear in mind in connection with this problem that a null observation at zero cost is typically among the choices available in real life. The generic act of this *compound problem* can conveniently be symbolized by $\Sigma\lambda(a)\mathbf{f}(\mathbf{x}_a)$, or sometimes simply by λ. Here, of course, $\lambda(a) \geq 0$, $\Sigma\lambda(a) = 1$; for choice of λ means choice, for each a, of the probability $\lambda(a)$ that the ath observation $\mathbf{x}_a$ will be made and also choice of the derived act $\mathbf{f}(\mathbf{x}_a)$ to be adopted in case $\mathbf{x}_a$ is made. It is intuitively evident, and follows easily from (1) below, that the mixture of several λ's is also a λ as far as income is concerned, so mixtures of λ's do not require explicit consideration. The income function can be written

$$(1) \qquad I(\lambda; i) = \Sigma\lambda(a)E(\mathbf{f}(\mathbf{x}_a) - \mathbf{c}_a \mid B_i).$$

Whence

$$(2) \qquad \max_{\lambda} I(\lambda; i) = \max_{\mathbf{f}} E(\mathbf{f} \mid B_i) - \min_{a} E(\mathbf{c}_a \mid B_i).$$

The loss function is accordingly

$$(3) \qquad L(\lambda; \beta) = \sum_{a} \lambda(a)\{L_a(\mathbf{f}(\mathbf{x}_a); \beta) + d_a(\beta)\},$$

where

$$(4) \qquad d_a(\beta) =_{\mathrm{Df}} \sum_{i} \{E(\mathbf{c}_a \mid B_i) - \min_{a'} E(\mathbf{c}_{a'} \mid B_i)\}\beta(i),$$

and $L_a(\mathbf{f}(\mathbf{x}_a); \beta)$ is the loss function of the observational problem derived from the ath observation.

The compound minimax problem is intimately related to the concave functions $h(\beta; \mathbf{x}_a)$ and the linear functions $d_a(\beta)$, as is explained by the following exercises.

[+] Some recent references appropriate to this title are Blackwell and Dubins (1962), Chao (1970), Fabius (1964), and Freedman (1965).

Exercises

1. Show that

(5) $h_\lambda(\beta) = \text{Df} \min_\lambda L(\lambda; \beta) = \min_a [h(\beta; \mathbf{x}_a) + d_a(\beta)].$

2. If $\lambda = 1 \cdot \mathbf{f}'(\mathbf{x}_{a'})$, then $L(\lambda; \beta) = h_\lambda(\beta)$; if and only if: first,

(6) $L_{a'}(\mathbf{f}'(\mathbf{x}_{a'}); \beta) = h(\beta; \mathbf{x}_{a'})$

(in which case $\mathbf{f}'(\mathbf{x}_{a'})$ will be called *well adapted* to $\mathbf{x}_{a'}$ and β); and, second,

(7) $h(\beta; \mathbf{x}_{a'}) + d_{a'}(\beta) = \min_a [h(\beta; \mathbf{x}_a) + d_a(\beta)]$

(in which case $\mathbf{x}_{a'}$ will be called *well adapted* to β).

3a. Show that

(8) $L_\lambda^* = \text{Df} \min_\lambda \max_\beta L(\lambda; \beta) = \max_\beta h_\lambda(\beta)$

$\leq \min_a \max_\beta [h(\beta; \mathbf{x}_a) + d_a(\beta)].$

3b. Under the important special condition that the $d_a(\beta)$ are equal to constants d_a, (8) specializes to

(9) $L_\lambda^* \leq \min_a [L^*(\mathbf{x}_a) + d_a].$

3c. When can equality hold in (8) and (9)?

3d. β' is maximin, if and only if $h_\lambda(\beta') = L_\lambda^*$.

4. A $\lambda = \Sigma \lambda(a) f(x_a)$ is minimax, if and only if:

(α) For every a for which $\lambda(a) > 0$, $\mathbf{x}_a$ is well adapted to every maximin β, and $\mathbf{f}(\mathbf{x}_a)$ is well adapted to $\mathbf{x}_a$ and every maximin β.

(β) $L(\lambda; i) \leq L_\lambda^*$ for every i. (Of course (β) is alone necessary and sufficient; the point of the exercise is that the necessary condition (α) may conveniently confine the search for minimax λ's to relatively few candidates.)

5. Suppose that: (α) r and i are confined to the values 1 and 2, and $L(\mathbf{f}_r; i) = |r - i|$; ($\beta$) $\mathbf{x}$ is confined to the values 1 and 2, and $P(1 \mid B_1) = 1/2$, $P(1 \mid B_2) = 1/4$; (γ) a is confined to the values 1 and 2, and the λ's of the compound problem attach weight $\lambda(1)$ to a basic act at zero cost and $\lambda(2)$ to an act derived from $\mathbf{x}$ at a non-negative constant cost d. Compute and graph: $h(\beta)$, $h(\beta; \mathbf{x})$, and (for various values of d) $h_\lambda(\beta)$. Graph L_λ^* as a function of d, and discuss the minimax λ's for various values of d.

7 Sequential probability ratio procedures

The type of decision problem that in § 7.7 led to the concept of a sequential probability ratio procedure has an intimate counterpart in an important type of compound objectivistic decision problem, for which the concept was in fact originally developed by Wald [W2]. The x_a's of a problem of this type range over the enormous variety of sequential observational programs associated with a sequence of (conditionally) identically distributed random variables $x(1)$, $x(2)$, $\cdots$. The technical assumption that the a's have a finite range is not fulfilled; but, as in § 7.7, I proceed with some lapse of rigor, referring to Wald's book [W3] or [A7] for the full details. Exercise 6.4 shows that attention may be confined to a's that are well adapted to at least one β, and that for those a's it may be confined to $f(x_a)$'s that are well adapted to x_a and the corresponding β. The way is paved by § 7.7, which states sharply restrictive properties of the x_a's and $f(x_a)$'s that are so adapted. In some cases, recognition of these properties contributes greatly to the possibility of actually computing minimax, or nearly minimax, procedures for sequential problems.

8 Randomization

Another important type of compound problem is illustrated by the second example of § 9.6. A generalization of part of that example is presented here to show how the minimax rule explains, or implies, the process called randomization, which is one of the most striking features of modern statistics, and one long antedating the minimax rule. Randomization represents the only important use of mixed acts that has thus far found favor with practicing statisticians, as will be discussed in the next section. The exact meaning of randomization seems a little elusive; no sharp definition is attempted here. But, roughly, **randomization** is the selection of an observation at random; that is, of a λ with more than one $\lambda(a)$ actually positive, the choice of the $\lambda(a)$'s and of the derived acts being governed largely by symmetry. The following example provides at least a fairly general illustration of the concept.

To set the stage and provide motivation for a formal statement, the example will first be stated in language that is suggestive though a little vague. The consequences of the basic acts in the example depend on the composition of a population of n objects, which may be thought of as numbered from 1 through n. It may be known of some compositions that they cannot occur; but, if a composition is considered possible, all populations having that composition (irrespective of ordering) are also considered possible. Each observation in the compound problem consists in the cost-free observation of some m of the objects, every subset of exactly m objects being available for observation.

Formally, the index i of the partition B_i runs over a certain set I of n-tuples, $\{i_1, \cdots, i_n\}$, of elements considered for definiteness to be integers. If $i = \{i_1, \cdots, i_n\}$ ε I, then any permutation Ti of i is also in I. It is assumed that

(1) $$E(\mathbf{f} \mid B_i) = E(\mathbf{f} \mid B_{Ti})$$

for every $\mathbf{f}$ ε F, i ε I, and permutation T.

To every subset A of m integers, $1 \le a_1(A) < a_2(A) < \cdots < a_{m-1}(A) < a_m(A) \le n$, there corresponds an observation $\mathbf{x}(A)$ the possible values of which are m-tuples $\{x_1(A), \cdots, x_m(A)\}$. The conditional distributions of the $x(A)$'s are defined thus: If $x_1(A) = i_{a_1(A)}$, etc., then $P(x_1(A), \cdots, x_m(A) \mid B_i) = 1$.

It is obvious that $L^*(\mathbf{x}(A))$ is the same for every A. In typical applications this common value is little, if at all, less than L^*.

If a compound act $\Sigma\lambda(A)\mathbf{f}(\mathbf{x}(A))$ is to be chosen, statistical common sense asserts that nothing is to be lost by:

(a) Letting $\lambda(A)$ be independent of A, and therefore equal to $\binom{n}{m}^{-1}$ for every A; that is, letting every sample of size m have the same probability of being chosen, or **randomizing,** as it is said.

(b) Letting $f(x_1(A), \cdots, x_m(A))$ be symmetric in its m arguments and independent of A.

It can in fact be shown, by the method illustrated in the second example of § 9.6 and discussed more generally in § 12.5, that there is at least one minimax satisfying (a) and (b), and even that there is an admissible one. Typically, if m is large, but small compared to n, L_λ^* is much smaller than the common value of the $L^*(\mathbf{x}(A))$'s.

The importance of randomization in applied statistics can scarcely be exaggerated. From the personalistic viewpoint it is one of the most important ways to bring groups of people into virtual unanimity; from the objectivistic viewpoint it not only makes possible great reductions in maximum loss, but it is seen as an invention by which the theory of probability is brought to bear on situations to which probability on first (objectivistic) sight would seem irrelevant.[+]

9 Mixed acts in statistics

Many have commented that modern applied statistics makes one, but only one, important use of mixed acts, namely in deciding, through the process of randomization, what to observe. Thus, for example, once the observation has been made, the derived act is in practice almost always chosen, without mixing, from a set of basic acts natural to the problem. This might seem to imply a sharp conflict between the minimax rule and ordinary statistical practice; but actually it reflects

+ I would express myself very differently today (Savage 1962, pp. 33-34).

agreement, for mixed acts greatly reduce the minimax loss in decision-problem interpretations of typical practical statistical situations, when and only when ordinary practice calls for mixed acts of the same sort, namely when randomization is called for.

There are certain mechanisms that systematically tend to make mixed acts have relatively little, or even absolutely no, advantage over unmixed acts. In the following discussion of these mechanisms, let $L(r; i)$ be the abstract game on which a bilinear game $L(\mathbf{f}; \mathbf{g})$ is based.

In the first place, supposing that $L(r; i)$ is non-negative for every r and i (as is appropriate to the context now at hand), (12.3.6) can be completed, so to speak, thus:

$$(1) \qquad L^* \min (R, I) \geq \min_r \max_i L(r; i),$$

where R and I denote for the moment the number of values of r and i, respectively, and min (R, I) is of course the minimum of the two integers R and I. An inequality stronger than (1) will actually be proved.

Consider a minimax $\mathbf{f}$ for which the smallest possible number R' of the $f(r)$'s are actually positive:

$$(2) \qquad R'L^* = \max_i R' \sum_r L(r; i)f(r)$$

$$\geq \max_i L(r'; i)$$

$$\geq \min_r \max_i L(r; i)$$

where r' is so chosen that $R'f(r') \geq 1$, as can obviously be done. It is known [B19] that $R' \leq \min (R, I)$.

The important lesson of (1) is that, unless R and I are both large, the introduction of mixed acts cannot reduce the minimax loss to a very small fraction of the value it would otherwise have.

To mention a different mechanism, Figure 12.4.1 suggests that, if there are many r's, the corners of the concave function emphasized in that figure may well be very blunt, in which case a minimax mixed act has almost as high a maximum loss as any one of its components. When the number of r's is infinite, the concave function may well be differentiable, in which case mixed acts have absolutely no advantage. The remark appended to Exercise 4.5a is pertinent here.

This mechanism can be related to a certain large class of infinite abstract (i.e., not necessarily bilinear) games, discovered by Kakutani [K1], for which $L^* = L_*$. Bilinear games are but a special case of these, and numerous others seem to arise frequently in applications.

If $L^* = L_*$ for an abstract game, nothing at all can be gained by adjoining mixed acts, as (12.3.5) shows.

Finally, it may be mentioned that in many cases where an observation $\mathbf{x}$ might be followed by a mixed derived act, the same, or nearly the same, consequences can often be realized by a pure act. Speaking a little loosely, this occurs whenever $\mathbf{x}$ has a continuous or nearly continuous contraction $\mathbf{y}$ that is irrelevant, or nearly irrelevant, for then $\mathbf{y}$ can play the role in selecting a basic derived act that would otherwise be assigned to a table of random numbers. If, for example, $\mathbf{x}$ is continuous, $\mathbf{y}(\mathbf{x})$ can be taken as the last few digits in the decimal expansion of $\mathbf{x}$ to an extravagant number of places. Again if, conditionally, $\mathbf{x} = \{\mathbf{x}_1, \cdots, \mathbf{x}_n\}$ is an n-tuple of continuously, identically, and independently distributed real random variables, $y(\mathbf{x})$ may be taken as the permutation that ranks the x's in ascending order, provided that $n!$ is fairly large: 10! should satisfy almost any need.

A recent technical reference on the superfluousness of mixed acts in the presence of continuous observations is [D13].

I have occasionally heard it conjectured that any mixed act made after the observation (in an observational decision problem) is wrong in principle. I would argue that the conjecture is mistaken thus: Any observational problem that calls for randomization can be simulated, so far as its loss function $L(r; i)$ is concerned, by a basic problem. A mixed act will be as appropriate to the basic problem as it was to the observational problem from which the basic one was derived. In this way a great variety of situations calling for mixed acts having nothing to do with choice of observation can be constructed, though they seem to be atypical in practice. Moreover, any basic problem can obviously occur as the decision problem remaining after some particular value x of an observation has been observed, so the situations just constructed lead to closely related ones calling for mixed acts *after* observation.

Less abstractly, consider a person choosing from a tray of assorted French pastries. Even after extensive visual observation and interrogation of the waiter, the person might justifiably introduce considerable mixture into his choice.

I think that the conjecture that mixed acts are necessarily inappropriate after observations stems partly from the mechanisms that do tend to make such acts inappropriate or unimportant in many typical cases and partly from justifiable dissatisfaction with specific mixed acts that have from time to time been suggested by statisticians. For example, the suggestion that ties in rank arising in non-parametric tests be removed by ranking the tied observations at random may in many, or perhaps all, cases fairly be regarded with suspicion.

CHAPTER 15

Point Estimation

1 Introduction

This chapter discusses point estimation, and the next two discuss the testing of hypotheses and interval estimation, respectively. Definitions of these processes must be sought in due course; but, for the moment, whatever notions about them you happen to have will afford sufficient background for certain introductory remarks applying equally well to both kinds of estimation and to testing.

Estimating and testing have been, and inertia alone would insure that they will long continue to be, cornerstones of practical statistics. Their development has until recently been almost exclusively in the verbalistic tradition, or outlook. For example, testing and interval estimation have often been expressed as problems of making assertions, on the basis of evidence, according to systems that lead, with high probability, to true assertions, and point estimation has even been decried as ill-conceived because it is not so expressible.

Wald's minimax theory has, as was explained in § 9.2, stimulated interest in the interpretation of problems of estimation and testing in behavioralistic terms; to objectivists this has, of course, meant interpretation as objectivistic decision problems. For reasons discussed in § 9.2, it does seem to me that any verbalistic concept in statistics owes whatever value it may have to the possibility of one or more behavioralistic interpretations.

The task of any such interpretation from one framework of ideas to another is necessarily delicate. In the present instance, there is a particular temptation to force the interpretation, namely, so that criteria proposed by the verbalistic outlook are translated into applications of the minimax theory, that is, of the minimax rule and the sure-thing principle (as expressed by the criterion of admissibility), for these are the only general criteria thus far proposed and seriously maintained for the solution of objectivistic decision problems. Of course it is to be expected, and I hope later sections of this chapter and the next demonstrate, that unforced interpretations do often translate verbalistic criteria into applications of the behavioralistic ones. In evaluating any such interpretations, it must be borne in mind that an analogy of great mathematical value may be valueless as an interpretation; correspondingly, what is put forward as mere analogy should not be taken to be an interpretation, much less branded as a forced one. For example, attention has already been called (in § 11.4) to the danger of regarding the analogy between the theory of two-person games and that of the minimax rule for objectivistic decision problems as an interpretation. In fact, minimax problems are of such mathematical generality that they arise, even within statistics, in contexts other than direct application of the minimax rule to objectivistic decision problems; a striking, though technical, example is Theorem 2.26 of Wald's book [W3].

The literature of estimation and testing is vast; indeed it has, I think, been seriously contended that statistics treats of no other subjects. This chapter and the next two cannot, therefore, pretend to present a complete digest of that literature, even so far as it pertains to the foundations of statistics. For further reading certain chapters of Kendall's treatise [K2] may be recommended as a key reference to the verbalistic tradition (Chapters 17 and 18 for point estimation; 19 and 20 for interval estimation; 21, 26, and 27 for testing). Many newer aspects are treated in Wald's book [W3]; and a recent review of testing by Lehmann [L4] is recommended.

2 The verbalistic concept of point estimation

Abstractly and very generally, but in verbalistic language (which is necessarily vague), the problem of **point estimation** is this: Knowing $P(x \mid B_i)$ for every i and having observed the value x, guess the value λ of a prescribed function, or **parameter** as it is often called, $\lambda(i)$ with values in a set Λ. Semi-behavioralistically this is, I think universally, understood to mean that a function 1 associating a value $l(x)$ ε Λ with each x (or possibly a mixture of such functions) is to be decided on, the function 1 being called an **estimate** (or, to be complete, a **point estimate**) of the parameter λ. A problem of point estimation has, thus, some of the structure of an objectivistic observational problem; but, since nothing has yet been said about the income, or consequence, resulting from the act l in case B_i obtains, it is at the moment impossible to advance criteria for the choice of 1.

3 Examples of problems of point estimation

It will now be well to present some examples after a few words of preparation. For simplicity, Λ will henceforth generally be supposed to be an interval (possibly unbounded) of real numbers. If $\lambda(i) =$

TABLE 1. SOME COMMON ESTIMATION PROBLEMS

Name	Conventional symbol for i	Conventional symbol for $\lambda(i)$	Range of x	Probability or density of x	Maximum-likelihood estimate of λ
(a) Binomial (of size n)	p	p	Integers 0 to n	$\binom{n}{x} p^x q^{n-x}$	x/n
(b) Poisson	μ	μ	Non-negative integers	$e^{-\mu}\mu^x/x!$	x
(c) Normal mean (for sample of n observations with variance one)	μ	μ	n-tuples x_i of real numbers	$\phi(x_1 - \mu) \cdots \phi(x_n - \mu)$, where $\phi(z) =_{\text{Df}} (2\pi)^{-\frac{1}{2}} e^{-\frac{1}{2}z^2}$	$\bar{x} =_{\text{Df}} \dfrac{1}{n}\sum x_i$
(d) Normal variance (for sample of n observations with mean zero)	σ^2	σ^2	n-tuples x_i of real numbers	$\sigma^{-n}\phi\left(\dfrac{x_1}{\sigma}\right) \cdots \phi\left(\dfrac{x_n}{\sigma}\right)$	$\dfrac{1}{n}\sum x_i^2$
(e) Normal mean (for sample of n observations with variance unknown)	(μ, σ^2)	μ	n-tuples x_i of real numbers	$\sigma^{-n}\phi\left(\dfrac{x_1 - \mu}{\sigma}\right) \cdots \phi\left(\dfrac{x_n - \mu}{\sigma}\right)$	$\bar{x}$
(f) Normal variance (for sample of n observations with mean unknown)	(μ, σ^2)	σ^2	n-tuples x_i of real numbers	$\sigma^{-n}\phi\left(\dfrac{x_1 - \mu}{\sigma}\right) \cdots \phi\left(\dfrac{x_n - \mu}{\sigma}\right)$	$\dfrac{1}{n}\sum x_i^2 - \bar{x}^2$

$\lambda(i')$ implies $i = i'$, then λ rather than i can be used to index the partition; such an estimation problem is said to be **free of nuisance parameters.** This usage corresponds to the fact that the i's can typically be represented as ordered couples (λ, θ), where λ is of course $\lambda(i)$ and θ is called the **nuisance parameter;** if θ in turn happens to be represented as an ordered n-tuple, ordinary usage calls θ an **n-tuple of nuisance parameters.** It must be recognized as atypical in estimation problems for i or λ to be confined to a finite set of values, and often x is not so confined either. It will therefore be necessary to proceed heuristically into domains where the mathematically limited theory developed in this book does not rigorously apply.

The specific estimation problems most commonly cited as examples, and most important in practice, are summarized in Table 1, together with their maximum-likelihood estimates, that is, estimates constructed in accordance with a rule to be defined in § 4. All but the last two examples of Table 1 are free of nuisance parameters.

4   Criteria that have been proposed for point estimates

As a matter of fact, verbalistic treatments typically do give some inkling of the consequence of the act l when B_i obtains. Thus, in the examples commonly cited, such as those in Table 3.1, Λ is a set of real numbers or a set of n-tuples of real numbers and, therefore, a set of objects between which the notion of proximity. has some meaning. Work in the verbalistic tradition has made it clear in connection with such examples that, if $l = \lambda(i)$ for the B_i that obtains, the guess is considered perfect and that, roughly speaking, it is considered rather poor if l is far from λ.

In spite of the apparently hopeless indefiniteness of estimation problems even as thus formulated, various criteria, or desiderata, for estimates have been suggested. A list of these criteria, intended to be essentially complete, is now presented. Each item is annotated and illustrated to make its meaning clear, and sometimes to call attention to related criteria not explicitly listed; motivation and criticism are, however, deferred until later sections, where they are treated in. connection with explicit hypotheses about the consequences of misestimation.

No attempt is made to include criteria like intellectual simplicity or facility of computation that depend not only on the estimate but also on the capabilities of the people who contemplate using it.   The list is in a sense logically inhomogeneous. For example, no one really considers it a virtue in itself for an estimate to be a maximum-likelihood estimate (Criterion 4); rather, it is believed that such estimates do typically have real virtues.

It has, to begin the list of criteria, been suggested by one person or another that:

1. If **y** is sufficient, nothing is to be lost by requiring the estimate l to be a contraction of **y**.

It will be instructive to bear in mind that necessary and sufficient statistics of the examples (a)–(f) in Table 3.1 are, respectively, x, x, $\bar{x}$, $\sum x^2$, $(\bar{x}, \sum x^2)$, $(\bar{x}, \sum x^2)$.

2. If, of two estimates l and l′,

$$(1) \qquad E([l - \lambda(i)]^2 \mid B_i) \leq E([l' - \lambda(i)]^2 \mid B_i)$$

for every i, with strict inequality for some i, then l is better than l′.

There are countless variants of this idea. In particular, the square of the difference may be replaced by any other positive power of the absolute difference. Again, (1) may be imposed at only one value of i, if l and l′ are subjected to some other condition, freedom from bias (Criterion 6 below) being the popular one.

Example (f) gives rise to a good illustration of this criterion, which is also interesting in a later connection. Letting $Q =_{\text{Df}} \sum x^2 - n\bar{x}^2$,[+] it is well known that $E(Q \mid \mu, \sigma^2) = (n - 1)\sigma^2$ and that $E(Q^2 \mid \mu, \sigma^2) = (n^2 - 1)\sigma^4$. Therefore

$$(2) \quad E([\alpha Q - \sigma^2]^2 \mid \mu, \sigma^2) = \{\alpha^2(n^2 - 1) - 2\alpha(n - 1) + 1\}\sigma^4$$

$$= \left\{\left(\alpha - \frac{1}{n + 1}\right)^2 (n^2 - 1) + \frac{2}{n + 1}\right\} \sigma^4$$

$$\geq \frac{2\sigma^4}{n + 1}$$

for all real α, with equality if and only if $\alpha = (n + 1)^{-1}$, omitting the pathological but trivial case that $n = 1$. By the criterion in question, $Q/(n + 1)$ is therefore better than any other estimate of the form αQ, including the maximum-likelihood estimate Q/n and the unbiased estimate $Q/(n - 1)$.

3. If, of two estimates l and l′,

$$(3) \quad P(-\epsilon_1 \leq l(x) - \lambda(i) \leq \epsilon_2 \mid B_i) \geq P(-\epsilon_1 \leq l'(x) - \lambda(i) \leq \epsilon_2 \mid B_i)$$

for every non-negative ϵ_1 and ϵ_2 and for every i, with strict inequality for some ϵ_1, ϵ_2, and some i, then l is better than l′.

[+] This example was given by Leo A. Goodman (1953).

Acceptance of this criterion is obviously implied by acceptance of Criterion 2, of which it may therefore be regarded as a skeptical counterpart; formal demonstration of a much more general assertion will be given in connection with (5.2—4). The criterion implies, for example, in connection with (c) of Table 3.1 that $\bar{x}$ is superior to any other weighted average of the x_i's. A more interesting example will be mentioned in connection with Criterion 5.

That modification of Criterion 3 in which it is concluded only that 1 is at least as good as l' is of some technical interest. Incidentally, if equality held identically in (3), there would presumably be nothing to choose between the two estimates by any reasonable criterion, for they would then both have the same system of conditional distributions.

4. A maximum-likelihood estimate is often a rather good estimate.

A **maximum-likelihood estimate** is an estimate 1 such that, for some function i of x, $l(x) = \lambda(i(x))$ and

$$(4) \qquad P(x \mid B_{i(x)}) \geq P(x \mid B_i)$$

for every i and x. In many natural problems there is only one maximum-likelihood estimate. Taking into account the analogy between probabilities and values of probability densities, the reader should verify that the estimates listed in Table 3.1 are indeed the unique maximum-likelihood estimates of the problems to which they refer. When there is a unique maximum-likelihood estimate, it is obviously a contraction of the likelihood ratios and, therefore, of any sufficient statistic; which fits neatly with Criterion 1.

5. A good estimate should have the same symmetry as the problem.

More precisely, if a permutation T of the i's and the x's is such that

$$(5) \qquad P(Tx \mid B_{Ti}) = P(x \mid B_i),$$

and such that $\lambda(i) = \lambda(i')$ implies $\lambda(Ti) = \lambda(Ti')$; then 1 should be such that, if $l(x) = \lambda(i)$, $l(Tx) = \lambda(Ti)$.

For example, adopting also Criterion 1, a good estimate for μ in (c) may be sought of the form $l(\bar{x})$. Symmetry then dictates $l(\bar{x} + \alpha) = l(\bar{x}) + \alpha$ and $l(-\bar{x}) = -l(\bar{x})$; in short, $l(\bar{x}) = \bar{x}$.

The same conclusion can be drawn for (e), though with a little more trouble. The criterion applied to (f) leads to estimates of the form αQ. The constant α might be fixed by appealing, for example, to Criterion 2, 4, or 6. These alone give three slightly different determinations— $\alpha^{-1} = (n + 1)$, n, and $(n - 1)$, respectively.

Again, it can be shown for Examples (c) and (e) that, among all estimates satisfying Criterion 5, $\bar{x}$ is best according to Criterion 3.

6. It is desirable that the estimate be unbiased.

An estimate l is called **unbiased,** if and only if

$$(6) \qquad E(1 \mid B_i) = \lambda(i)$$

for every i.

It is easy to verify that the maximum-likelihood estimates of (a)–(e) in Table 3.1 are all unbiased; that of (f), however, is not, for $E(Q/n \mid \mu, \sigma^2) = (1 - 1/n)\sigma^2$ instead of σ^2. Again, if l is a maximum-likelihood estimate of λ, e^1 is a maximum-likelihood estimate of e^λ. But, if l is not definitive, and l is an unbiased estimate of λ, e^1 is not an unbiased estimate of e^λ, as Theorem 1 of Appendix 2 implies.

7. If $P(\mid 1 - \lambda(i) \mid < \mid 1' - \lambda(i) \mid \mid B_i) > 1/2$ for every i, then l is better than $1'$.

Any resemblance between this criterion and Criterion 3 seems to be dispelled by the following example. Suppose that, for every i, $P(1 - \lambda(i) = a, 1' - \lambda(i) = b \mid B_i)$ equals 2/11 if a and b are integers such that $0 \leq a < b \leq 2$, equals 5/11 if a and b are 2 and 0 respectively, and equals 0 otherwise. According to Criterion 7, 1 is better than $1'$, because $6/11 > 1/2$; but, according to Criterion 3, $1'$ is better than 1, because $5/11 > 4/11$ and $7/11 > 6/11$. The example can easily be modified to suit any taste for symmetry and continuity. But, if 1 and $1'$ are conditionally independent (which is not a natural assumption), and 1 is better than $1'$ according to Criterion 7; then, as may easily be shown, $1'$ cannot be better than 1 by Criterion 3.

The list of criteria is here interrupted by several paragraphs of explanation in preparation for two concluding criteria.

The approach to certainty treated in §§ 3.6 and 7.6 has its counterpart in the theory of estimation. In particular, if $\mathbf{x}(n) = \{\mathbf{x}_1, \cdots, \mathbf{x}_n\}$ is an n-tuple of conditionally independent and identically distributed observations, there will typically exist sequences of estimates $1(n)$ based on $\mathbf{x}(n)$, such that

$$(7) \qquad \lim_{n \to \infty} P(\mid l(x(n), n) - \lambda(i) \mid \leq \epsilon \mid B_i) = 1$$

for every positive ϵ and every i. A sequence of estimates satisfying (7) relative to any sequence of observations $\mathbf{x}(n)$ (not necessarily n-tuples of conditionally independent observations) is called **consistent.**

The condition of consistency is often realized in a very special way, namely that the **error** $[l(x(n); n) - \lambda(i)]$ is, for every B_i and for large n, practically normally distributed about zero with variance inversely proportional to n. More formally, a sequence of estimates may be such that

$$(8) \quad \lim_{n \to \infty} P\left(\frac{n^{1/2}[l(x(n); n) - \lambda(i)]}{\sigma(i)} \leq \alpha \mid B_i\right) = \frac{1}{(2\pi)^{-1/2}} \int_{-\infty}^{\alpha} e^{-1/2 z^2} dz$$

for every i and α, where $\sigma(i)$ is some positive function of i; it is then said that $n^{1/2}[l(x(n); n) - \lambda(i)]$ is **asymptotically normal** about zero with **asymptotic variance** $\sigma^2(i)$. If, in addition, for every i, $\sigma^{-2}(i)$ is not less than a certain function, the differential information, to be defined in § 6, then the sequence l_n is called **efficient**.

There is a possible pitfall in connection with the idea of asymptotic normality. Though (8) implies that, for large n, the distribution of the error is, in a sense, almost the normal distribution with zero mean and variance $\sigma^2(i)/n$, it does not imply that the mean of the error is close to zero, or even finite or well defined. Similarly, the variance of the error may be much larger than $\sigma^2(i)/n$, infinite, or ill defined; but it cannot, for large n, be smaller than $\sigma^2(i)/n$ by a fixed fraction or less.

Much literature on estimation has concentrated on sequences of estimation problems in which $\mathbf{x}(n)$ is an n-tuple consisting of the first n elements of an infinite sequence of conditionally independent and conditionally identically distributed random variables or, as it will be called in the present chapter, a *standard* sequence; because these are the simplest examples of sequences of increasingly informative observations. Examples (c)–(f) in Table 3.1 refer directly to standard sequences; the binomial distributions (a) can be regarded as the distribution of the sufficient statistic $\sum \mathbf{x}_i$ of the standard sequence $\mathbf{x}(n)$ in which each $\mathbf{x}_i$ takes the values 1 and 0 with probabilities p and $1 - p$, respectively (cf. Exercise 7.4.1); again, if each $\mathbf{x}_i$ is Poisson-distributed with parameter μ, then $\sum \mathbf{x}_i$ is sufficient for $\mathbf{x}(n)$ and is itself Poisson-distributed with parameter $n\mu$. Thus, all the examples in Table 3.1 give rise more or less directly to examples of standard sequences.

In speaking of standard, and occasionally of other, sequences the ellipsis of referring to a sequence of estimates simply as "an estimate" has been widely adopted, so one reads recommendations that "an estimate" should be consistent or efficient. This ellipsis, though often convenient, sometimes proves dangerous. It distracts from the fact that a person is called upon to make *an* estimate, not a sequence of estimates; so that the question of what constitutes a good sequence does not arise. Again, it makes one feel that if an estimate, say l_{13}, has been defined for $\mathbf{x}(13)$, then the definition of l_{14} is thereby implied. One forgets, for example, that "the" average of n observations is a whole sequence of statistics, a sequence singled out by human tastes and interests, rather than by any mathematical necessity. In short, the ellipsis establishes the atmosphere of the logically nonsensical (though perhaps psychologically revealing) questions on intelligence tests such as: "What are the two missing terms in the sequence __ __ 1 8 2 8 1 8 2 8?" †

The recommendations of consistency and efficiency quoted above can be added to the numbered list of suggestions, in a form that avoids the ellipsis:

8. If each $l(n)$ is a good estimate for the corresponding $\mathbf{x}(n)$ of a standard sequence, then the sequence $l(n)$ is consistent.

The sequence of maximum-likelihood estimates of the sequences of problems (a), (c)–(f) are consistent; and, for the sequence of problems of estimating from an observation $\mathbf{y}_n$ Poisson-distributed with parameter $n\mu$, the maximum-likelihood estimates $\mathbf{y}_n/n$ are consistent.

If there is one consistent sequence of estimates, for a sequence of problems there is a plethora. Each term of a consistent sequence can, for example, be multiplied by $(1 + n^{-\frac{1}{2}})$ without destroying consistency. Again, the sample medians ‡ are in (c) a consistent sequence different from the sequence of maximum-likelihood estimates.

9. Under the hypothesis of Criterion 8, the sequence $l(n)$ is efficient, at least if any efficient sequence of estimates exists.

The six sequences of maximum-likelihood estimates mentioned under Criterion 8 are all well known to be efficient, as sequences of maximum-likelihood estimates for standard sequences typically are. The asymptotic variances and certain other interesting quantities associated with these six sequences are presented in Table 1. It is remarkable that, for each of the examples in Table 1, the expected values of the estimates approach the estimated parameter; n times the variance of the estimate, and n times the expected squared error, both approach the asymptotic variance of $n^{\frac{1}{2}}$ times the error. For the first five examples the relations mentioned hold, indeed, not only in the limit, but exactly, for all n. All six examples are rather special, or magical, but the limiting relations just mentioned may fairly be expected to hold in some generality, though they are not (as has already been mentioned) really implied by the asymptotic normality of the sequence of errors times $n^{\frac{1}{2}}$. To illustrate the exceptions that can occur, $|\bar{x}|^{-1}$ is, in (c), the

---

† $e = 2.7182818285$ to eleven significant figures.

‡ See any statistics text for definition, if necessary.

maximum-likelihood estimate of $|\mu|^{-1}$ for $\mu \neq 0$; this sequence of estimates is efficient; and $n^{1/2}(|\bar{x}|^{-1} - |\mu|^{-1})$ is asymptotically normal about zero with asymptotic variance μ^{-4}; but the other three entries for Table 1 are infinite in this example.

TABLE 1.   EXAMPLES OF BEHAVIOR OF MAXIMUM-LIKELIHOOD ESTIMATES

Sequence	Mean	$n \times$ variance	$n \times$ expected square of error	Asymptotic variance of $n^{1/2} \times$ error
(a) Poisson μn	p	pq	pq	pq
	μ	μ	μ	μ
(c)	μ	1	1	1
(d)	σ^2	$2\sigma^4$	$2\sigma^4$	$2\sigma^4$
(e)	μ	σ^2	σ^2	σ^2
(f)	$\left(1 - \dfrac{1}{n}\right)\sigma^2$	$2\left(1 - \dfrac{1}{n}\right)\sigma^4$	$\left(2 - \dfrac{1}{n}\right)\sigma^4$	$2\sigma^4$

As in the case of consistency, where there is one efficient sequence, there are many, but efficiency is, of course, a much more restrictive property than consistency. For example, multiplication by $(1 + n^{-1/2})$ typically destroys efficiency, though multiplication by $(1 + n^{-1})$ never does. Again, the consistent sequence of medians mentioned under Criterion 8 is not efficient. Indeed, it is well known of that sequence that the sequence of errors times $n^{1/2}$ is asymptotically normal about zero with asymptotic variance $\pi/2$ rather than 1.

5   A behavioralistic review of the criteria for point estimation

It is time now to introduce the notion of consequences, or (equivalently, I believe) of loss, thereby interpreting estimation problems as decision problems. Let it be said then that an *estimation decision problem* is an observational decision problem with the following distinguishing feature. There is a one-to-one correspondence between the basic acts **f** and the values attained by a real-valued function $\lambda(i)$, such that $L(\mathbf{f}; i) = 0$, if **f** is the act that corresponds with $\lambda(i)$. It is simpler, more suggestive, and harmless to let the number l that corresponds to **f** replace **f** itself in all further discussion of estimation decision problems. To illustrate the new notation, it may be said that $L(l; i) = 0$, if $l = \lambda(i)$.

I believe that any situation ordinarily said to call for (point) estimation can be analyzed as an estimation decision problem. For example, estimating how much paint will cover a wall may, depending on circumstances, mean deciding: how much paint to buy, what to bid for a contract, or what number to enter in a guessing pool. Under each of those interpretations there will be zero loss, if and, typically, only if the estimate is "correct," as one says.

The consequences of an estimate may, like those of many real life decisions, be difficult to appraise. It is hard to say even in relatively concrete situations what it will cost to misestimate the speed of light, a particular mortality rate, or the national income. If, to revert to an example already discussed, the estimate is to be published somewhere for the use of whoever has a use for it, the consequences of publication may seem beyond all reckoning. None the less, I reaffirm the conviction that the concept of consequence measured in income or loss is valuable in dealing with such situations, as I hope the present treatment of estimation will illustrate.[+] Incidentally, it seems indifferent, as I have already said, whether loss or income is taken as the starting point. It is easily shown that the decisions of the idealized person of the personalistic probability theory will be the same in two problems having possibly different income, but the same loss, functions. This feature I would expect to be acceptable even to objectivists, and I also think it appropriate to theories of group decision.

I know of nothing interesting that distinguishes estimation decision problems as a class from observational decision problems generally. But actual estimation situations suggest certain relatively wide classes of estimation decision problems about which interesting and valuable conclusions can be drawn. Indeed, it will be shown in this and the next two sections that seven of the nine listed criteria for estimation can be justified to some extent as flowing from application of the principle of admissibility and the minimax rule to such classes of estimation decision problems.

Before making any real specialization, it may be most systematic to mention that Criterion 1 is simply an instance of the general principle, which we have now studied from several points of view, that nothing is lost by confining attention to sufficient statistics, at least if mixtures are allowed.

It is clear in almost any estimation situation, even in those for which the notion of loss is vaguest, that if two errors have the same sign the larger entails at least as great a loss as the smaller. Analytically,

$$(1) \qquad L(l; i) \leq L(l'; i)$$

for $\lambda(i) \leq l < l'$ and for $\lambda(i) \geq l > l'$. Situations to which (1) fails to apply can readily be imagined. William Tell, for example, in esti-

[+] This idea was expressed by Gauss (1821, Section 6).

mating the angle by which to elevate his cross-bow for the apple shot might have preferred a downward error of 10° to one of 1°; but such circumstances seem exceptional. Furthermore, it is usually justifiable to assume that strict inequality holds in (1), though there are many exceptions in which, for example, "a miss is as good as a mile" or one hit is as good as another.

As is, I think, intuitively evident, when strict inequality holds in (1), Criterion 3 is simply an application of the principle of admissibility. That conclusion can be shown in complete generality without serious difficulty, but, in compliance with the usual mathematical limitations of this book, it will here be shown only under the assumption that x is confined to a finite number of values.

What is to be shown is this: If l and l' are a pair of estimates satisfying the hypothesis of Criterion 3, and if (1) holds with strict inequality; then $L(l; i) - L(l'; i) \leq 0$ for every i, with strict inequality for some i. To begin the proof calculate thus:

$$(2) \quad L(l; i) - L(l'; i) = \sum_l L(l; i)[P(l(x) = l \mid B_i) - P(l'(x) = l \mid B_i)]$$

$$= \sum_l L(l; i)Q(l; i)$$

$$= \sum_{l < \lambda(i)} L(l; i)Q(l; i) + \sum_{l > \lambda(i)} L(l; i)Q(l; i),$$

where the definition of $Q(l; i)$ is clear from the context, and where it has been taken into account that $L(\lambda(i); i) = 0$. It will be shown that both sums in the last part of (2) are non-positive and that for some i at least one of them is negative. Focus, for definiteness, on the second sum. Let $l_0 = \lambda(i)$ and $l_1, l_2, \cdots$ be, in order of increasing magnitude, the values of $l > \lambda(i)$ for which $Q(l; i) \neq 0$. With the abbreviations $L(k) =_{\text{Df}} L(l_k; i)$, $\Delta(k) =_{\text{Df}} L(k) - L(k-1)$, and $Q(k) =_{\text{Df}} Q(l_k; i)$, the sum to be investigated is

$$(3) \quad \sum_{0 < k} L(k)Q(k) = \sum_{0 < k} Q(k) \sum_{0 < k' \leq k} \Delta(k')$$

$$= \sum_{0 < k'} \Delta(k') \sum_{k \geq k'} Q(k).$$

(This rearrangement may seem bizarre on first encounter, but it is widely used in mathematics generally and is in fact an exact analogue, for sums, of the more familiar integration by parts, for integrals.) It follows from (1) read with strict inequality that $\Delta(k) > 0$; and it follows from the hypothesis of Criterion 3 that $Q(k) \leq 0$, and that some $Q(k)$—or an analogous term associated with the first sum in the last line of (2)—is strictly negative for some i. This completes the deduction of Criterion 3 from the strict form of (1) and the principle of admissibility. Essentially the same argument leads from (1) as actually written to the modification mentioned in the note under Criterion 3.

A very slight strengthening of (1), together with the minimax rule, provides a widely applicable justification of Criterion 8 (consistency), as will now be explained. Suppose that (1) not only holds but also is strict, if $l = \lambda(i)$; that is, in addition to (1) suppose only that $L(l'; i) > 0$ for all $l' \neq \lambda(i)$. In this context, let $\mathbf{x}(n)$ be a sequence of observations such that the minimax $L^*(n)$ of the corresponding estimation problems approaches zero with increasing n; then any sequence of minimax estimates $1(n)$ is consistent. Indeed, if the sequence $1(n)$ is not consistent, then, for some i, and some positive ϵ and δ,

$$(4) \qquad P(|\, l(x_n; n) - \lambda(i)\,| > \epsilon \,|\, B_i) > \delta$$

for some arbitrarily large values of n. This implies

$$(5) \quad L^*(n) \geq L(1(n); i) \geq \delta \min \{L(\lambda(i) + \epsilon; i), L(\lambda(i) - \epsilon; i)\} > 0,$$

which contradicts the hypothesis.

Turn next to Criterion 5 (symmetry). Suppose that the estimation decision problem has symmetry in the sense defined under Criterion 5. That does not in itself really call for estimates with the same symmetry. But, if L also has the symmetry, that is, if $L(\lambda(i'); i) = L(\lambda(Ti'); Ti)$ for all appropriate T, then the discussion of symmetry in § 12.5 suggests that typically there is, at any rate, a symmetrical, admissible, minimax estimate. Whether L has the requisite symmetry is a question that can often be answered without detailed knowledge of L.

It is often justifiable to suppose that the function $L(l; i)$ is smooth enough to be differentiated twice with respect to l, at least when l is near $\lambda(i)$. This condition, though very often met, is not quite so devoid of content as it may seem to a reader brought up in the tradition that it makes no practical difference whether a function has a few sharp corners because they can always be rounded off with almost no change in the function. If, for example, $L(l; i)$ is for all practicable purposes equal to $|\, l - \lambda\,|$; then L cannot be regarded as differentiable even once when $l = \lambda$, and the theory to be developed here for twice differentiable $L(l; i)$'s in the presence of extensive observation does not apply. It will therefore be useful to digress to the consideration of an example, illustrating how corners can arise and the phenomena that tend to round them off.

Suppose that a person must estimate the amount λ of shelving for books, priced at \$1.00 per foot, to be ordered for some purpose. It is

possible that the following economic analysis of the situation would be
sufficiently realistic. The person holds every foot of shelving less than
the number of feet, λ, of books to be shelved to be worth \$$\alpha$, $\alpha > 1$,
but superfluous shelving he holds to be worthless. Formally,

$$(6) \qquad L(l; \lambda) = (\alpha - 1)(\lambda - l) \qquad \text{for } l \leq \lambda$$

$$= (l - \lambda) \qquad \text{for } l \geq \lambda.$$

There is then a corner, or kink, at $l = \lambda$; so differentiation, even once, is
impossible.

But the following analysis is much more likely to be sufficiently real-
istic. The urgency of the shelving of the books is variable. Some would
be worth shelving, even if the cost of shelving were very high; at the
other extreme, there are some that would not be worth shelving unless
the cost were very low. More fully, the value of l feet of shelving is a
function $i(l)$ that presumably has the following features. It is mono-
tonically increasing, strictly concave, and twice differentiable in l;
$i(0) = 0$; $i(\infty) < \infty$; $i'(0) > 1$. The income attached to ordering L
feet of shelving, at the price \$1.00 per foot, is clearly

$$(7) \qquad I(l; i) = i(l) - l.$$

It is maximized at the one and only value λ for which $di(\lambda)/d\lambda = 1$, so
that

$$(8) \qquad L(l; i) = [i(\lambda) - \lambda] - [i(l) - l],$$

which is of course twice differentiable in l.

The moral of these two possible economic analyses of one example is
of wide applicability, as is well known among economists. Where a
superficial analysis suggests a kink, or even a discontinuity, in an in-
come function, deeper analysis will often show that the function is
smoothed out by various economic phenomena such as the inhomo-
geneity and the mutual substitutability of commodities.

To return from the digression, if L is twice differentiable in l (at
least when l is close to λ), L can be expanded in a Taylor series thus:

$$(9) \quad L(l; i) = L(\lambda; i) + (l - \lambda) \frac{\partial}{\partial l} L(l; i) \bigg|_{l=\lambda(i)}$$

$$+ \frac{1}{2}(l - \lambda)^2 \frac{\partial^2}{\partial l^2} L(l; i) \bigg|_{l=\lambda(i)} + o((l - \lambda)^2),$$

where, following standard usage, $o((l - \lambda)^2)$ is a function of l and i, not
necessarily the same from one context to another, such that $o((l - \lambda)^2) \div$

$(l - \lambda)^2$ approaches zero as l approaches $\lambda(i)$ for fixed i. The first term on the right side of (9) vanishes by the definition of estimation; the second must vanish also, for otherwise L could be negative. Therefore,

$$(10) \qquad L(l; i) = \frac{1}{2}(l - \lambda)^2 \frac{\partial^2}{\partial l^2} L(l; i) \bigg|_{l=\lambda} + o((l - \lambda)^2)$$

$$= (l - \lambda(i))^2 \alpha(i) + o((l - \lambda)^2),$$

where $\alpha(i)$ is defined by the context.

In view of (10), it is plausible that L may, in many problems where estimates of great accuracy are possible, be supposed to be practically of the form

$$(11) \qquad L(l; i) = (l - \lambda(i))^2 \alpha(i),$$

where $\alpha(i) > 0$ for every i. This does not exactly mean that a reasonable L can be closely approximated by functions of the form (11) for all l. In particular, the absurd assumption that L is unbounded (which such approximation would typically imply) is not to be made. It means, rather, that under favorable circumstances (11) may lead to a reasonably good evaluation of $L(1; i)$. In so far as the form (11) can be supposed adequately to represent L, Criterion 2 is obviously an application of the principle of admissibility. An interesting discussion and application of (11) is given by Yates [Y2].

6 A behavioralistic review, continued

Thus far, Criteria 1, 2, 3, 5, and 8 have been discussed in behavioralistic terms. In fact, under suitable hypotheses, each has been found to have considerable behavioralistic justification. Criteria 4 and 9 also have such justification, but my discussion of them is so bulky it had better be isolated in a special section. As for Criteria 6 and 7, the only ones remaining, they do not seem to me to have any serious justification at all, as will be discussed in still another section.

Criterion 4, the recommendation of maximum-likelihood estimates, is of extraordinary interest, for, of all the criteria of the verbalistic tradition, it is essentially the only one that selects a unique estimate in almost every estimation situation of practical importance. The present section demonstrates that, in the presence of extensive observation, maximum-likelihood estimates are often almost minimax estimates; it also gives some analysis of Criterion 9, which refers to efficiency. The way to these goals is roundabout; it begins with a study of information in the technical sense mentioned in § 3.6. In this section it will be assumed for mathematical simplicity that each observation under discussion is confined to a finite number of values, each having positive probability for every element of whatever partition is under discussion.

If B_i and B_j are elements of a partition, not necessarily finite, and $\mathbf{x}$ is an observation, say, in the spirit of (3.6.11), that the *information of j relative to i for the observation* $\mathbf{x}$ is

$$(1) \quad J(i, j; \mathbf{x}) =_{\mathrm{Df}} -E\left(\log \frac{P(x \mid B_j)}{P(x \mid B_i)} \,\middle|\, B_i\right) = -E\left(\log \frac{r_j}{r_i} \,\middle|\, B_i\right).$$

The expression of J in terms of likelihood ratios is important, especially for the extension of the discussion to more general observations than those contemplated here. The reader should, therefore, try to bear in mind that the whole discussion could be carried on in terms of likelihood ratios; I refrain from so doing only for momentary reasons of notational convenience. The theory of $\mathbf{J}$ can conveniently be presented in a series of exercises.

Exercises

1a. If $\mathbf{y}$ is a contraction of $\mathbf{x}$, then $J(i, j; \mathbf{x}) \geq J(i, j; \mathbf{y})$. With equality when? Hint:

$$(2) \qquad -E\left(\log \frac{P(x \mid B_j)}{P(x \mid B_i)} \,\middle|\, B_i, y\right) \geq -\log \frac{P(y \mid B_j)}{P(y \mid B_i)}.$$

1b. $J(i, j; \mathbf{x}) \geq 0$. With equality when?

2a. If $\mathbf{x}_1, \cdots, \mathbf{x}_n$ are conditionally independent, then

$$(3) \qquad\qquad J(i, j; \mathbf{x}_1, \cdots, \mathbf{x}_n) = \sum_s J(i, j; \mathbf{x}_s)$$

2b. If in addition the $\mathbf{x}_s$'s are conditionally identically distributed, then

$$(4) \qquad\qquad J(i, j; \mathbf{x}_1, \cdots, \mathbf{x}_n) = nJ(i, j; \mathbf{x}_1).$$

It is interesting to evaluate the information $J(\lambda, \lambda + \Delta\lambda; \mathbf{x})$ where λ and $\lambda + \Delta\lambda$ are two closely neighboring values of the parameter of an estimation problem, supposed, for simplicity, to be free of nuisance parameters. If $P(x \mid \lambda)$ is continuous in λ, it is almost obvious that $J(\lambda, \lambda + \Delta\lambda; \mathbf{x})$ approaches zero as $\Delta\lambda$ approaches zero. If $P(x \mid \lambda)$ is differentiable in λ, it is easy to show further (considering that J is nonnegative) that even $J(\lambda, \lambda + \Delta\lambda; \mathbf{x})/\Delta\lambda$ approaches zero as $\Delta\lambda$ approaches zero. But in this case much more can and will be shown, namely,

$$(5) \qquad \lim_{\Delta\lambda \to 0} \frac{J(\lambda, \lambda + \Delta\lambda; \mathbf{x})}{\Delta\lambda^2} = \frac{1}{2} H(\lambda; \mathbf{x})$$

$$=_{\mathrm{Df}} \frac{1}{2} E\left[\left(\frac{\partial \log P(x \mid \lambda)}{\partial \lambda}\right)^2 \Big| \lambda\right].$$

The function H is generally, following Fisher, called information, but here we had better call it *differential information*. Chronologically, as explained at the end of § 3.6, the concept of differential information is older than that here called simply information and of which it is, according to (5), a limiting case.

The demonstration of (5) begins with the consideration that

$$(6) \qquad \log(1 + t) = t - \tfrac{1}{2}t^2 + o(t^2).$$

Therefore,

$$(7) \quad \log \frac{P(x \mid \lambda + \Delta\lambda)}{P(x \mid \lambda)} = \log\left\{1 + \frac{P(x \mid \lambda + \Delta\lambda) - P(x \mid \lambda)}{P(x \mid \lambda)}\right\}$$

$$= \left\{\frac{P(x \mid \lambda + \Delta\lambda) - P(x \mid \lambda)}{P(x \mid \lambda)}\right\}$$

$$- \frac{1}{2}\left\{\frac{P(x \mid \lambda + \Delta\lambda) - P(x \mid \lambda)}{P(x \mid \lambda)}\right\}^2 + o(\Delta\lambda^2).$$

Since the expected value given λ of the term in the second line of (7) is easily seen to be exactly zero, it will be tactful to leave that term alone; but the second may be approximated thus:

$$(8) \quad \left\{\frac{P(x \mid \lambda + \Delta\lambda) - P(x \mid \lambda)}{P(x \mid \lambda)}\right\}^2 = \left\{\frac{\Delta\lambda \, \partial P(x \mid \lambda)}{P(x \mid \lambda) \, \partial \lambda} + o(\Delta\lambda)\right\}^2$$

$$= \Delta\lambda^2 \left\{\frac{\partial \log P(x \mid \lambda)}{\partial \lambda}\right\}^2 + o(\Delta\lambda^2).$$

Therefore,

$$(9) \qquad J(\lambda, \lambda + \Delta\lambda; \mathbf{x}) = \tfrac{1}{2} H(\lambda; \mathbf{x})\Delta\lambda^2 + o(\Delta\lambda^2),$$

which establishes (5).

More exercises

3. If the kth derivative ($k > 0$) with respect to λ of $P(x \mid \lambda)$ exists for every x, then

$$(10) \qquad E\left(\frac{1}{P(x \mid \lambda)} \frac{\partial^k}{\partial \lambda^k} P(x \mid \lambda) \Big| \lambda\right) = \frac{\partial^k}{\partial \lambda^k}\left(\sum_x P(x \mid \lambda)\right) = 0.$$

4. If the requisite second derivative exists, then

$$(11) \qquad H(\lambda; \mathbf{x}) = -E\left(\frac{\partial^2}{\partial\lambda^2} \log P(x \mid \lambda) \Big| \lambda\right).$$

5. If $\mathbf{y}$ is a contraction of $\mathbf{x}$ (and $H(\lambda; \mathbf{x})$ is well defined), then $H(\lambda; \mathbf{y})$ $\leq H(\lambda; \mathbf{x})$.

Remark: The inequality is obvious in the light of Exercise 1a and the first part of (5). But it can also be derived from the following application of Theorem 1 of Appendix 2, which is useful in the next exercise.

$$(12) \qquad \left\{\frac{1}{P(y \mid \lambda)} \frac{\partial P(y \mid \lambda)}{\partial\lambda}\right\}^2 = E^2\left(\frac{1}{P(x \mid \lambda)} \frac{\partial P(x \mid \lambda)}{\partial\lambda} \Big| y, \lambda\right)$$

$$\leq E\left(\left\{\frac{1}{P(x \mid \lambda)} \frac{\partial P(x \mid \lambda)}{\partial\lambda}\right\}^2 \Big| y, \lambda\right),$$

with equality for every y and λ, if and only if $\dfrac{\partial}{\partial\lambda} \log P(x \mid \lambda)$ can be expressed as a function of y and λ alone.

6a. If $\mathbf{y}$ is a contraction of $\mathbf{x}$, $H(\lambda; \mathbf{x}) = H(\lambda; \mathbf{y})$ for every λ; if and only if $\mathbf{y}$ is sufficient for $\mathbf{x}$.

6b. $H(\lambda; \mathbf{x}) = 0$ for every λ, if and only if $\mathbf{x}$ is utterly irrelevant.

7a. If $\mathbf{x}_1, \cdots, \mathbf{x}_n$ are independent given λ, then

$$(13) \qquad H(\lambda; \mathbf{x}_1, \cdots, \mathbf{x}_n) = \sum_s H(\lambda; \mathbf{x}_s).$$

7b. If, in addition, the $\mathbf{x}_s$'s are identically distributed given λ, then

$$(14) \qquad H(\lambda; \mathbf{x}_1, \cdots, \mathbf{x}_n) = nH(\lambda; \mathbf{x}_1).$$

8. If $\mathbf{l}$ is a real-valued contraction of $\mathbf{x}$, and $H(\lambda; \mathbf{x})$ is well defined, then

(a)

$$(15) \qquad \frac{d}{d\lambda} E(\mathbf{l} \mid \lambda) = E\left(l(\mathbf{x}) \frac{\partial \log P(l(\mathbf{x}) \mid \lambda)}{\partial\lambda} \Big| \lambda\right).$$

(b)

$$(16) \qquad E([\mathbf{l} - \lambda]^2 \mid \lambda)H(\lambda; \mathbf{l}) \geq \left\{\frac{d}{d\lambda} E(\mathbf{l} \mid \lambda)\right\}^2,$$

with equality if and only if

$$(17) \qquad \frac{\partial}{\partial\lambda} \log P(l \mid \lambda) = (l - \lambda)k$$

for some constant k. Hint: Use Exercise 3 and apply the Schwartz inequality to (15).

(c) If $H(\lambda; \mathbf{x}) > 0$, then

(18)
$$E([1 - \lambda]^2 \mid \lambda) \geq \left\{ \frac{d}{d\lambda} E(1 \mid \lambda) \right\}^2 / H(\lambda; \mathbf{x}).$$

Exercise 8c is an important, and now famous, inequality. It, together with its n-dimensional generalization, has been called the Cramér-Rao inequality because of its independent publication by Rao and Cramér in 1945 and 1946 respectively (see [H6]). But the name is not at all well justified historically. Fréchet presented the inequality in 1943 [F8], and Darmois extended Fréchet's inequality to n dimensions, at least for unbiased estimates, in a publication [D1] not later than Rao's. The inequality has also, though I think erroneously, been attributed to an early paper by Aitken and Silverstone [A1], and to one by Doob [D10]. My point is, of course, not to give a definitive history of the inequality, but merely to suggest that for the time being an impersonal name would be better. I tentatively propose calling it the *information inequality*. Some recent references pertinent to the information inequality and other topics treated thus far in this section are [W15], [M5], [C6], and [H6]. The techniques used in the remainder of this section, which revolve around the information inequality, were published posthumously by Wald [W5].

The information inequality has an important bearing on application of the minimax rule to estimation, of which the following theorem may, in view of (5.11) be taken as a first illustration.

THEOREM 1

HYP. 1. For every λ in a closed interval of length δ, $H(\lambda; \mathbf{x}) \leq H$, where H is a constant.

2. 1 is a real-valued contraction of $\mathbf{x}$.

CONCL. For some λ in the interval, $E((1 - \lambda)^2 \mid \lambda) \geq \left(H^{1/2} + \frac{2}{\delta} \right)^{-2}$.

PROOF. Suppose that the theorem is false. Then according to Exercise 8c,

(19)
$$1 > H^{1/2} \left(H^{1/2} + \frac{2}{\delta} \right)^{-1} > \left| \frac{d}{d\lambda} E(1 \mid \lambda) \right|$$

for every λ in the interval. Therefore,

(20)
$$\frac{d}{d\lambda} [\lambda - E(1 \mid \lambda)] > 1 - H^{1/2} \left(H^{1/2} + \frac{2}{\delta} \right)^{-1} = \frac{2}{(\delta H^{1/2} + 2)}$$

for every λ in the interval. Therefore, at one end of the interval or the other,

$$(21) \qquad |\lambda - E(1\,|\,\lambda)| > \frac{2}{(\delta H^{1/2} + 2)} \cdot \frac{\delta}{2} = \left(H^{1/2} + \frac{2}{\delta}\right)^{-1}.$$

This leads to a contradiction through the well-known inequality

$$(22) \qquad E([1 - \lambda]^2\,|\,\lambda) \geq \{E(1 - \lambda\,|\,\lambda)\}^2 = |\lambda - E(1\,|\,\lambda)|^2,$$

which can be derived as a direct application of Theorem 1 of Appendix 2, or of the Schwartz inequality, or of the useful identity

$$(23) \qquad E([1 - \lambda]^2\,|\,\lambda) = V(1\,|\,\lambda) + \{E(1 - \lambda\,|\,\lambda)\}^2. \;\blacklozenge$$

In the remaining portion of this section, let it be understood that:

1. The x_s's are an infinite sequence of observations that are, given λ, identically distributed and independent.
2. $x(n) = \{x_1, \cdots, x_n\}$ for $n = 1, 2, \cdots$.
3. $1(n)$ is a real-valued contraction of $x(n)$.

The contraction $1(n)$ is to be thought of as an estimate of λ based on observation of $x(n)$. In the spirit of the minimax theory it is really mixed, rather than ordinary, estimates that should be treated here. But this entails no essential change in the following discussion once it is recognized that a mixed estimate is, in effect, an ordinary estimate based on observation of $y(n) =_{\text{Df}} (1(n), x(n))$, where $x(n)$ is sufficient for $y(n)$, so that $H(\lambda; y(n)) = H(\lambda; x(n))$ for all λ.

4. ϵ and δ are positive numbers.
5. Λ_0 is a closed interval of length δ contained in the range of λ and including a given value λ_0.

The next theorem shows that, if $L(l; \lambda)$ is of the form (5.11), $L(1(n); \lambda)$ cannot ordinarily be kept much smaller than $\alpha(\lambda_0)/nH(\lambda_0; x_1)$ for large n, even in a small interval about λ_0.

THEOREM 2   If $H(\lambda; x_1)$ is continuous and positive at λ_0, and if $\alpha(\lambda)$ is a non-negative function continuous at λ_0, then, for sufficiently large n, $E((1(n) - \lambda)^2\alpha(\lambda)\,|\,\lambda) \geq (1 - \epsilon)\alpha(\lambda_0)/nH(\lambda_0; x_1)$ for some $\lambda \,\epsilon\, \Lambda_0$.

PROOF.   There is no loss of generality in supposing that $\epsilon < 1$ and Λ_0 such that, for $\lambda \,\epsilon\, \Lambda_0$, $\alpha(\lambda) \geq \alpha(\lambda_0)(1 - \epsilon)^{1/2}$ and $H(\lambda; x_1)^{1/2} \leq H(\lambda_0; x_1)^{1/2} [1 + (1 - \epsilon)^{-1/4}]/2$. Using Exercise 7b,

$$(24) \quad H(\lambda; x(n))^{1/2} = n^{1/2}H(\lambda; x_1)^{1/2} \leq \frac{n^{1/2}}{2} H(\lambda_0; x_1)^{1/2}[1 + (1 - \epsilon)^{-1/4}]$$

for $\lambda \, \varepsilon \, \Lambda_0$. By Theorem 1, if $n \geq 16/\delta^2 H(\lambda_0; \mathbf{x}_1)[(1 - \epsilon)^{-\frac{1}{4}} - 1]^2$, then

$$(25) \quad E((\mathbf{l}(n) - \lambda)^2 \mid \lambda) \geq \left\{ \frac{n^{\frac{1}{2}}}{2} H(\lambda_0; \mathbf{x}_1)^{\frac{1}{2}}[1 + (1 - \epsilon)^{-\frac{1}{4}}] + \frac{2}{\delta} \right\}^{-2}$$

$$\geq \frac{(1 - \epsilon)^{\frac{1}{2}}}{nH(\lambda_0; \mathbf{x}_1)}$$

for some $\lambda \, \varepsilon \, \Lambda_0$. ◆

The next theorem extends Theorem 2 to practically any loss function that is twice differentiable in l for l and λ close to λ_0.

THEOREM 3

HYP. 1. $H(\lambda; \mathbf{x}_1)$ is positive and continuous at λ_0.

2. $\alpha(\lambda) =_{\mathrm{Df}} \dfrac{1}{2} \dfrac{\partial^2}{\partial l^2} L(l; \lambda) \Big|_{l=\lambda}$ is continuous at λ_0.

3. Inequality (5.1) holds for λ in Λ_0.

CONCL. For sufficiently large n, $L(\mathbf{l}(n); \lambda) \geq (1 - \epsilon)\alpha(\lambda_0)/nH(\lambda_0; \mathbf{x}_1)$ for some $\lambda \, \varepsilon \, \Lambda_0$.

PROOF. It may be supposed without loss of generality that $\epsilon < 1$; and that, for $l, \lambda \, \varepsilon \, \Lambda_0$, $L(l; \lambda) \geq (1 - \epsilon)^{\frac{1}{2}}\alpha(\lambda)(l - \lambda)^2$.

It may also be supposed that $l(x; n) \, \varepsilon \, \Lambda_0$. This is so, because it would suffice to prove the theorem for a new estimate $\mathbf{l}'(n)$, where $l'(x; n)$ is defined to be the number in Λ_0 closest to $l(x; n)$, which in turn follows from the fact that $L(\mathbf{l}'(n); \lambda) \leq L(\mathbf{l}(n); \lambda)$ for $\lambda \, \varepsilon \, \Lambda_0$.

These suppositions having been made, the theorem is a direct consequence of Theorem 2. ◆

COROLLARY 1   If $L(l; \lambda)$ satisfies (5.1) and has two derivatives with respect to l continuous in λ for every λ and for every l sufficiently close to λ, and if $H(\lambda; \mathbf{x}_1)$ is continuous and positive, then, for sufficiently large n,

$$(26) \qquad L^*(n) \geq (1 - \epsilon) \sup_{\lambda} \alpha(\lambda)/nH(\lambda; \mathbf{x}_n),$$

where $L^*(n)$ is the minimax value of the estimation decision problem derived from $L(l; \lambda)$ and $\mathbf{x}(n)$, unless the supremum in question is infinite, in which case $nL^*(n)$ approaches infinity.

Of course, it would be enough to assume only that $L(l; \lambda)$ and $H(\lambda; \mathbf{x}_1)$ are well behaved at some sequence of values of λ on which the supremum in question is approached. In particular, if the supremum is actually attained at some λ, they need only be well behaved there.

Now, turning to the sequence of maximum-likelihood estimates, let them be denoted for the moment by $\tilde{1}(n)$. It is known that under rather general hypotheses $n^{1/2}(\tilde{1}(n) - \lambda)$ is asymptotically normal about zero with asymptotic variance $1/H(\lambda; x_1)$.† This suggests, and examples tend to confirm, that, under some supplementary conditions,

$$(27) \qquad \lim_{n \to \infty} nE((\tilde{1}(n) - \lambda)^2) = \frac{1}{H(\lambda; x_1)}.$$

Indeed, one set of conditions implying (27) is stated in [W5], but one that seems difficult to apply. It can be shown that (27), together with the usual asymptotic behavior of $\tilde{1}(n)$, implies

$$(28) \qquad \lim_{n \to \infty} nL(\tilde{1}(n); \lambda) = \frac{\alpha(\lambda)}{H(\lambda; x_1)},$$

provided, for example, that $L(l; \lambda)$ is bounded for each λ and that the second derivative of $L(l; \lambda)$ with respect to l exists when $l = \lambda$. Easily applied rigorous theorems implying (28) much less (27) do not seem to have been formulated yet; but examples suggest that, under conditions general enough for many applications, (28) actually does hold uniformly, in the sense that, for n sufficiently large,

$$(29) \qquad \frac{(1 - \epsilon)\alpha(\lambda)}{nH(\lambda; x_1)} \le L(\tilde{1}(n); \lambda) \le \frac{(1 + \epsilon)\alpha(\lambda)}{nH(\lambda; x_1)}$$

for all λ simultaneously. If (29) holds, then, in view of Corollary 1, $\tilde{1}(n)$ is nearly minimax for large n, in the sense that

$$(30) \qquad L^*(n) \ge (1 - \epsilon) \sup_{\lambda} L(\tilde{1}(n); \lambda).$$

Good examples can be based on (a) of Tables 3.1 and 4.1, letting $L(l; p)$ be any loss function having two continuous derivatives in l throughout $0 \le l, p \le 1$. In particular, the example discussed in § 13.4 arises, if $L(l; p) = (l - p)^2$. It can be argued that the phenomenon discussed in connection with that example is probably not rare;

† Some key references for the asymptotic behavior of $\tilde{1}(n)$ are [K2], [C9], [L3], [W16], [N4]. The literature on this subject is extraordinarily complicated. There are acknowledged mathematical mistakes in some of its most sophisticated publications; others prove much less than any but the most attentive reader would be led to suppose; few give an adequate statement of their relations to their predecessors; and those that make serious pretentions to rigor involve complicated hypotheses. For documentation of this lament see [N4], [W4], and [L3];

because, for minimax $1(n)$, $L(1(n); \lambda)$ is, judging from examples, often constant and, therefore, nearly equal to $\sup_{\lambda} \alpha(\lambda)/nH(\lambda; x_1)$, but $L(\tilde{1}; \lambda)$ closely follows the rise and fall of $\alpha(\lambda)/nH(\lambda; \mathbf{x}_1)$.

Turn now to Criterion 9, efficiency. It seems difficult to defend the criterion as it has been defined in connection with (4.8); for what virtue is there in the asymptotic normality required by (4.8)? It is perhaps noteworthy that the sequence of minimax estimates, $\hat{\mathbf{p}}_1(n)$, arising in connection with § 13.4 does not satisfy (4.8). Indeed, (13.4.3) implies that $n^{1/2}(\hat{\mathbf{p}}_1(n) - p)$ is asymptotically normal not about zero, but about $(\frac{1}{2} - p)$.

It is my impression that the essence of the efficiency concept resides not in asymptotic normality, but in the overall behavior of the mean square error of a sequence of estimates. I therefore propose tentatively to modify the definition and to call a sequence of estimates $1(n)$ efficient, if and only if its mean square error behaves at least as well as can typically be expected for a sequence of maximum-likelihood estimates.

Formally, I propose to call $1(n)$ efficient, if and only if, for n sufficiently large,

$$(31) \qquad E([1(n) - \lambda]^2) \leq \frac{(1 + \epsilon)}{nH(\lambda; \mathbf{x}_1)}$$

for every λ simultaneously.

I think the main objection that is likely to be raised to this proposed definition is associated with the possibility that in some problems of theoretical, and perhaps also of practical, importance (31) is not satisfied by any sequence of estimates whatsoever, though the maximum-likelihood sequence is efficient in the "official" sense. In such a problem, are the maximum-likelihood estimates not as good for all practical purposes for sufficiently large n as though their variances were actually equal to those of the normal distributions to which they approximate? It is natural to think so by analogy with other contexts in the theory of probability, but approximate normality is actually no substitute for (31) in the present context. The next paragraph is devoted to an example illustrating the inadequacy of asymptotic variance as a measure of asymptotic loss. It can be skipped without loss by anyone not interested in such technicalities.

The best example I have been able to construct is derived from a sequence of observations that is not a standard sequence. Whether the interesting features that it exhibits can actually be realized by standard sequences, I do not know; but the example will do to illustrate the issue. Let $\mathbf{y}(n)$ be any real random variable subject to the density

$n^{1/2}\phi((y - \lambda)n^{1/2}; n)$, defined thus: $\phi(z; n)$ is the standard normal density inside the interval $[-\delta(n), \delta(n)]$, $\delta(n)$ being such that the standard normal probability of this interval is $(1 - n^{-1})$; $\phi(z; n) = z^{-2}\delta(2n)/4$ for $\delta(2n) \leq |z| \leq n^{1/2}$; $\phi(z; n)$ is so defined elsewhere as to be a symmetric positive probability density with the first two moments finite, with a bounded derivative approaching zero like z^{-4} with increasing z, and with unique absolute maximum at $z = 0$. It is evident that $n^{1/2}$ $(\mathbf{y}(n) - \lambda)$ is asymptotically normal about zero with unit variance. The information $H(\lambda; \mathbf{y}(n))$ is well defined (even according to the strict conditions imposed by Cramér, Lemma 1, Section 32.2 of [C9]). The maximum-likelihood estimates of λ are $\mathbf{y}(n)$, and these are also (according to Theorem 3.3 of [G1]) minimax for the simple quadratic loss function $(l - \lambda)^2$. But

$$(32) \qquad E([\mathbf{y}(n) - \lambda]^2 \mid \lambda) = E(\mathbf{y}(n)^2 \mid 0)$$

$$\geq 2n^{1/2} \int_{\delta(2n)n^{-1/2}}^{1} y^2\phi(yn^{1/2}; n) \, dy$$

$$= \tfrac{1}{2}n^{-1/2}[1 - \delta(2n)n^{-1/2}] \, \delta(2n),$$

which does not satisfy (31). Even for the bounded, and therefore more realistic, loss function,

$$(33) \qquad L(l; \lambda) = \min \{1, [l - \lambda]^2\},$$

it follows easily from Theorem 3.3 of [G1] that every estimate must somewhere incur a loss at least as great as the lower bound established by (32). To summarize, there are no estimates efficient in the sense of (31), nor even in the sense that would arise from (31) on replacing the simple quadratic loss function by a bounded loss function; the sequence of estimates $\mathbf{y}(n)$ is efficient in the official sense, so to speak, but does not, of course, result in losses of the order of n^{-1}.

What can be said in positive justification of the criterion of efficiency as defined by (31) or the like? Roughly, the elements of such a sequence nearly dominate every estimate for every smooth loss function. A little more precisely, for large n, the loss associated with an element of a sequence efficient in the sense of (31) is at most larger by a small fraction than that of any other estimate, except possibly in some short intervals.† The maximum loss of such an element is at most larger by a small fraction than the minimax loss, so the elements of the sequence are typically nearly minimax. Moreover, they typically have consid-

---

† It has actually been demonstrated that the total length of these exceptional intervals (within any fixed interval) is small [L3].

erably smaller losses than any minimax estimate, except in short intervals that are typically very improbable a priori in the personal sense. Thus the principle of admissibility, the minimax rule, and the personalistic concept of probability combine to suggest that efficiency as defined by (31) is a promising guide in the search for good estimates.

An extensive critique of the concept of efficiency, including much material on its history, has been given by LeCam in [L3], which unfortunately was not available to me in its entirety as I wrote this section.

R. A. Fisher's name is the most prominent in the history of maximum-likelihood estimation and efficiency. Some historical details are given in [N4] and on p. 45 of Vol. II of [K2].

7 A behavioralistic review, concluded

Criteria 6 (unbiasedness) and 7 are now the only ones in the list for which I have not suggested some justification in terms of the theory of decision problems, and, indeed, I cannot. Unbiased estimates fascinate many theoretical statisticians, including myself, and the study of them undoubtedly has certain valuable by-products. Yet it is now widely agreed that a serious reason to prefer unbiased estimates seems never to have been proposed.

Three weak defenses are sometimes heard. First, unbiasedness is asserted to have an intuitive appeal; whether it does or not depends, of course, on the experience of the intuiter. Second, averages of increasingly many unbiased estimates are typically consistent. If this is a virtue, it is a limited one and pertains to the unbiased estimate not as an estimate, but as a step in the definition of other estimates. Third, an allusion is made to equity. If, for example, it has been agreed that one party will buy a sack of sugar from another at so much per pound, it seems fair that the nominal weight of the sack be determined by unbiased estimate. This ethical conclusion could perhaps be given some justification in terms of approximately linear utility functions or a long-run argument, though there is danger of falling into such pitfalls as the conclusion that accuracy is unimportant for equity; and it might find some application in the theory of barter; but it seems, at best, tangential to estimation in the sense of the present chapter.

For a proper appraisal of the criterion of unbiasedness it should be realized that, even if λ admits an unbiased estimate, many not-at-all pathological functions of λ (which can in turn be regarded as parameters), may fail to do so and that such unbiased estimates as λ does admit may be preposterous. These phenomena are both illustrated by the following simple example. Let x be confined to two values, say 1 and 2; let $P(1 \mid \lambda) = 1 - P(2 \mid \lambda) = \lambda$; and let λ be confined to the interval

[1/3, 2/3]. Then, by definition, 1 is an unbiased estimate of $\phi(\lambda)$, if and only if $l(1)\lambda + l(2)(1 - \lambda) = l(2) + (l(1) - l(2))\lambda = \phi(\lambda)$—a condition that can be met, if and only if ϕ is linear. Suppose, for example, $\phi(\lambda) = \lambda$ for every λ, then $l(1) = 1$, $l(2) = 0$ defines the only unbiased estimate of $\phi(\lambda)$. This estimate is worse, according to an emphatic variant of Criterion 3, than the biased estimate $1'$ such that $l'(1) = 2/3$ and $l'(2) = 1/3$; for $1'$ (when it errs at all) errs in the same direction as 1, but never nearly as far.

As for Criterion 7, it is on first encounter appealing to postulate that, if 1 is usually closer to λ than $1'$ is, then 1 is better than $1'$. But, speaking at least for myself, the initial appeal of Criterion 7 seems to have been bound up with the conjecture that Criterion 7 is in some sense of the same sort as Criterion 3. The example given under Criterion 7 almost entirely evaporates the conjecture, and with it the appeal.

In the paper [P5] in which the criterion is put forward for consideration and exploration, Pitman mentions that the criterion seems acceptable in contexts where "the devil takes the hindmost." This allusion to the devil seems to offer no justification for the criterion as a criterion of estimation, for I understand the allusion to refer only to the following kind of decision problem, which is quite remote from estimation as ordinarily understood and is hardly ever encountered: A person must choose between 1 and $1'$, winning a prize if the estimate of his choice falls closer to λ than does the other one.

According to Pitman, the relationship of "better than," or "closer than" as he calls it, defined by Criterion 7, is not necessarily transitive. He argues, I think with some justice, that this breakdown of transitivity does not in itself invalidate the criterion when the criterion is applied to select the "best" from some prescribed class of estimates; but "best" cannot here be taken literally.

Criterion 7 is unusual in that it depends on the joint conditional distributions of pairs of estimates rather than on the distributions of each estimate considered separately. On any ordinary interpretation of estimation known to me, it can be argued (as it was under Criterion 3) that no criterion need depend on more than the separate distributions.

CHAPTER 16

Testing

1 Introduction

In principle, this chapter on the statistical process of testing (often referred to more fully as making tests of hypotheses or significance tests) might have been organized on the pattern of the preceding chapter on point estimation: a statement of verbalistic ideas, followed by motivation and criticism in terms of behavioralistic ideas. But I am dissuaded from repeating that pattern by several considerations. It would, in the first place, be needlessly repetitious. Thus, in the presence of the preceding chapter I need mention only in passing that sufficient statistics and symmetry play the same role in testing as in other observational decision problems, and that a certain scheme of testing, closely related to maximum-likelihood estimation, has asymptotic, or large sample, virtues. Again, the pattern of the preceding chapter is less attractive here, because the criteria for tests developed in the verbalistic tradition do not on the whole seem to have such satisfying behavioralistic motivation as do their counterparts in the theory of point estimation. Finally, it is inappropriate to attempt anything like a complete list of verbalistic criteria for tests here, especially in view of the availability of two excellent and mutually complementary key references (Chapters 21, 26, and 27 of [K2]; and [L4]).

The organization actually adopted is this: First, testing and criteria for tests are discussed from a frankly behavioralistic viewpoint. In this discussion ideas stemming from the verbalistic tradition are used freely, and some criteria of the verbalistic tradition are criticized. Second, an attempt is made to analyze some of the important statistical situations to which the theory of testing is ordinarily applied. It is becoming increasingly recognized that many of these applications are very crude, and that their replacement by sounder procedures constitutes some of the most important and provocative statistical problems of today.

Terms introduced in boldface in this chapter are among the most frequent in ordinary statistical usage. The definitions given are intended to be in reasonable accord with that usage, but some small concessions are made to the particular form in which the theory of testing is expressed here.

2 A theory of testing

Verbalistically, the problem of testing means to guess, on the basis of observation, which of two disjoint and mutually exhaustive hypotheses obtains. Behavioralistically, this would generally be agreed to point to the definition: A *testing problem* is an observational decision problem derived from exactly two basic acts f_0 and f_1. These two basic acts are called (for a reason that will soon be clear) **accepting** and **rejecting the null hypothesis,** respectively.

Considered abstractly as bilinear games, testing problems may, so far as I know, have no special feature beyond the uninteresting one that one of two f's is appropriate to each i. But, considered as observational problems, testing problems do present some interesting special features. In the first place, since at least one of the two basic acts is appropriate to each i, the set I of all i's can be partitioned into three sets, H_0, H_1, and N, defined thus:

$$L(f_0; i) = 0 \quad \text{and} \quad L(f_1; i) > 0 \quad \text{for } i \in H_0,$$

(1) $$L(f_0; i) > 0 \quad \text{and} \quad L(f_1; i) = 0 \quad \text{for } i \in H_1,$$

$$L(f_0; i) = 0 \quad \text{and} \quad L(f_1; i) = 0 \quad \text{for } i \in N.$$

When it is recalled that the i's correspond to a partition B_i of S, the sets H_0, H_1, and N may, with a slight clash of logical gears, be regarded as three events partitioning S. The traditional names of H_0 and H_1 are the **null** and the **alternative hypothesis,** respectively; N, being quite unimportant and often either ignored or made vacuous by some trick of definition, has no such name. Rejecting the null hypothesis when it does in fact obtain and accepting it when it does not obtain are called **errors,** more specifically **errors of the first** and **second kind,** respectively.

A **test** is a derived act of a testing problem. A test may conveniently be identified with the real-valued contraction z of the observation x, such that $z(x)$ is the probability prescribed by the test for rejection of the null hypothesis in case x is observed. An unmixed test (which was until recently the only kind contemplated) corresponds to a z confined to the two values 0 and 1, which respectively imply outright acceptance and rejection of the null hypothesis.

The loss associated with the test z when i obtains is clearly

(2)
$$L(\mathbf{z}; i) = L(\mathbf{f}_0; i)E(1 - \mathbf{z} \mid i) + L(\mathbf{f}_1; i)E(\mathbf{z} \mid i)$$

$$= L(\mathbf{f}_1; i)E(\mathbf{z} \mid i) \qquad \text{for } i \in H_0$$

$$= L(\mathbf{f}_0; i)[1 - E(\mathbf{z} \mid i)] \qquad \text{for } i \in H_1$$

$$= 0 \qquad \text{for } i \in N.$$

The functions $E(\mathbf{z} \mid i)$ and $[1 - E(\mathbf{z} \mid i)]$ are, respectively, the probability of rejecting and accepting the null hypothesis with the test $\mathbf{z}$ when i obtains. There is obviously nothing to choose between them in importance or convenience, each being equivalent to the other. They are commonly called the **power function**, and **operating characteristic**, respectively.

In view of (2), one test $\mathbf{z}$ dominates another $\mathbf{z}'$, if and only if

(3)
$$E(\mathbf{z} \mid i) \le E(\mathbf{z}' \mid i) \qquad \text{for } i \in H_0$$

$$E(\mathbf{z} \mid i) \ge E(\mathbf{z}' \mid i) \qquad \text{for } i \in H_1;$$

or, again, if and only if the probability of error with $\mathbf{z}'$ is at least as great as with $\mathbf{z}$ for every i. Thus, dominance, admissibility, and equivalence depend on the basic loss function, $L(\mathbf{f}_r; i)$, only in so far as that function determines H_0 and H_1. This is not only remarkable but also useful; for H_0 and H_1 may well be clearly defined in contexts where the basic loss is vague, or otherwise ill determined.

If $\mathbf{z}$ is admissible in the spirit of (3) relative to a pair of sets H_0 and H_1, then (if ∞ is for the moment admitted as a possible value for a loss) there exists a basic loss function leading to H_0 and H_1 and having $\mathbf{z}$ as its essentially unique minimax. Indeed, let

(4)
$$L(\mathbf{f}_0; i) = [1 - E(\mathbf{z} \mid i)]^{-1} \qquad \text{for } i \in H_1$$

$$= 0 \qquad \text{elsewhere;}$$

$$L(\mathbf{f}_1; i) = E(\mathbf{z} \mid i)^{-1} \qquad \text{for } i \in H_0$$

$$= 0 \qquad \text{elsewhere.}$$

With this loss and reckoning $0 \cdot \infty = 0$ (as is appropriate here), $L(\mathbf{z} \mid i)$ $= 1$ or 0, according as there is or is not positive probability of making an error at i with $\mathbf{z}$. In view of (2) and (4), any minimax $\mathbf{z}'$ not equivalent to $\mathbf{z}$ would strictly dominate $\mathbf{z}$, contrary to the assumption that $\mathbf{z}$ is admissible. The moral of that conclusion can be put thus: Without special assumptions about the basic loss, the principle of admissibility and the minimax rule lead to no criteria expressible solely in terms of H_0, H_1, and the conditional distributions of the observation $\mathbf{x}$ other than that of admissibility itself. Whether some other objectivistic principle could justify such criteria may be considered an open question, but, as I have already said (in § 15.1), no other general objectivistic principles have been seriously maintained.

It is natural, for example, to demand that $\mathbf{z}$ have the same symmetry as $P(x \mid i)$ and H_0 and H_1; but that criterion can surely not be justified at all, unless the basic loss is also assumed to have the same symmetry, the justifiability of which in turn depends on the case.

To take another important example, it is often proposed that a satisfactory test must be unbiased,† that is, its power function must never be higher in H_0 than in H_1. More formally, the test $\mathbf{z}$ is **unbiased,** if and only if

(5) $$E(\mathbf{z} \mid i_0) \leq E(\mathbf{z} \mid i_1)$$

for every $i_0 \in H_0$ and every $i_1 \in H_1$.

Assuming that $L(\mathbf{f}_0; i)$ and $L(\mathbf{f}_1; i)$ are constant in H_1 and H_0, respectively, it will be shown that any minimax must be unbiased. As a step toward that demonstration, consider a testing problem as a minimax problem, without any special assumption about the basic loss. It is possible that $L^* = 0$, in which case the minimax tests are all equivalent and all unbiased. Putting that possibility aside, I assert, and will show, that (under the usual mathematical simplifications)

(6) $$\max_{i \in H_0} L(\mathbf{z}; i) = \max_{i \in H_1} L(\mathbf{z}; i) = L^*$$

for any minimax $\mathbf{z}$. It is obvious that neither maximum exceeds L^*, and also that one or the other must equal L^*. But suppose, for example, that the second maximum were actually less than L^*, and consider $\mathbf{z}' = \alpha \mathbf{z}$ with $0 < \alpha < 1$. According to (2), if $\mathbf{z}'$ is substituted for $\mathbf{z}$, the first maximum in (6) will be depressed, and, for α sufficiently close to 1, the second would remain actually less than L^*, which contradicts the assumption that $\mathbf{z}$ is minimax, establishing (6).

Now make the special assumption that

(7)
$$L(\mathbf{f}_0; i) = A \qquad \text{for } i \in H_1$$
$$L(\mathbf{f}_1; i) = B \qquad \text{for } i \in H_0,$$

and suppose that $\mathbf{z}$ could be minimax but biased. There would then

---

† A definition unifying the various concepts of unbiasedness in statistics is put forward in [L5].

exist $i_0 \varepsilon H_0$ and $i_1 \varepsilon H_1$ such that

$$(8) \qquad L^* = L(\mathbf{z}; i_0) = BE(\mathbf{z} \mid i_0) = A - AE(\mathbf{z} \mid i_1) = L(\mathbf{z}; i_1),$$

and such that $E(\mathbf{z}; i_0) > E(\mathbf{z}; i_1)$. But consideration of the test that simply assigns to every x the number β midway between $E(\mathbf{z}; i_0)$ and $E(\mathbf{z}; i_1)$ shows that $\mathbf{z}$ could not be minimax.

The condition (7) is a reasonable assumption in some testing problems, and, where (7) is satisfied, the criterion of unbiasedness has such support as the minimax rule can give. In many other typical testing problems, however, there are borderline errors that hardly matter at all but can scarcely be prevented, and serious errors that can largely be prevented. The following example, which can be varied to suit diverse tastes, shows that it can be folly to insist on unbiasedness in such problems.

Let i take the three values 0, 1, 2, and let $\mathbf{x}$ take the values 0 and 1 with conditional probabilities defined thus:

$$(9) \qquad P(0 \mid 0) = 99/100, \qquad P(0 \mid 1) = 0, \qquad P(0 \mid 2) = 1.$$

Let the basic loss be defined by the condition that $i \varepsilon H_0$ or $i \varepsilon H_1$, according as $i = 0$ or not, and by

$$(10) \qquad L(\mathbf{f}_1; 0) = 1, \qquad L(\mathbf{f}_0; 1) = 1, \qquad L(\mathbf{f}_0; 2) = 1/101.$$

Then

$$L(\mathbf{z}; 0) = [99z(0) + z(1)]/100$$

$$(11) \qquad L(\mathbf{z}; 1) = 1 - z(1)$$

$$L(\mathbf{z}; 2) = [1 - z(0)]/101.$$

It is easily verified that the only minimax $\mathbf{z}^*$ is defined by $z^*(0) = 0$, $z^*(1) = 100/101$, and that $L(\mathbf{z}^*; i) = L^* = 1/101$ for every i. But it is also easily verified that the only unbiased tests are absurd in that they ignore the observation $\mathbf{x}$; they are in fact just those for which $z(0) = z(1)$.

It has until quite recently been said by many that attention should be confined to tests such that there is a fixed probability α (called the size of the test) of making an error of the first kind for every $i \varepsilon H_0$. Indeed, the criterion of size has often been taken so seriously as to be incorporated into the very definition of a test. Though many important tests happen to have a size, others equally important do not; so it now seems to be recognized [L4] that the possession of a size cannot be taken seriously as a criterion.† To take an everyday example, consider the binomial distributions

(12) $$P(x \mid p) = \binom{101}{x} p^x (1 - p)^{101-x},$$

where the parameter p confined to $[0, 1]$ plays the role of i and $x = 0,$ $\cdots, 101$; and suppose that H_0 is the hypothesis that $p < 1/2$. A test of size α is a test for which

(13) $$\sum_x z(x) \binom{100}{x} p^x (1 - p)^{101-x} = \alpha$$

for all $p < 1/2$. This obviously implies

(14) $$\sum_x [z(x) - \alpha] \binom{101}{x} \left(\frac{p}{1 - p}\right)^x = 0$$

for all $p < 1/2$, whence $z(x) = \alpha$ for every x. So only absurd tests have size, in this example, though there are clearly tests here that are quite satisfactory for many applications, for example, let $z(x)$ equal 0 or 1 according as $x \leq 50$ or $x > 50$.

In view of the criticism just made, there is a tendency to redefine size so that any test has a *size* α, namely,

(15) $$\alpha = _{\text{Df}} \max_{i \, \varepsilon \, H_0} E(\mathbf{z} \mid i).$$

In terms of this definition of size, a concept of testing somewhat different from that proposed in this section has been defined and defended (Wald, p. 21 of [W3], and Lehmann, pp. 17–18 of [L4]; namely, it is postulated that a test is to be chosen not from among all possible tests, but only from among those having a size α (in the sense of (15)) given as part of the testing problem.‡ This concept of testing is not defended to the exclusion of the one proposed here, but it is asserted by the authors cited to be more realistic for some problems. The arguments of both authors on this point are similar and, I think, quite weak in two crucial places, for the advantage is supposed to flow in some *unspecified* way from the *undemonstrated* impossibility of comparing preferences for consequences of qualitatively different kinds. It seems, if I may be allowed such a conjecture, that the concept of testing under a

† Statisticians interested in the Behrens-Fisher problem may be interested in pp. **35.**173a–b of [F6], which hinge on the question of size as a criterion.

‡ The constraint actually imposed, especially by Lehmann [L4], is that the size be at most α. But, as Lehmann explains, this difference is more apparent than real.

constraint of size represents a Procrustean attempt to fit the (older) Neyman-Pearson theory of testing hypotheses too closely with the (newer) minimax theory. It is not to be denied, of course, that there may sometimes be a mathematical advantage in studying and comparing tests of given size.

It should be mentioned, before concluding the subject, that any theory taking size seriously introduces an asymmetry of the theory with respect to H_0 and H_1, an asymmetry that is surely not always appropriate.

Significance level, or **level of significance,** is a synonym (neglecting a slight distinction made in [L4]) of size, probably more widely used than size itself.

3 Testing in practice

The theory of testing admits some fairly realistic applications, but the present state of statistics is such that the theory of testing is invoked more often than not in problems on which it does not bear squarely. This section discusses typical applications of the theory, pointing out the shortcomings I am aware of.

The development of the theory of testing has been much influenced by the special problem of **simple dichotomy,** that is, testing problems in which H_0 and H_1 have exactly one element each. Simple dichotomy is susceptible of neat and full analysis (as in Exercise 7.5.2 and in § 14.4), likelihood-ratio tests here being the only admissible tests; and simple dichotomy often gives insight into more complicated problems, though the point is not explicitly illustrated in this book.

Coin and ball examples of simple dichotomy are easy to construct, but instances seem rare in real life. The astronomical observations made to distinguish between the Newtonian and Einsteinian hypotheses are a good, but not perfect, example, and I suppose that research in Mendelian genetics sometimes leads to others. There is, however, a tradition of applying the concept of simple dichotomy to some situations to which it is, to say the best, only crudely adapted. Consider, for example, the decision problem of a person who must buy, f_0, or refuse to buy, f_1, a lot of manufactured articles on the basis of an observation x. Suppose that i is the difference between the value of the lot to the person and the price at which the lot is offered for sale, and that $P(x \mid i)$ is known to the person. Clearly, H_0, H_1, and N are sets characterized respectively by $i > 0$, $i < 0$, $i = 0$. This analysis of this, and similar, problems has recently been explored in terms of the minimax rule, for example by Sprowls [S16] and a little more fully by Rudy [R4], and by Allen [A3]. It seems to me natural and promising for many fields of application, but it is not a traditional analysis. On the contrary, much literature recommends, in effect, that the person pretend that only two values of i, $i_0 > 0$ and $i_1 < 0$, are possible and that the person then choose a test for the resulting simple dichotomy. The selection of the two values i_0 and i_1 is left to the person, though they are sometimes supposed to correspond to the person's judgment of what constitutes good quality and poor quality—terms really quite without definition. The emphasis on simple dichotomy is tempered in some acceptance-sampling literature, where it is recommended that the person choose among available tests by some largely unspecified overall consideration of operating characteristics and costs, and that he facilitate his survey of the available tests by focusing on a pair of points that happen to interest him and considering the test whose operating characteristic passes (economically, in the case of sequential testing) through the pair of points. These traditional analyses are certainly inferior in the theoretical framework of the present discussion, and I think they will be found inferior in practice.

To make a small digression, there is a complication in connection with testing whether to buy that is not ordinarily envisaged by statistical theory; namely, the economic reaction between the buyer and the supplier. If, for example, the supplier knows the test the buyer is going to apply, that knowledge will influence the quality of the lot supplied. There seems to be little, if any, successful work on the economic problem thus raised about the game-like behavior of the two people involved (cf. pp. 331, 340, and 346 of [W6]).

The problem whether to buy a lot obviously has many formal counterparts in other domains. In some of them it is particularly clear that purely objectivistic methods do not suffice. To illustrate, imagine two experiments: one designed to determine whether it is advantageous to add a certain small amount of sodium fluoride to the drinking water of children, the other to determine whether the same amount of oil of peppermint is advantageous. Granting that each of the two additions can be made at the same cash cost for labor and material and that the designs of the two hypothetical experiments differ only in the interchange of the roles of sodium fluoride and oil of peppermint, the corresponding testing problems are objectivistically completely parallel, that is, the same with regard to loss function and conditional probability of the observations. But it must be acknowledged, I think, that the people actually charged with the decision in either of these two cases would and should take into account opinions they had before the observation. For example, they might originally have considered it nearly impossible that the oil of peppermint could result in any hygienic advantage large enough to compensate for even the small cost of its administration, but, in view of recent dental researches on the subject, they might not have considered it at all unlikely that the sodium fluoride should have an overall advantage. In that case, parallel observations in the two experiments would not always lead to parallel decisions. Objectivists typically admit such a possibility but go on to say that it is unreasonable to isolate the experiment and that it is the totality of information bearing on the subject that should be treated objectivistically. If objectivists could give a more detailed discussion of how to deal with such a totality of information, it might do much to clarify their position.

I turn now to a different and, at least for me, delicate topic in connection with applications of the theory of testing. Much attention is given in the literature of statistics to what purport to be tests of hypotheses, in which the null hypothesis is such that it would not really be accepted by anyone. The following three propositions, though playful in content, are typical in form of these *extreme* null hypotheses, as I shall call them for the moment.

A   The mean noise output of the cereal Krakl is a linear function of the atmospheric pressure, in the range from 900 to 1,100 millibars.

B   The basal metabolic consumption of sperm whales is normally distributed [W11].

C   New York taxi drivers of Irish, Jewish, and Scandinavian extraction are equally proficient in abusive language.

Literally to test such hypotheses as these is preposterous. If, for example, the loss associated with f_1 is zero, except in case Hypothesis A is exactly satisfied, what possible experience with Krakl could dissuade you from adopting f_1?

The unacceptability of extreme null hypotheses is perfectly well known; it is closely related to the often heard maxim that science disproves, but never proves, hypotheses. The role of extreme hypotheses in science and other statistical activities seems to be important but obscure. In particular, though I, like everyone who practices statistics, have often "tested" extreme hypotheses, I cannot give a very satisfactory analysis of the process, nor say clearly how it is related to testing as defined in this chapter and other theoretical discussions. None the less, it seems worth while to explore the subject tentatively; I will do so largely in terms of two examples.

Consider first the problem of a cereal dynamicist who must estimate the noise output of Krakl at each of ten atmospheric pressures between 900 and 1,100 millibars. It may well be that he can properly regard the problem as that of estimating the ten parameters in question, in which case there is no question of testing. But suppose, for example, that one or both of the following considerations apply. First, the engineer and his colleagues may attach considerable personal probability to the possibility that A is very nearly satisfied—very nearly, that is, in terms of the dispersion of his measurements. Second, the administrative, computational, and other incidental costs of using ten individual estimates might be considerably greater than that of using a linear formula. It might be impractical to deal with either of these considerations very rigorously. One rough attack is for the engineer first to examine the observed data x and then to proceed either as though he actually believed Hypothesis A or else in some other way. The other way might be to make the estimate according to the objectivistic formulae that would have been used had there been no complicating considerations, or it might take into account different but related complicating considerations not explicitly mentioned here, such as the advantage of using a quadratic approximation. It is artificial and inadequate to regard this decision between one class of basic acts or another as a test, but that is what in current practice we seem to do. The choice of which test to adopt in such a context is at least partly motivated by the vague idea that the test should readily accept, that is, result in acting as though the extreme null hypotheses were true, in the farfetched case that the null hypothesis is indeed true, and that the worse the approximation of the null hypotheses to the truth the less probable should be the acceptance.

The method just outlined is crude, to say the best. It is often modified in accordance with common sense, especially so far as the second consideration is concerned. Thus, if the measurements are sufficiently precise, no ordinary test might accept the null hypotheses, for the experiment will lead to a clear and sure idea of just what the departures from the null hypotheses actually are. But, if the engineer considers those departures unimportant for the context at hand, he will justifiably decide to neglect them.

Rejection of an extreme null hypothesis, in the sense of the foregoing discussion, typically gives rise to a complicated subsidiary decision problem. Some aspects of this situation have recently been explored, for example by Paulson [P3], [P4]; Duncan [D11], [D12]; Tukey [T4], [T5]; Scheffé [S7]; and W. D. Fisher [F7].

To summarize abstractly, I would say that, in current practice, so-called tests of extreme hypotheses are resorted to when at least a little credence is attached to the possibility that the null hypothesis is very nearly true and when there is some special advantage to behaving as though it were true. One other illustration will make it clear that point estimation is not essential to the situation and that belief in the approximate truth of the null hypothesis alone does not always justify testing.

Consider the personnel manager of a great New York taxi company. Wishing, of course, that his drivers should be as proficient as possible, he would, under simple circumstances, hire exclusively from the national-extraction group that had obtained the highest mean scores in a standard proficiency examination; for why should he not be guided by a positive indication, however slight? A statistical test of the extreme Hypothesis C would not, therefore, be called for, as has been pointed out in general terms by Bahadur and Robbins [B3]. Even strong belief that ethnic differences are extremely small in the respect in question would not alone be any reason for departing from this simple policy, dictated by the principle of admissibility—quite in contrast to the example framed around Hypothesis A. If, however, public opinion, a shortage of labor, or administrative difficulty militates against any discrimination at all, the manager may resort to a test based on the examination scores.

In practice, tests of extreme hypotheses are typically chosen from a relatively small arsenal of standard types, or families, each family consisting of one unmixed test at every significance level (as size is always called in this context). In publications, it is standard practice not simply to report the result of a test, but rather to report that level of significance for which the corresponding test of the relevant family would be on the borderline between acceptance and rejection. The rationale usually given for this procedure is that it enables each user of the publication to make his own test at the significance level he deems appropriate to his particular problem. Thus the significance level is supposed to play much the same practical role as a sufficient statistic.

An interesting contribution to the theory of extreme hypotheses is given by Bahadur [B1] in the special context of the two-sided t-test.

CHAPTER 17

Interval Estimation
and Related Topics

1 Estimates of the accuracy of estimates

The doctrine is often expressed that a point estimate is of little, or no, value unless accompanied by an estimate of its own accuracy. This doctrine, which for the moment I will call the *doctrine of accuracy estimation*, may be a little old-fashioned, but I think some critical discussion of it here is in order for two reasons. In the first place, the doctrine is still widely considered to contain more than a grain of truth. For example, many readers will think it strange, and even remiss, that I have written a long chapter (Chapter 15) on estimation without even suggesting that an estimate should be accompanied by an estimate of its accuracy. In the second place, it seems to me that the concept of interval estimation, which is the subject of the next section, has largely evolved from the doctrine of accuracy estimation and that discussion of the doctrine will, for some, pave the way for discussion of interval estimation.

The doctrine of accuracy estimation is vague, even by the standards of the verbalistic tradition, for it does not say what should be taken as a measure of accuracy, that is, what an estimate of accuracy should estimate. Any measure would be rather arbitrary; a typical one, here adopted for definiteness, is the **root-mean-square error,**

(1) $\qquad E^{\frac{1}{2}}([1 - \lambda(i)]^2 \mid B_i) = \{V(1 \mid B_i) + [E(1 \mid B_i) - \lambda(i)]^2\}^{\frac{1}{2}},$

using (15.6.23). The root-mean-square error reduces to the **standard deviation,** $V^{\frac{1}{2}}(1 \mid B_i)$, in case the estimate 1 is unbiased.

Taking the doctrine literally, it evidently leads to endless regression, for an estimate of the accuracy of an estimate should presumably be accompanied by an estimate of its own accuracy, and so on forever.

Even supposing that the doctrine were somehow purged of vagueness and endless regression, it would still be in clear conflict with the behavioralistic concept of estimation studied in Chapter 15. If a decision problem consists in deciding on a number in the light of an observation, the person concerned wants to adopt an 1 that is, in some sense or other, as good as possible; but, since he must make some decision, it could at most satisfy idle curiosity to know how good the best is— idle, I say, because, his decision once made, there is no way to use knowledge of its accuracy.

Since it seems to me that the kind of problem envisaged in Chapter 15 is of frequent occurrence and may properly be called estimation, I am inclined to say that the doctrine of accuracy estimation is erroneous. However, it is possible that someone should point out a different class of problems, also properly called problems of estimation, with respect to which the doctrine has some validity; though, so far as I know, this has not yet occurred.

One sort of situation that might, through what I would consider faulty analysis, seem to support the doctrine of accuracy estimation is illustrated by the following, highly schematized example. A person has to estimate the number n of replacement parts of a certain sort that should be carried by an expedition. He can conduct a trial the outcome of which will, let us say, be an observation x distributed in the Poisson distribution with mean equal to $\alpha c n$; that is,

$$(2) \qquad P(x \mid n) = e^{-\alpha cn}(\alpha cn)^{x}/x!,$$

where α is a known constant and c, which the person can choose, is the cost (beyond overhead) of the trial. Under reasonable hypotheses, once c has been chosen and the value x observed, $\hat{n}(x) = x/\alpha c$ is a good estimate of n; and in so far as the problem is of the type envisaged in Chapter 15, that is the end of the matter.

But there may be features of the problem that have not yet been stated, though in principle they should have been. In particular, it may be that the person is free to conduct a second trial, though there will typically be a high penalty for doing so. One rough, but sometimes natural and practical, step toward deciding whether a second trial is called for is to remark that $(\hat{n}/\alpha c)^{\frac{1}{2}}$ is a good estimate of the root-mean-square error of $\hat{n}$ and may give a fairly good basis on which to judge whether the risk of misestimation warrants the expense of a second trial.

My own conviction is that we should frankly regard such a problem as has just been described as a special problem in sequential analysis and treat it as an organic whole. Viewed thus, c is to be chosen in the light of the possibility of making a second trial. The decision to be based on x is the complex one of whether to go to the expense of a second trial; if so, of what magnitude; and, if not, what estimate of n to adopt.

Another sort of situation that seems to have stimulated the doctrine of accuracy estimation is the following. Suppose that a research worker has observed $x_1, \cdots, x_n$, which are independent and normally distributed about the mean μ with variance σ^2 given μ and σ. If he wishes to publish the results of his investigation for all concerned to use as their own needs and opinions may dictate, he should, ideally, publish a sufficient statistic of his observation, stating how it is distributed given μ and σ. Any other course may deprive some reader of some information he might be able to put to use. So far as the primary aim is concerned, all sufficient statistics are equivalent, but secondary considerations greatly narrow the research worker's choice. To illustrate, consider the five sufficient statistics the values of which for $\{x_1, \cdots, x_n\}$ are:

(a) $\{x_1, \cdots, x_n\}$.
(b) The n order statistics of $\{x_1, \cdots, x_n\}$.
(c) $\sum x_i$ and $\sum x_i^2$.
(d) $\bar{x} =_{\mathrm{Df}} \sum x_i/n$ and $s^2 =_{\mathrm{Df}} (\sum x_i^2 - \bar{x} \sum x_i)/n - 1$.
(e) $\bar{x}$ and $s/n^{1/2}$.

If n is at all large, (c), (d), and (e) are cheaper to publish than (a) and (b). Moreover, for almost any use to which a reader might wish to put the data, (c), (d), and (e) will save him a considerable amount of computation. In so far as it is true that almost any reader who has a use for the data at all will use $\bar{x}$, but not necessarily $\sum x_i$, statistics like (d) and (e) are slightly preferable to (c). There is something to be said both for (d) and for (e), in view of the ready availability of certain tables; but, at least when n is very large, there is a slight advantage to (e) for those calculations a reader is most likely to perform. In particular, a reader using (e) can, when n is large, often ignore the actual value of n. Even if the distributions of the $x_1, \cdots, x_n$ are not exactly normal, (c), (d), and (e) often can play almost the same role as sufficient statistics. It is no wonder then that (e) is often chosen as a convenient way to present data. But, in my opinion, it is a mistake to lay great theoretical emphasis on the fact that (e) happens to consist of what is ordinarily a good estimate of μ, namely $\bar{x}$, together with what is ordinarily a good estimate of the root-mean-square error of that estimate, namely $s/n^{1/2}$.

2   Interval estimation and confidence intervals

The verbalistic tradition has suggested a procedure different from point estimation but somehow related to it. This other procedure, here called *interval estimation*, can be defined as follows, though the definition is necessarily vague. Where x is an observation subject to the conditional distributions $P(x \mid B_i)$ and $\lambda(i)$ is a function of i, guess that $\lambda(i)$ lies in some set $M(x)$ (to be called an **interval estimate**) determined for each value of **x**. It is almost a part of the definition to say that the function $M(x)$ is to be so chosen that $P(\lambda(i) \; \varepsilon \, M(x) \mid B_i)$ shall be nearly 1 for every i and that $M(x)$ should tend to be small and "close knit" in a geometrical sense, some compromise being effected between these two conflicting desiderata. The parameter $\lambda(i)$ could in principle be a very general function, but it will here be enough to suppose for definiteness and simplicity that $\lambda(i)$ is real. Though more general possibilities are contemplated in principle, the set $M(x)$ is in practice typically a bounded interval, which corresponds with what I meant in saying that $M(x)$ is supposed to be "close knit."

The idea of interval estimation is complicated; an example is in order. Suppose that, for each λ, **x** is a real random variable normally distributed about λ with unit variance; then, as is very easy to see with the aid of a table of the normal distribution, if $M(x)$ is taken to be the interval $[x - 1.9600, x + 1.9600]$, then

$$(1) \qquad\qquad P(\lambda \; \varepsilon \, M(x) \mid \lambda) = \alpha,$$

where α is constant and almost equal to 0.95.

It is usually thought necessary to warn the novice that such an equation as (1) does not concern the probability that a random variable λ lies in a fixed set $M(x)$. Of course, λ is given and therefore not random in the context at hand; and, given λ, α is the probability that $M(\mathbf{x})$, which is a contraction of **x**, has as its value an interval that contains λ.

Why seek an interval estimate? One sort of verbalistic answer runs like this: At first glance, the problem of estimation seems to require that a person guess, on observing that **x** takes the value x, that $\lambda(i)$ has some particular value $l(x)$; but, since it is virtually impossible that such a guess should be correct, it seems better to try something else. In particular, it is often possible to assert that $\lambda(i)$ is in a comparatively narrow interval $M(x)$, chosen according to such a system that it is very improbable for each i that the assertion will be false. Less extreme verbalistic explanations tend to give the impression that point estimation need not be altogether rejected, but that interval estimation satisfies a parallel need.

The first part of the explanation just cited is specious, since no one really expects a point estimate to be correct, and since, when one really is obliged by circumstances to make a point estimate in the behavioralistic sense, there is no escaping it. None the less, that part of the explanation does seem to give some insight into the appeal of interval estimation. The second part of the explanation is a sort of fiction; for it will be found that whenever its advocates talk of making assertions that have high probability, whether in connection with testing or estimation, they do not actually make such assertions themselves, but endlessly pass the buck, saying in effect, "This assertion has arisen according to a system that will seldom lead you to make false assertions, if you adopt it. As for myself, I assert nothing but the properties of the system."

From the behavioralistic point of view, I maintain that point estimation fulfils an important function. On the other hand, I can cite no important behavioralistic interpretation of interval estimation. Moreover, in such direct and indirect contact as I have had with actual statistical practice, I have—with but one extraordinary exception, which will soon be discussed—encountered no applications of interval estimation that seemed convincing to me as anything more than an informal device for exploring data or crudely summarizing it for others. In short, not being convinced myself, I am in no position to present convincing evidence for the usefulness of interval estimation as a direct step in decision. The reader should know, however, that few are as pessimistic as I am about interval estimation and that most leaders in statistical theory have a long-standing enthusiasm for the idea, which may have more solid grounds than I now know.

The following is a schematized example of one sort of decision problem that does call for something like interval estimation. An observation x bears on the position λ of a lifeboat, the occupants of which will be saved or lost according as the boat is or is not sighted by a searching aircraft before nightfall. The decision problem is, therefore, to choose, from all the domains that the airplane could search in time, one domain $M(x)$; and the loss must, in effect, be reckoned as 0 or 1 according as $M(x)$ does or does not contain λ. This type of problem seems, however, too rare and too special to be taken as representative of those for which interval estimation is so widely advocated.

Many criteria have been put forward for interval estimation, but I am of course in no position to discuss them critically. J. Neyman has gone about the search for criteria systematically, setting up a parallelism between the theory of interval estimation and of testing. In particular, paralleling the criterion of fixed size for tests, he has emphasized interval estimates such that

$$(2) \qquad\qquad P(\lambda(i) \; \varepsilon \; M(x) \mid B_i) = \alpha$$

for a fixed α (typically close to 1) and for every i. Such interval estimates are called **confidence intervals** at the **confidence level** α. The interval estimate mentioned in connection with (1) is obviously a confidence interval. Wald [W3] sought to include the theory of confidence intervals in the minimax theory, but in my opinion he did not succeed in giving interval estimation a behavioralistic interpretation.

Though I am in no position to criticize any criterion of interval estimation, I venture to ask whether (2) is not gratuitous, as I have more positively asserted of its analogue in the theory of testing.

Chapters 19 and 20 of [K2] will serve as key references for interval estimation.

3 Tolerance intervals

There has recently been considerable study of what are called **tolerance intervals** (or **limits**). They are related to the problem of guessing the actual value of a real random variable **y**, on the basis of an observation of **x**. A **tolerance interval** for **y** at **tolerance level** α and **confidence level** β is an interval-valued function $Y(x)$ such that

$$(1) \qquad P[P(y \ \varepsilon \ Y(x) \mid B_i, x) \geq \alpha \mid B_i] = \beta$$

for every i.

The concept expressed by (1) is a slippery one; perhaps it will help to express it in words thus: For every B_i, there is probability β that x is such that y will fall in $Y(x)$ with probability at least α, given B_i and x. In typical applications y is independent of x; this permits a slight simplification of the definition. The notion of tolerance interval seems to me at least as unamenable to behavioralistic interpretation as that of confidence interval, and I therefore venture no discussion of it here. Key references are [B22] and [W7].

4 Fiducial probability

This is not really a section on fiducial probability, but rather an apology for not having such a section. The concept of fiducial probability put forward and stressed by R. A. Fisher is the most disputed technical concept of modern statistics, and, since the concept is largely concerned with interval estimation, I wanted to discuss it here. I have, however, been privileged to see certain as yet unpublished manuscripts of R. M. Williams [W12] and J. W. Tukey which convince me that such discussion by me now would be premature.

Some key references to fiducial probability and to the Behrens-Fisher problem, which is the most disputed field of application of fiducial probability, are Fisher's own papers, especially [F5], and Papers 22, 25, 26, 27, and 35 of the collection [F6]; Kendall [K2], Chapter 20; Yates [Y1]; Owen [O1]; Segal [S9]; Bartlett [B6]; Scheffé [S6], [S5]; Walsh [W9]; and Chand [C5].[+]

[+] And I can now add Barnard (1963), Dempster (1964), Fisher (1956, Sections III 3, IV 6, V 5, V 8, VI 8, VI 12), Linnik (1968, Chapters VIII-X), Patil (1965), Scheffé (1970), Tukey (1957), and Williams (1966).

APPENDIX 1

Expected Value

This appendix, a brief account of some relatively elementary aspects of the badly named mathematical concept, expected value, is presented for those who might otherwise be handicapped in reading this book. No proofs are given here, but the reader who needs this appendix will probably be willing and able to accept the facts cited without proof, especially if he acquires intuition for the subject by working the suggested exercises. The requisite proofs are, however, given implicitly in any standard work on integration or measure (e.g., Chapters I–V of [H2]).

Throughout this appendix, let S be a set with elements s and subsets A, B, C, $\cdots$ on which a (finitely additive) probability measure P is defined. Bounded real **random variables**, that is, bounded real-valued functions, defined for each $s \in S$, will here be denoted by $\mathbf{x}$, $\mathbf{y}$, $\cdots$, and real numbers by x, y, z, and lower-case Greek letters.

The **expected value** of $\mathbf{x}$, generally written $E(\mathbf{x})$, is characterized as the one and only function attaching a real number to every bounded random variable $\mathbf{x}$, subject to the following three conditions for every $\mathbf{x}$, $\mathbf{y}$, ρ, σ, and B:

(1) $$E(\rho\mathbf{x} + \sigma\mathbf{y}) = \rho E(\mathbf{x}) + \sigma E(\mathbf{y}).$$

(2) $$E(\mathbf{x}) \geq 0 \quad \text{whenever} \quad P(x(s) < 0) = 0.$$

(3) $$E(\mathbf{c}(\mid B)) = P(B).$$

In (3), $\mathbf{c}(\mid B)$ is the **characteristic function** of B, that is, $c(s \mid B) = 1$, if $s \in B$, and $c(s \mid B) = 0$, if $s \in \sim B$. In mathematical contexts remote from the topics in this book, the term "characteristic function" has at least two other meanings virtually unconnected with the one at hand, one in connection with linear operators on function spaces, and another in connection with the Fourier analysis of distributions.

Often the expected value of $\mathbf{x}$ is referred to as the **integral** of $\mathbf{x}$ over S, in which case it is generally written $\int x(s) \, dP(s)$.

Exercises

1. If $\mathbf{x}$ takes only a finite number of values, $x_1, \cdots, x_n$, except on a set of probability zero; then

(4)
$$E(\mathbf{x}) = \sum_{i=1}^{n} x_i P(x(s) = x_i),$$

that is, the average of the x_i's, each weighted by the probability of its occurrence.

2. If $P(x(s) < y(s)) = 0$, $E(\mathbf{x}) \geq E(\mathbf{y})$; and if, in addition, $P(x(s) > y(s) + \epsilon) > 0$ for some $\epsilon > 0$, then $E(\mathbf{x}) > E(\mathbf{y})$.†

3. If $\mathbf{x}$ is a real random variable, B_i a partition, ρ_i and σ_i real numbers such that $\rho_i \leq x(s) \leq \sigma_i$ for $s \varepsilon B_i$, then

(5)
$$\Sigma \rho_i P(B_i) \leq E(\mathbf{x}) \leq \Sigma \sigma_i P(B_i).$$

4.
$$\mathbf{c}(|\ A \cap B) = \mathbf{c}(|\ A)\mathbf{c}(|\ B),$$

$$\mathbf{c}(|\ {\sim}A) = 1 - \mathbf{c}(A),$$

$$\mathbf{c}(|\ A \cup B) = \mathbf{c}(|\ A) + \mathbf{c}(|\ B) - \mathbf{c}(|\ A)\mathbf{c}(|\ B).$$

As is explained in texts on measure theory, the expected value can (at least for countably additive measures), and in practice must, be extended to many unbounded random variables.

Since, provided $P(B) > 0$, the conditional probability, defined by $P(C \mid B) = P(C \cap B)/P(B)$, is itself a probability measure, the expectation of $\mathbf{x}$ with respect to a conditional probability is a meaningful concept. This **conditional expectation** is written $E(\mathbf{x} \mid B)$ and read "the expected value of $\mathbf{x}$ given B."

More exercises

5. $E(\mathbf{x} \mid B) = E(\mathbf{xc}(|\ B))/P(B)$. Hint: It suffices to verify that the expression on the right satisfies the three conditions parallel to (1–3) that define $E(\mathbf{x} \mid B)$.

6. If B_i is a partition of S, then

(6)
$$\sum_i \mathbf{c}(s \mid B_i) = 1 \qquad \text{for every } s.$$

7. $E(\mathbf{x}) = \sum_i E(\mathbf{x} \mid B_i)P(B_i)$. Hint: Use $\mathbf{x} = 1\mathbf{x}$.

---

† Technical note: In the event that P is countably additive, $P(x(s) > y(s)) > 0$ implies the existence of a suitable ϵ, so then ϵ need not be mentioned at all.

Suppose **y** is a (not necessarily real) random variable that takes on only a finite number of values. It will be understood that $E(\mathbf{x} \mid y)$ is the expected value of **x** given that $y(s) = y$, provided y is such that this event has positive probability. Furthermore, it will be understood that $E(\mathbf{x} \mid \mathbf{y})$ is a bounded real random variable that for each s takes the value $E(\mathbf{x} \mid y(s))$. The definition leaves $E(\mathbf{x} \mid \mathbf{y})$ undefined on the null set of those points s where $y(s)$ is a value that **y** takes on with probability zero. It is immaterial how this blemish is removed; in particular $E(\mathbf{x} \mid \mathbf{y})$ may as well be set equal to 0, where it has not already been defined.

Still more exercises

8. $E(E(\mathbf{h} \mid \mathbf{y})) = E(\mathbf{h})$.

9. If **f** is a real-valued function defined on the values of **y**; then $f(\mathbf{y})$ is a bounded real variable, and

(7) $$E(f(\mathbf{y})\mathbf{x}) = E(f(\mathbf{y})E(\mathbf{x} \mid \mathbf{y})).$$

10. If $h(\mathbf{x})$ is such that, for all **f**,

(8) $$E(f(\mathbf{y})\mathbf{x}) = E(f(\mathbf{y})h(\mathbf{y})),$$

then $h(y(s)) = E(\mathbf{x} \mid y(s))$, except possibly on a set of s's of probability zero.

Exercise 9 and its corollary, 8, present the most frequently used properties of conditional expectation. Exercise 10 shows that the property presented in 9 characterizes conditional expectation. Through this characterization Kolmogoroff [K7] extends the ideas of conditional expectation and also of conditional probability (for countably additive measures) to random variables **y** not necessarily confined to a finite or even denumerable set of values; though the definition in terms of ordinary conditional probability then breaks down completely, the probability that $y(s) = y$ often being 0 for every y.

APPENDIX 2

Convex Functions

This appendix gives a brief account of convex functions in the same spirit as the preceding one gives an account of expected value. Reasonable facsimiles of the proofs omitted here are scattered through [H4], where they may be found by anyone not content to skip them.

An **interval** is a set I of real numbers; such that, if x, z ε I and $x \leq y \leq z$, then y ε I. It is not difficult to see that intervals can be classified according to Table 1, where it is to be understood that $x < z$.

TABLE 1. THE VARIOUS TYPES OF INTERVALS

Symbolic designation	The set of real y's such that	Verbal description
$(-\infty, +\infty)$	$y = y$	The infinite interval (the set of all real numbers)
$(x, +\infty)$ $(-\infty, x)$	$x < y$ $x > y$	Open } half-infinite intervals
$[x, +\infty)$ $(-\infty, x]$	$x \leq y$ $x \geq y$	Closed }
(x, z)	$x < y < z$	Open } bounded intervals
$[x, z)$ $(x, z]$	$x \leq y < z$ $x < y \leq z$	Half-open }
$[x, z]$	$x \leq y \leq z$	Closed }
$[x, x]$	$x = y$	One-point intervals
	$y < y$	The vacuous interval (the vacuous set)

A real-valued function **t** defined for z in an interval I is **convex**, if and only if the graph of the function never rises above any chord of itself. Analytically, if ρ and σ are positive, $\rho + \sigma = 1$, and x, y ε I; then

(1) $$t(\rho x + \sigma y) \leq \rho t(x) + \sigma t(y).$$

If equality holds in (1) for some ρ; then, as is easily verified, it holds for every ρ, and t is **linear**, i.e., of the form $\alpha z + \beta$, in the closed interval $[x, y]$. An interval in which t is linear will here be called an **interval of linearity**. If and only if there are no intervals of linearity other than the one-point and vacuous intervals, t is **strictly convex**.

Exercises

1. Verify, at least graphically, that the following functions are convex in the indicated intervals; discuss their intervals of linearity; and say which are strictly convex.

$I = (-\infty, +\infty)$:

(a) $e^{\rho x}$ for every ρ,

(b) $x^2 + \rho x + \sigma$ for every ρ and σ,

(c) $|x|$,

(d) $|x|^{\rho}$ for $\rho \geq 1$,

(e) x.

$I = (0, \infty)$:

(f) $-\log x$,

(g) x^{ρ} for $-\infty < \rho < 0$.

$I = (-1, +1)$:

(h) $(1 - x^2)^{-\frac{1}{2}}$,

(i) $1 - \cos (\pi x/2)$.

2. In an interval where t is convex, if $d^2t(x)/dx^2$ exists at x, then $d^2t(x)/dx^2 \geq 0$; and if, for every x in an interval I, $d^2t(x)/dx^2$ exists and is non-negative, then t is convex in I.

3. Re-explore Exercise 1 in the light of 2.

4. Let **T** be a non-vacuous set of functions, t, t', $\cdots$, convex in I, and let

(2) $$t^*(s) = \sup_{t} t(s).$$

In (2), as always in mathematics, the **sup**, or **supremum**, of a set of numbers is the least number, possibly ∞, that is not less than any element of the set. If $t^*(s) < \infty$ for every $s \in I$, then **t*** is convex in I. Explore the proposition just stated, first graphically, especially for a finite set of linear t's, and then analytically. What if the elements of **T** are all strictly convex?

5. In an open interval where t is convex, it is also continuous. What are the facts for closed and half-closed intervals?

6. If t is convex in I, $x_k \, \varepsilon \, I$, $\rho_k > 0$, and $\Sigma\rho_k = 1$, where $k = 1, \cdots,$ r; then

(3)
$$\sum_k \rho_k t(x_k) \geq t\left(\sum_k \rho_k x_k\right).$$

Equality obtains, if and only if all the x_k's are in a single interval of linearity of t.

  (a) Interpret the propositions above in terms of probability.
  (b) Prove them by arithmetic induction on r.
  (c) What if t is strictly convex?

Exercise 6 suggests, and indeed proves a special case of, the following well-known and most useful theorem, which cannot be proved here in full generality.

THEOREM 1   If t is convex and bounded in the interval I, and $x(s) \, \varepsilon \, I$ for all $s \, \varepsilon \, S$, then

(4)
$$E(t(\mathbf{x})) \geq t(E(\mathbf{x})).$$

Equality obtains, if and only if the values of x are with probability one contained in a single interval of linearity of t. Here and throughout this appendix, such conditions for equality are to be understood to apply only in the event that either P is countably additive or the random variable is with probability one confined to a finite set of values; the general situation for finitely additive measures is a little more complicated.

More exercises

7. The **variance** of x, often written $V(\mathbf{x})$, is defined thus:

(5)
$$V(\mathbf{x}) = E([\mathbf{x} - E(\mathbf{x})]^2).$$

Show that

(6)
$$V(\mathbf{x}) = E(\mathbf{x}^2) - E^2(\mathbf{x}) \geq 0,$$

with equality if and only if $P(x(s) = E(\mathbf{x})) = 1$.
  8. Show that, if x is never smaller than some positive number,

(7)
$$\log E^{-1}(\mathbf{x}^{-1}) \leq E(\log \mathbf{x}) \leq \log E(\mathbf{x}).$$

When can either equality obtain? Write the analogue of (7) suggested by (3), and show thereby that (7) is a generalization of the familiar fact that the arithmetic mean (of positive numbers) is at least as great as the geometric mean and the geometric mean is at least as great as the harmonic mean.

One of the most famous of all inequalities is the **Schwartz inequality,** which can, though not quite obviously, be derived from Theorem 1, and which can be stated in terms of expected values thus:

$$(8) \qquad E^2(\mathbf{xy}) \le E(\mathbf{x}^2)E(\mathbf{y}^2),$$

with equality obtaining if and only if for some numbers ρ and σ not both zero

$$(9) \qquad P(\rho x(s) = \sigma y(s)) = 1.$$

Note that (9) expresses (perhaps too compactly) that, except on some set of probability zero, either $\mathbf{x}$ or $\mathbf{y}$ vanishes identically or else each is a fixed multiple of the other.

Statistically speaking, the Schwartz inequality expresses, in effect, the familiar fact that any correlation coefficient must lie between $+1$ and -1, one of the extremes occurring if and only if at least one of the two random variables involved is a linear function of the other.

The concept of convex functions and its implications can easily be extended to real-valued functions defined on vectors in an n-dimensional vector space, the role of intervals there being replaced by convex subsets of the vector space; but an understanding of this extension, though desirable, is not absolutely essential in reading this book.

One good introduction to convex subsets of vector spaces is Sections 16.1–2 of [V4], and another especially adapted to statistical applications is incorporated in [B18]. The standard treatise on the topic is that of Bonnessen and Fenchel [B20].

APPENDIX 3

Bibliographic Material

The bibliography of about 170 items that terminates this appendix lists not only all works referred to in this book but also some others, for it is intended to serve not only as a mechanical aid to reference but also as a briefly and informally annotated list of suggested readings in the foundations of statistics. In addition to the notes incorporated into the bibliography, information about many of the works listed there is given in other parts of the book, where it can be found by referring to the author's name in the author index. References that have come to my attention since the first edition are in Appendix 4: Bibliographic Supplement. They are cited by the convention according to which the first of them is called (Aczél 1966).

Todhunter has abundant references scattered in chronological order through [T3], emphasizing the mathematical aspects of probability up through the period of Laplace. Keynes, in [K4], gives a formal bibliography which purposely does not overlap Todhunter's material very extensively, the emphasis being on more philosophical aspects of probability and on the period between Laplace and Keynes. Carnap in [C1] also gives a formal bibliography, which emphasizes publications since Keynes. Carnap promises an even fuller bibliography in the projected second volume of his work, and he recommends the bibliography of Georg Henrik von Wright in [V5].

Bibliographies of statistics proper are of some, though diluted, relevance. Of these, the most useful is that of M. G. Kendall in Vol. II of [K2]. Carnap at the beginning of his bibliography gives reference to some other statistical bibliographies. The enormous work of O. K. Buros in statistical bibliography, [B23], [B24], and [B25], should also be mentioned. His volumes bring together pointed excerpts from reviews of statistical books. Buros also directed a bibliographic department, entitled "Statistical Methodology," in the *Journal of the American Statistical Association* from September 1945 to September 1948, listing current articles, books, theses, and chapters dealing with statistics. In

Volume 20 (1949) of the *Annals of Mathematical Statistics*, an important journal of statistical theory, there are two cumulative indexes of Volumes 1–20, one arranged by author, the other by subject.

Bibliography

Aitken, A. C., and H. Silverstone
  [A1] "On the estimation of statistical parameters," *Proceedings of the Royal Society of Edinburgh*, 61 (1941–43), 186–194 (issued separately April 2, 1942).
Allais, Maurice
  [A2] "Le comportement de l'homme rationnel devant le risque: Critique des postulats et axioms de l'école Americaine," *Econometrica*, 21 (1953), 503–546.
Allen, S. G., Jr.
  [A3] "A class of minimax tests for one-sided composite hypotheses," *Annals of Mathematical Statistics*, 24 (1953), 295–298.
Anscombe, F. J.
  [A4] "Mr. Kneale on probability and induction," *Mind*, 60 (1951), 299–309.
    Says much of general interest on the foundations of statistics, in the course of comments on [K5].
Arrow, Kenneth J.
  [A5] *Social Choice and Individual Values*, Cowles Commission Monograph No. 12, New York, John Wiley & Sons, 1951. (Second edition, 1963.)
  [A6] "Alternative approaches to the theory of choice in risk-taking situations," *Econometrica*, 19 (1951), 404–437.
Arrow, K. J., David Blackwell, and M. A. Girshick
  [A7] "Bayes and minimax solutions of sequential decision problems," *Econometrica*, 17 (1949), 213–243.
Bahadur, Raghu Raj
  [B1] "A property of the *t*- statistic," *Sankhyā*, 12 (1952), 79–88.
  [B2] "Sufficiency and statistical decision functions," *Annals of Mathematical Statistics*, 25 (1954), (to appear).
Bahadur, Raghu Raj, and Herbert Robbins
  [B3] "The problem of the greater mean," *Annals of Mathematical Statistics*, 21 (1950), 469–487.
Banach, S.
  [B4] *Théorie des opérations linéaires*, Warsaw, Fundusz Kultury Narodowej, 1932.
Banach, S., and A. Tarski
  [B5] "Sur la décomposition des ensembles de points en parties respectivement congruentes," *Fundamenta Mathematicae*, 6 (1924), 244–277.
Bartlett, M. S.
  [B6] "Completely simultaneous fiducial distributions," *Annals of Mathematical Statistics*, 10 (1939), 128–138.
Baumol, William J.
  [B7] "The Neumann-Morgenstern utility index—an ordinalist view," *Journal of Political Economy*, 59 (1951), 61–66.
Bayes, Thomas
  [B8] *Facsimiles of Two Papers by Bayes: i. An Essay Toward Solving a Problem in the Doctrine of Chances, With Richard Price's Foreword and Discussion; Phil. Trans. Royal Soc., pp. 370–418, 1763. With a Commentary by Edward C. Molina.*

ii. A Letter on Asymptotic Series from Bayes to John Canton; pp. 269–271 of the Same Volume. With a Commentary by W. Edwards Deming, ed. W. Edwards Deming, Washington, D. C., The Graduate School, The Department of Agriculture, 1940; republished as (Bayes 1958).

The first of these two papers, in which a special case of what is now called Bayes' rule is introduced, figures prominently in controversies about the foundations of probability, for this paper first put several of the major issues in the limelight.

Bell, E. T.

[B9] *Men of Mathematics*, New York, Simon and Schuster, 1937.

Bernoulli, Daniel

[B10] "Specimen theoriae novae de mensura sortis," *Commentarii academiae scientiarum imperialis Petropolitanae* (for 1730 and 1731), 5 (1738), 175–192.

[B11] *Die Grundlage der modernen Wertlehre. Versuch einer neuen Theorie der Wertbestimmung von Glücksfällen* (German translation of [B10] by Alfred Pringsheim, with introduction by Ludwig Frick), Leipzig, Duncker V. Humblot, 1896.

[B11a] "Exposition of a new theory on the measurement of risk" (English translation of [B10] by Louise Sommer), *Econometrica*, 22 (1954), 23–26.

Bernoulli, Jacob (=James)

[B12] *Ars conjectandi*, Basel, 1713.

[B13] *Wahrscheinlichkeitsrechnung* (German translation of [B12] by R. Haussner), Ostwald's Klassiker der Exakten Wissenschaften, Nos. 107 and 108, Leipzig, W. Engelmann, 1899.

Contains, besides much of primary mathematical interest, what I understand to be the first extended discussion of the application of probability to the problem of inference. Unfortunately, the German translation is said to be incomplete.

Birkhoff, G., and S. MacLane

[B14] *A Survey of Modern Algebra*, New York, The Macmillan Co., 1941.

Bizley, M. T. L.

[B15] "Some notes on probability," *Journal of the Institute of Actuaries Students' Society*, 10 (1951), 161–203.

Blackwell, David

[B16] "Comparison of experiments," pp. 93–102 of *Proceedings of the Second [1950] Berkeley Symposium on Mathematical Statistics and Probability*, ed. Jerzy Neyman, Berkeley, University of California Press, 1951.

[B17] "On the translation parameter problem for discrete variables," *Annals of Mathematical Statistics*, 22 (1951), 393–399.

Blackwell, David, and M. A. Girshick

[B18] *The Theory of Games and Statistical Decisions*, New York, John Wiley & Sons, 1954.

Bohnenblust, H. F., S. Karlin, and L. S. Shapley

[B19] "Solutions of discrete two-person games," pp. 51–72 of [K13].

Bonnessen, T., and W. Fenchel

[B20] *Theorie der konvexen Körper*, Ergebnisse der Mathematik und ihrer Grenzgebiete, Vol. III, Part I, Berlin, J. Springer, 1934; reprinted, New York, Chelsea Publishing Co., 1948.

Borel, Emile

[B21] "The theory of play and integral equations with skew symmetric kernels; On games that involve chance and the skill of the players; On systems of linear forms of skew symmetric determinant and the general theory of play (translated by Leonard J. Savage)," *Econometrica*, 21 (1953), 97–124.

Bowker, A. H.

[B22] "Tolerance limits for normal distributions," Chapter 2, pp. 95–110 of *Techniques of Statistical Analysis*, by the Statistical Research Group, Columbia University, New York, McGraw-Hill Book Co., 1947.

Buros, O. K. (ed.)

[B23] *Research and Statistical Methodology, Books and Reviews (1933–38)*, New Brunswick, New Jersey, Rutgers University Press, 1938.

[B24] *The Second Yearbook in Research and Methodology, Books and Reviews*, Highland Park, New Jersey, The Griffin Press, 1941.

[B25] *Statistical Methodology Reviews 1941–1950*, New York, John Wiley & Sons, 1951.

Carnap, Rudolf

[C1] *Logical Foundations of Probability*, Chicago, University of Chicago Press, 1950.

This is the first of a projected pair of volumes designed to demonstrate meticulously the author's contention that a certain almost necessary view of probability is essential to science—not denying the meaningfulness of the objectivistic concept. Reviewed by me in [S4].

[C2] *The Nature and Application of Inductive Logic*, Chicago, University of Chicago Press, 1951.

A reprint of selected sections of [C1].

[C3] *The Continuum of Inductive Methods*, Chicago, University of Chicago Press, 1952.

Essentially a chapter of the second volume of the projected pair referred to under [C1].

Centre National de Recherche Scientifique

[C4] *Fondements et applications de la théorie du risque en économetrie*, Paris, Centre National de la Recherche Scientifique, 1954.

Report of an international econometric colloquium on risk, in which there was much discussion of utility, held in Paris, May 12–17, 1952.

Chand, Uttam

[C5] "Distributions related to comparison of two means and two regression coefficients," *Annals of Mathematical Statistics*, 21 (1950), 507–522.

Chapman, Douglas G., and Herbert Robbins

[C6] "Minimum variance estimation without regularity assumptions," *Annals of Mathematical Statistics*, 22 (1951), 581–586.

Chernoff, Herman

[C7] "Remarks on a Rational Selection of a Decision Function," Cowles Commission Discussion Paper, Statistics, No. 326 (January 10, 1949). Unpublished.

Churchman, C. West

[C8] *Theory of Experimental Inference*, New York, The Macmillan Co., 1948.

A discussion of current statistics from the viewpoint of technical philosophy.

Cramér, Harald

[C9] *Mathematical Methods of Statistics*, Princeton, Princeton University Press, 1946.

By far the most comprehensive rigorous book on mathematical methods of statistics.

Darmois, G.

[D1] "Sur les limites de la dispersion de certains estimations," *Revue de l'Institut International de Statistique*, 13 (1945), 9–15.

de Finetti, Bruno

[D2] "La prévision: ses lois logiques, ses sources subjectives," *Annales de l'Institut Henri Poincaré*, 7 (1937), 1–68. (Translated in (Kyburg and Smokler 1964).)

The first two and the final chapters of this paper give a statement, which I have found particularly stimulating, of the author's view (a personalistic one) of the foundations of probability. The three intervening chapters are mathematically rather technical. A bibliography of the author's work on the foundations of probability up to sometime in 1937 is included.

[D3] "Le vrai et le probable," *Dialectica*, 3 (1949), 78–93.

[D4] "Sull' impostazione assiomatica del calcolo delle probabilità," *Annali Triestini*, Series 2, 19 (1949), 29–81.

[D5] "La 'logica del plausibile' secondo la concezione di Polya," *Atti della XLII Riunione della Società Italiana per il Progresso delle Scienze (Novembre 1949)*, Rome, Società Italiana per il Progresso delle Scienze, 1951 (10 pages).

[D6] "Recent suggestions for the reconciliations of theories of probability," pp. 217–226 of *Proceedings of the Second [1950] Berkeley Symposium on Mathematical Statistics and Probability*, ed. Jerzy Neyman, Berkeley, University of California Press, 1951.

Especially through the suggestions it makes about multipersonal problems, an early manuscript of [D6] has had much influence on this book.

[D7] "Sulla preferibilità," *Giornale degli Economisti e Annali di Economia*, 11 (1952), 685–709.

[D7a] "La notion de 'distribution d'opinion' comme base d'un essai d'interprétation de la statistique," *Publications de l'Institut de Statistique de l'Université de Paris*, 1 (1952), 1–19.

Delorme, S. (ed.)

[D8] *Colloque de calcul des probabilités (Actualités scientifiques et industrielles 1146)*, Paris, Hermann et Cie., 1951.

A collection of papers by several authors, mostly on the philosophy of probability, read in a colloquium held under the 13th International Congress of the Philosophy of Science, in Paris, 1949. There is an overall review by M. Fréchet, president of the colloquium. All papers are in French, except one in English.

Dialectica

[D9] *Dialectica*, Vol. 3 (1949), Nos. 9–10.

This issue of *Dialectica*, a quarterly "international review of the philosophy of knowledge," is devoted to probability, and mainly to its foundations. It is composed of papers by several authors, each in English, French, or German.

Doob, J.

[D10] "Statistical estimation," *Transactions of the American Mathematical Society*, 39 (1936), 410–421.

Duncan, D. B.

[D11] "A significance test for differences between ranked treatments in an analysis of variance," *Virginia Journal of Science*, 2 (1951), 171–189.

[D12] "On the properties of the multiple comparisons test," *Virginia Journal of Science*, 3 (1952), 49–57.

Dvoretzky, A., A. Wald, and J. Wolfowitz
[D13] "Elimination of randomization in certain statistical decision procedures and zero-sum two-person games," *Annals of Mathematical Statistics*, 22 (1951), 1–21.

Elfving, G.
[E1] "Sufficiency and completeness in decision function theory," *Annales Academiae Scientarum Fennicae, Series A.I.*, 135 (1952), 9 pages.

Feller, William
[F1] *An Introduction to Probability Theory and Its Applications*, Vol. 1, New York, John Wiley & Sons, 1950. (Third edition, 1968; Vol. 2, 1966 and 1971.)

    A sophisticated introduction to the mathematics of probability. Certain relatively advanced mathematical techniques are avoided by a severe restriction of the material treated, which is none the less extensive and varied. A second volume, removing the restriction, is promised.

Féraud, D.
[F2] "Induction amplifiante et inférence statistique," *Dialectica*, 3 (1949), 127–152.

Fisher, R. A.
[F3] *Statistical Methods for Research Workers*, Edinburgh and London, Oliver and Boyd, 1925; and later editions.

    The author is the outstanding member of the British-American School, and this book of his has had far more influence on the development of statistics in the current century than any other publication.

[F4] *The Design of Experiments*, Edinburgh and London, Oliver and Boyd, 1935; and later editions.

    Second only to [F3] in the extent of its influence.

[F5] "A note on fiducial inference," *Annals of Mathematical Statistics*, 10 (1939), 383–388.

[F6] *Contributions to Mathematical Statistics*, New York, John Wiley & Sons, 1950.

    A collection of Fisher's papers selected and annotated by himself. With a biography of Fisher by P. C. Mahalanobis. Reviewed in [N4].

Fisher, Walter D.
[F7] "On a pooling problem from the statistical decision viewpoint," *Econometrica* 21 (1953), 567–585.

Fréchet, Maurice
[F8] "Sur l'extension de certains évaluations statistiques au cas de petits échantillons," *Revue de l'Institut International de Statistique*, 11 (1943), 183–205.

[F9] "Emile Borel, initiator of the theory of psychological games and its application," *Econometrica* 21 (1953), 95–96.

Fréchet, Maurice, and J. von Neumann
[F10] "Commentary on the Borel notes," *Econometrica*, 21 (1953), 118–127.

Friedman, Milton
[F11] "Choice, chance, and personal distribution of income," *Journal of Political Economy*, 61 (1953), 277–290.

Friedman, Milton, and L. J. Savage
[F12] "The utility analysis of choices involving risk," *Journal of Political Economy*, 56 (1948), 279–304; reprinted, with a correction, in [S19].

[F13] "The expected-utility hypothesis and the measurability of utility," *Journal of Political Economy*, 60 (1952), 463–474.

Girshick, M. A., and L. J. Savage

[G1] "Bayes and minimax estimates for quadratic loss functions," pp. 53–74 in *Proceedings of the Second [1950] Berkeley Symposium on Mathematical Statistics and Probability,* ed. Jerzy Neyman, Berkeley, University of California Press, 1951.

Good, I. J.

[G2] *Probability and the Weighing of Evidence,* London, Charles Griffin and Co., and New York, Hafner Publishing Co., 1950.

Presents, with many interesting examples and arguments, Good's view, a personalistic one, on the foundations of probability. Reviewed by me in [S3].

Graves, Lawrence M.

[G3] *The Theory of Functions of Real Variables,* New York, McGraw-Hill Book Co., 1946.

Halmos, Paul R.

[H1] "The foundations of probability," *American Mathematical Monthly,* 51 (1944), 493–510.

Short, easy exposition of the Kolmogoroff probability concept.

[H2] *Measure Theory,* New York, Van Nostrand Co., 1950.

Halmos, Paul R., and L. J. Savage

[H3] "Application of the Radon-Nikodym theorem to the theory of sufficient statistics," *Annals of Mathematical Statistics,* 20 (1949), 225–241.

Hardy, G. H., J. E. Littlewood, and G. Polya

[H4] *Inequalities,* Cambridge, Cambridge University Press, 1934.

Hildreth, Clifford

[H4a] "Alternative conditions for social orderings," *Econometrica,* 21 (1953), 81–94.

Hodges, J. L., Jr., and E. L. Lehmann

[H5] "Some problems in minimax point estimation," *Annals of Mathematical Statistics,* 21 (1950), 182–197.

[H6] "Some applications of the Cramér-Rao inequality," pp. 13–22 in *Proceedings of the Second [1950] Berkeley Symposium on Mathematical Statistics and Probability,* ed. Jerzy Neyman, Berkeley, University of California Press, 1951.

Hume, David

[H7] *An Enquiry Concerning Human Understanding,* London, 1748; and later editions.

An early and famous presentation of the philosophical problem of inductive inference, around which almost all later discussion of the problem pivots.

Jeffreys, Harold

[J1] *Theory of Probability* (Second edition), Oxford, Clarendon Press, 1948.

An ingenious and vigorous defense of a necessary view, similar to, but more sophisticated than, Laplace's. (Second edition, 1961.)

Kakutani, S.

[K1] "A generalization of Brouwer's fixed-point theorem," *Duke Mathematical Journal,* 8 (1941), 457–459.

Kendall, Maurice G.

[K2] *The Advanced Theory of Statistics,* Vol. I, 1947, Vol. II, 1948, London, Charles Griffin and Co.

Virtually an encyclopedia of statistical theory, history, and bibliography (as of 1943).

[K3] "On the reconciliation of the theories of probability," *Biometrika,* 36 (1949), 101–116.

Keynes, John Maynard

[K4] *A Treatise on Probability*, London and New York, Macmillan & Co., 1921; second edition, 1929.

A long, but often entertaining, account of Keynes' view, a necessary one. The historical passages and bibliography are of special interest.

Kneale, William

[K5] *Probability and Induction*, Oxford, Clarendon Press, 1949.

An ingenious presentation, philosophical in spirit and background, of a new view with objectivistic and necessary aspects. Reviewed in [A4].

[K6] "Probability and induction," *Mind*, 60 (1951), 310–317.

A reply to [A4].

Kolmogoroff, A. N.

[K7] *Grundbegriffe der Wahrscheinlichkeitsrechnung*, Berlin, J. Springer, 1933.

[K8] *Foundations of the Theory of Probability* (English translation of [K7] edited by Nathan Morrison), New York, Chelsea Publishing Co., 1950.

Statement and very compact development of the Kolmogoroff concept of mathematical probability. Excellent reading, though mathematically rather advanced.

Koopman, B. O.

[K9] "The axioms and algebra of intuitive probability," *Annals of Mathematics*, Ser. 2, 41 (1940), 269–292.

[K10] "The bases of probability," *Bulletin of the American Mathematical Society*, 46 (1940), 763–774.

[K11] "Intuitive probabilities and sequences," *Annals of Mathematics*, Ser. 2, 42 (1941), 169–187.

These three papers present the personalistic view that Koopman holds along with an objectivistic one.

[K12] Reviews of eleven papers, *Mathematical Reviews*, 7 (1946), 186–193; and 8 (1947), 245–247.

A connected sequence of reviews of papers, by several authors, that were published as a symposium in *Philosophy and Phenomenological Research*, Vols. 5 and 6 (1945–46).

Kuhn, H. W., and A. W. Tucker (eds.)

[K13] *Contributions to the Theory of Games*, Vol. I (Annals of Mathematics Study No. 24), Princeton, Princeton University Press, 1950.

[K14] *Contributions to the Theory of Games*, Vol. II (Annals of Mathematics Study No. 28), Princeton, Princeton University Press, 1953.

Loosely coordinated collection of articles on the theory of games by several authors, with bibliographies.

Kullback, S., and R. A. Leibler

[K15] "On information and sufficiency," *Annals of Mathematical Statistics*, 22 (1951), 79–86.

Laplace, Pierre Simon de

[L1] *Essai philosophique sur les probabilités* (First edition), Paris, 1814; and several subsequent editions, of which the Fifth, 1825, was the last to be revised by Laplace.

[L2] *A Philosophical Essay on Probabilities* (English translation of [L1], Second edition), New York, John Wiley & Sons, 1917; reprinted, New York, Dover Publications, 1952.

This essay, published both separately and as the preface of the author's great technical treatise, *Théorie analytique des probabilités*, is, save for a few dull spots, one of the most delightful and stimulating classics of probability. Laplace's view is a naive necessary one. His concept of the domain of applicability of the theory of probability is inspiring, if not always realistic.

LeCam, Lucien

[L3] "On some asymptotic properties of maximum likelihood estimates and related Bayes' estimates," pp. 277–329 of *University of California Publications in Statistics*, Vol. 1, No. 11, Berkeley and Los Angeles, University of California Press, 1953.

Lehmann, E. L.

[L4] "Some principles of the theory of testing hypotheses," *Annals of Mathematical Statistics*, 21 (1950), 1–26.

[L5] "A general concept of unbiasedness," *Annals of Mathematical Statistics*, 22 (1951), 587–597.

Lehmann, E. L., and Henry Scheffé

[L6] "Completeness, similar regions, and unbiased estimation, Part I," *Sankhyā*, 10 (1950), 305–340.

Lewis, Clarence Irving, and Cooper Harold Langford

[L7] *Symbolic Logic*, New York and London, The Century Company, 1932.

Lindley, D. V.

[L8] "Statistical inference," *Journal of the Royal Statistical Society, Series B*, 15 (1953), 30–76.

Excellent reading in connection with Chapters 14–17 of this book. Unfortunately, I did not see the paper in time to reflect its contents in those chapters.

Markowitz, Harry

[M1] "The utility of wealth," *Journal of Political Economy*, 60 (1952), 151–158.

Marshall, Alfred

[M2] *Principles of Economics* (First edition), London, Macmillan & Co., 1890; and many subsequent editions of which the Eighth (1927) is standard.

McKinsey, J. C. C.

[M3] *Introduction to the Theory of Games*, New York, McGraw-Hill Book Co., 1952.

Mosteller, Frederick, and Philip Nogee

[M4] "An experimental measurement of utility," *Journal of Political Economy*, 59 (1951), 371–404.

Mourier, Edith

[M5] "Tests de choix entre divers lois de probabilité," *Trabajos de estadística*, 2 (1951), 234–259.

Munroe, M. E.

[M6] *Theory of Probability*, New York, McGraw-Hill Book Co., 1951.

An elementary modern text on the mathematics of probability. Easier and more general than [F1], but not so penetrating.

Nagel, Ernest

[N1] "Principles of the theory of probability," *International Encyclopedia of Unified Science*, Vol. I, No. 6, Chicago, University of Chicago Press, 1939.

Neyman, Jerzy

[N2] "Outline of a theory of statistical estimation based on the classical theory of probability," *Philosophical Transactions of the Royal Society*, Ser. A, 236 (1937), 333–380.

[N3] "L'estimation statistique, traitée comme un problème classique de probabilité," pp. 25–57 of *Actualités scientifiques et industrielles No. 739*, Paris, Hermann et Cie., 1938.

Representative works of the Neyman-Pearson School, a major branch of what in this book is called the British-American School.

[N4] "Fisher's collected papers," *Scientific Monthly*, 42 (1951), 406–408.

Nunke, R. J., and L. J. Savage

[N5] "On the set of values of a nonatomic, finitely additive, finite measure," *Proceedings of the American Mathematical Society*, 3 (1952), 217–218.

Owen, A. R. G.

[O1] "Ancillary statistics and fiducial distributions," *Sankhyā*, 9 (1948–49), 1–18.

Pareto, Vilfredo

[P1] *Manuel d'économie politique* (Second edition), Paris, Giard, 1927. (First edition 1909. Based on a still earlier book in Italian.)

Pascal, Blaise

[P2] *Pensées*, with introduction and notes by Louis Lafuma, Paris, Delmas, 1947.

Paulson, Edward

[P3] "A multiple decision procedure for certain problems in the analysis of variance," *Annals of Mathematical Statistics*, 20 (1949), 95–98.

[P4] "On comparison of several experimental categories with a control," *Annals of Mathematical Statistics*, 23 (1952), 239–246.

Pitman, E. J. G.

[P5] "The 'closest' estimate of statistical parameters," *Proceedings of the Cambridge Philosophical Society*, 33 (1937), 212–222.

Polya, G.

[P6] "Preliminary remarks on a logic of plausible inference," *Dialectica*, 3 (1949), 28–35.

Ramsey, Frank P.

[R1] "Truth and probability" (1926), and "Further considerations" (1928), in *The Foundations of Mathematics and Other Logical Essays*, London, Kegan Paul, and New York, Harcourt, Brace and Co., 1931.

Penetrating development of a personalistic view of probability and utility. Ramsey's concepts of probability and utility are essentially the same as those presented in this book, but his logical development of them is an interesting alternative to the one used here, his definitions of probability and utility being simultaneous and interdependent.

Reichenbach, Hans

[R2] *The Theory of Probability*, Berkeley and Los Angeles, University of California Press, 1949.

The most recent and complete statement of Reichenbach's elaborately worked out objectivistic view.

Richter, Hans

[R2a] "Zur Grundlegung der Wahrscheinlichkeitstheorie, I, II, III, IV," *Mathematische Annalen*, 125 (1952), 129–139; 125 (1953), 223–224, 335–343; and 126 (1953), 362–374.

A personalistic theory of probability with a physical orientation, pertaining not to behavior but to "feeling of expectation."

Rousseas, Stephen W., and Albert G. Hart

[R3] "Experimental verification of a composite indifference map," *Journal of Political Economy*, 59 (1951), 288–318.

Rudy, Norman

[R4] "Some problems in the economics of industrial sampling inspection," an as yet unpublished dissertation submitted to the University of Chicago in 1952.

Samuelson, Paul A.

[S1] "Probability, utility, and the independence axiom," *Econometrica*, 20 (1952), 670–678.

Savage, L. J.

[S2] "The theory of statistical decision," *Journal of the American Statistical Association*, 46 (1951), 55–67.

[S3] Review of I. J. Good's *Probability and the Weighing of Evidence*, in *Journal of the American Statistical Association*, 46 (1951), 383–384.

[S4] Review of Rudolf Carnap's *Logical Foundations of Probability*, in *Econometrica*, 20 (1952), 688–690.

Scheffé, Henry

[S5] "On solutions of the Behrens-Fisher problem, based on the t-distribution," *Annals of Mathematical Statistics*, 14 (1943), 35–44.

[S6] "A note on the Behrens-Fisher problem," *Annals of Mathematical Statistics*, 15 (1944), 430–432.

[S7] "A method of judging all contrasts in the analysis of variance," *Biometrika*, 40 (1953), 1–18.

Searle, S. R.

[S8] "Probability—the difficulties of definition," *Journal of the Institute of Actuaries Students' Society*, 10 (1951), 204–212.

Segal, I. E.

[S9] "Fiducial distributions of several parameters with application to a normal system," *Proceedings of the Cambridge Philosophical Society*, 34 (1938), 41–47.

Shackle, G. L. S.

[S10] *Expectation in Economics*, Cambridge, Cambridge University Press, 1949.

Shannon, Claude E., and Warren Weaver

[S11] *The Mathematical Theory of Communication*, Urbana, University of Illinois Press, 1949.

Shapley, L. S., and R. N. Snow

[S12] "Basic solutions of discrete games," pp. 27–36 of [K13].

Shohat, J. A., and J. D. Tamarkin

[S13] *The Problem of Moments* (Mathematical Surveys, No. 1) American Mathematical Society, New York, 1943; reprinted with small changes in 1950.

Smith, Cedric A. B.

[S14] "Some examples of discrimination," *Annals of Eugenics*, 13 (1947), 272–282.

Sobczyk, A., and P. C. Hammer

[S15] "The ranges of additive set functions," *Duke Mathematical Journal*, 11 (1944), 847–851.

Sprowls, R. Clay

[S16] "Statistical decision by the method of minimum risk: an application," *Journal of the American Statistical Association*, 45 (1950), 238–248.

Statistical Research Group, Columbia University

[S17] *Sequential Analysis of Statistical Data: Applications*, New York, Columbia University Press, 1945.

Stigler, George J.

[S18] "The development of utility theory," *Journal of Political Economy*, Part I, 58 (1950), 307–327; Part II, 58 (1950), 373–396.

Stigler, George J., and Kenneth E. Boulding (eds.)

[S19] *Readings in Price Theory*, Chicago, Richard D. Irwin, 1952.

Thrall, Robert M., Cyde H. Coombs, and Robert L. Davis (eds.)
[T1] *Decision Processes*, John Wiley & Sons, New York, 1954.

Tintner, Gerhard
[T1a] "A contribution to the non-static theory of choice," *Quarterly Journal of Economics*, 56 (1942), 274–306.

Tippett, L. H. C.
[T2] *Statistics*, London, Oxford University Press, 1947.
   A short, easy introduction to the ideas of applied statistics, with emphasis on the social sciences.

Todhunter, I.
[T3] *A History of the Mathematical Theory of Probability from the Time of Pascal to that of Laplace*, Cambridge and London, Macmillan & Co., 1865; reprinted, New York, G. E. Stechert, 1931; New York, Chelsea Publishing Co., 1949.

Tukey, John W.
[T4] "Comparing individual means in the analysis of variance," *Biometrics*, 5 (1949), 99–114.
[T5] "Quick and dirty methods in statistics. Part II. Simple analyses for standard designs," *Proceedings of the Fifth Annual Convention of the American Society for Quality Control*, 1951, pp. 189–197.

van Dantzig, D.
[V1] "Sur l'analyse logique des relations entre le calcul des probabilités et ses applications," pp. 49–66 of [D8].

von Mises, Richard
[V2] *Probability, Statistics and Truth*, London, William Hodge and Co., 1939.
   English translation of the original German. Interesting for its presumably authoritative statement of von Mises' unusual view on the *mathematical* foundations of probability. His view on the intellectual foundations is objectivistic. (Second edition, 1957.)

von Neumann, John
[V3] "Zur Theorie der Gesellschaftsspiele," *Mathematische Annalen*, 100 (1928), 295–320.

von Neumann, John, and Oskar Morgenstern
[V4] *Theory of Games and Economic Behavior* (Second edition), Princeton, Princeton University Press, 1947.
   Contains, as a digression, the important new treatment of utility from which the treatment of utility in this book derives. The second edition contains more than the first on this subject, especially a technical appendix. Also, the idea of regarding multiple choices as single overall choices is discussed in great detail. Finally, the chapters on "zero-sum two-person" games are mathematically intimately connected with the statistical minimax theory.

von Wright, Georg Henrik
[V5] *The Logical Problem of Induction* (Acta Philosophica Fennica, Fasc. 3), Helsinki, 1941. (Second edition, 1957.)

Wald, Abraham
[W1] *On the Principles of Statistical Inference*, Notre Dame, Indiana, 1942.
   A compact exposition of the dominant formal ideas of the British-American School.
[W2] *Sequential Analysis*, New York, John Wiley & Sons, 1947.
[W3] *Statistical Decision Functions*, New York, John Wiley & Sons, 1950.
   Wald's great contributions to statistical theory are well represented by this book, but it is not for beginners.
[W4] "Note on the consistency of the maximum likelihood estimate," *Annals of Mathematical Statistics*, 20 (1949), 595–601.

[W5] "Asymptotic minimax solutions of sequential point estimation problems," pp. 1–12 of *Proceedings of the Second* [1950] *Berkeley Symposium on Mathematical Statistics and Probability,* ed. Jerzy Neyman, Berkeley, University of California Press, 1951.

Wallis, W. Allen

[W6] "Standard sampling-inspection procedures," *Proceedings of the International Statistical Conferences,* 3 (1947), 331–350.

[W7] "Tolerance intervals for linear regression," pp. 43–52 of *Proceedings of the Second* [1950] *Berkeley Symposium on Mathematical Statistics and Probability,* ed. Jerzy Neyman, Berkeley, University of California Press, 1951.

Wallis, W. Allen, and Milton Friedman

[W8] "The empirical derivation of indifference functions," pp. 175–189 in *Studies in Mathematical Economics and Econometrics,* ed. O. Lange *et al.,* Chicago, University of Chicago Press, 1942.

Walsh, John E.

[W9] "On the power function of the 'best' *t*-test solution of the Behrens-Fisher problem," *Annals of Mathematical Statistics,* 20 (1949), 616–618.

Wiener, Norbert

[W10] *Cybernetics,* New York, John Wiley & Sons, 1948.

White, Paul D., Robert L. King, and James Jenks, Jr.

[W11] "The relation of heart size to the time intervals of the heart beat, with particular reference to the elephant and the whale," *The New England Journal of Medicine,* 248 (1953), 69–70.

Williams, R. M.

[W12] "The use of fiducial distributions with special reference to the Behrens-Fisher problem," Part II of an unpublished dissertation submitted to the University of Cambridge and filed in 1949 in the University of Cambridge Library as *Ph.D. 1671.*

Wisdom, John Oulton

[W13] *Foundations of Inference in Natural Science,* London, Methuen, 1952.

A new book in the philosophical tradition, but motivated by the idea of examining how inductive inference is actually used in science. Though almost entirely verbalistic in outlook, behavioralistic ideas including the sure-thing principle play an important role in the final and culminating chapter.

Wold, H.

[W14] "Ordinal preferences or cardinal utility," *Econometrica,* 20 (1952), 661–664.

Wolfowitz, J.

[W15] "The efficiency of sequential estimates and Wald's equation for sequential processes," *Annals of Mathematical Statistics,* 18 (1947), 215–230.

[W16] "On Wald's proof of the consistency of the maximum likelihood estimate," *Annals of Mathematical Statistics,* 20 (1949), 601–602.

[W17] "On ϵ-complete classes of decision functions," *Annals of Mathematical Statistics,* 22 (1951), 461–464.

Yates, F.

[Y1] "An apparent inconsistency arising from tests of significance based on fiducial distributions of unknown parameters," *Proceedings of the Cambridge Philosophical Society,* 35 (1939), 579–591.

[Y2] "Principles governing the amount of experimentation in developmental work," *Nature,* 170 (1952), 138–140.

APPENDIX 4

Bibliographic Supplement

Since the publication of the first edition of this book, the literature of the foundations of statistics, like that of all science, has been growing with awesome rapidity. The relatively short list of about 180 items below includes a few older references overlooked in the first edition, but most are more recent. They are chosen in the spirit of those in the first edition, Appendix 3: Bibliographic Material. Some support new assertions made in this edition, some bring up to date reading lists and key references for certain topics, and some are selected for their quality and originality.

Pages in this edition that cite a given entry in the list below are shown by italic numbers following the entry—a neglected invention going back at least to (Coolidge 1940). Where there is neither such a page number nor a comment, the entry is supposed to speak for itself.

Some of the new entries are special bibliographies (Edwards 1969; George 1968; Greenwood et al. 1962; Joiner et al. 1970; Lancaster 1968, 1969, 1970; Miller 1969; Savage 1970; Wasserman and Silander 1958, 1964).

Bibliographies of statistics itself, not to mention those of related fields, have so proliferated that Lancaster (1968, 1969, 1970) has already published three bibliographies of statistical bibliographies, one of book length. Several important journals have published cumulative indexes as shown by the table below.

Journal	Years (and Volumes)	Principal types of coverage
Annals of Mathematical Statistics	1930-1960 (1-31)	Citation, author, subject, tables
Biometrika	1901-1950 (1-37)	Subject
Biometrika	1901-1961 (1-48)	Author

Journal	Years (and Volumes)	Principal types of coverage
Econometrica	1932-1952 (1-20)	Author, subject, book reviews indexed by author of book
Journal of the American Statistical Association	1888-1939 (1-34)	Author, subject
Journal of the American Statistical Association	1940-1955 (35-50)	Author, subject, book reviews indexed by author of book and by subject
Revue de Statistique Appliquée	1953-1969 (1-17)	Author, subject, book reviews indexed by subject
Statistical Theory and Method Abstracts	1959-1966 (1-7)	Each year separately, author, subject, book reviews indexed by author of book

Of these, the thirty-year index of the *Annals of Mathematical Statistics* (Greenwood et al. 1962) is a landmark in bibliographic technique and is still very useful though no longer recent. The index of several journals compiled by Joiner et al. (1970) is timely and of a very useful type, probably much cheaper to compile than that exemplified by (Greenwood et al. 1962).

Since 1964 the *Science Citation Index* (the Institute for Scientific Information, Philadelphia) has been published with ever increasing coverage. This is an enormous enterprise showing who has cited whom in about 3,000 different journals, which makes it relatively easy to find recent literature on any scientific topic for which one or two older references are known. Though the coverage for statistics and related fields may not yet be very complete, this facility is already useful.

Additional bibliography

Aczél, János
  1966   *Lectures on Functional Equations and Their Applications*, New York, Academic Press.
        Section 7.1.4 is a key reference for a mathematical approach that investigates the consequences of functional equations that somehow suggest themselves as axioms for probability but does not seek to interpret probability.
Anscombe, Francis J.
  1961   "Bayesian statistics," *The American Statistician*, 15, No. 1, 21-24.
Archibald, G. C.
  1959   "Utility, risk, and linearity," *Journal of Political Economy*, 67, 437-450.
Barnard, George A.
  1947   "A review of 'Sequential Analysis' by Abraham Wald," *Journal of the American Statistical Association*, 42, 658-664.

Barnard, George A.
1963 "Some logical aspects of the fiducial argument," *Journal of the Royal Statistical Society, Series B*, 25, 111-114. 262

Barnard, George A.
1965 "The use of the likelihood function in statistical practice," *Proceedings of the Fifth Berkeley Symposium on Mathematical Statistics and Probability*, 1, 27-40. iv

Barnard, G. A., G. M. Jenkins, and C. B. Winsten
1962 "Likelihood, inference, and time series," *Journal of the Royal Statistical Society, Series A*, 125, 321-372. iv

Bayes, Thomas
1958 "Essay toward solving a problem in the doctrine of chances: with a biographical note by G. A. Barnard," *Biometrika*, 45, 293-315. (Also published separately by the Biometrika Office, University College, London.)
New edition of [B8].

Birnbaum, Allan
1962 "On the foundations of statistical inference," *Journal of the American Statistical Association*, 57, 269-306. iv

Birnbaum, Allan
1969 "Concepts of statistical evidence," pp. 112-143 in *Essays in Honor of Ernest Nagel*, eds. Sidney Morgenbesser, Patrick Suppes, and Morton White, New York, St. Martin's Press.

Blackwell, D., and L. Dubins
1962 "Merging of opinions with increasing information," *Annals of Mathematical Statistics*, 33, 882-887. 214

Blum, Julius, and Judah Rosenblatt
1967 "On partial a priori information in statistical inference," *Annals of Mathematical Statistics*, 38, 1671-1678.
Seeks a compromise between the personalistic and frequentistic approaches.

Borel, Emile
1924 "À propos d'un traité de probabilités," *Revue Philosophique*, 98, 321-336; reprinted in *Pratique et Philosophie des Probabilités* by Borel, 1939, Paris, Gauthier-Villars; translated in (Kyburg and Smokler 1964).
This review of [K4] contains the earliest account of the modern concept of personal probability known to me.

Box, George E. P., and G. C. Tiao
1962 "A further look at robustness via Bayes' theorem," *Biometrika*, 49, 419-433.
A personalistic account of an important general problem in statistics.

Broad, C. D.
1969 *Induction, Probability, and Causation*, Dordrecht, Holland, D. Reidel Publishing Company.
Collected papers of a famous philosopher on the title themes.

Bross, Irwin D. J.
1963 "Linguistic analysis of a statistical controversy," *The American Statistician*, 17, 18-21.
Antipersonalistic commentary.

Bühlmann, H.
1960 "Austauschbare stochastische Variablen und ihre Grenzwert-
      saetze," *University of California Publications in Statistics*, 3, 1-36.   *53*
Carnap, Rudolph
1962 "The aim of inductive logic," pp. 303-318 in (Nagel, Suppes,
      and Tarski 1962).
Cérésole, Pierre
1915 "L'irriductibilité de l'intuition des probabilités et l'existance de
      propositions mathématique indémontrables," *Archives de Psycho-
      logie*, 15, 255-305.   *7*
      Remarkable early ideas about the subjectivity of probability
      and about the elusive concept of the probability of mathematical
      propositions.
Chambers, Michael L.
1970 "A simple problem with strikingly different frequentistic and
      Bayesian solutions," *Journal of the Royal Statistical Society,
      Series B*, 32, 278-282.
Chao, M. T.
1970 "The asymptotic behavior of Bayes' estimators," *Annals of
      Mathematical Statistics*, 41, 601-608.   *214*
Clarke, R. D.
1954 "The concept of probability," *Journal of the Institute of Actu-
      aries*, 80, 1-13 (followed by abstract of discussion, 14-31).
      A personalistic view from an actuarial standpoint.
Coolidge, Julian L.
1940 *A History of Geometrical Methods*, New York, Oxford Univer-
      sity Press. (Reprinted: Dover Publications, New York, 1963.   *283*
Cornfield, Jerome
1966 "Sequential trials, sequential analysis, and the likelihood prin-
      ciple," *The American Statistician*, 20, No. 2, 18-23.   *iv*
Cornfield, Jerome
1969 "The Bayesian outlook and its applications," *Biometrics*, 25,
      617-642.
      A significant statement of the personalistic position in statistics.
Costantini, Domenico
1970 *Fondamenti del Calcolo delle Probabilità*, Milano, Giangiacomo
      Feltrinelli Editore.
      A recent survey of the foundations of probability by an eclec-
      tically inclined author.
de Finetti, Bruno
1954 "Media di decisioni e media di opinioni," *Bulletin de l'Institut
      International de Statistique*, 28th session, 34, 144-157.   *177*
de Finetti, Bruno
1961 "The Bayesian approach to the rejection of outliers," pp. 199-
      210 in Vol. 1 of *Proceedings of the Fourth [1960] Berkeley Sym-
      posium on Mathematical Statistics and Probability*, ed. Jerzy
      Neyman, Berkeley and Los Angeles, University of California Press.
de Finetti, Bruno
1968 "Probability: Interpretations," pp. 496-505 in the *International
      Encyclopedia of the Social Sciences*, New York, Macmillan.
      A penetrating overview of the philosophy of probability.

de Finetti, Bruno, and Leonard J. Savage
1962 "Sul modo di scegliere le probabilità iniziali," *Sui fondamenti della statistica, Biblioteca del Metron, Series C*, 1, 81-147. (A fairly extensive English summary, 148-151.) Discusses at length many facets of the interpretation and application of the personalistic position.

de Jouvenel, Bertrand
1967 *The Art of Conjecture*, New York, Basic Books. A historical and literary approach to intelligent and imaginative guessing.

De Zeeuw, G., C. A. J. Vlek, and W. A. Wagenaar
1970 *Subjective Probability: Theory, Experiments, Applications* (consists of issue No. 2/3, Vol. 34 of *Acta Psychologica*), Amsterdam, North-Holland Publishing Co.

Dempster, Arthur P.
1964 "On the difficulties inherent in Fisher's fiducial argument," *Journal of the American Statistical Association*, 59, 67-88.   *262*

Dempster, Arthur P.
1968 "A generalization of Bayesian inference," *Journal of the Royal Statistical Society, Series B*, 30, 205-247.   *58*

Dickey, James M.
1971 "The weighted likelihood ratio, linear hypotheses on normal location parameters," *Annals of Mathematical Statistics*, 42, 204-223. A key reference for the personalistic approach in multivariate statistics.

Dreze, Jacques
1961 "Fondements logiques de la probabilité subjective et de l'utilité," pp. 73-87 in *La Décision*, Paris, Centre National de la Recherche Scientifique. Flees the tyranny of consequences that do not depend on states and states that do not depend on acts.

Dubins, Lester E.
1969 "An elementary proof of Bochner's finitely additive Radon-Nikodym theorem," *American Mathematical Monthly*, 76, No. 5, 520-523.   *35*

Dubins, Lester E., and Leonard J. Savage
1965 *How to Gamble If you Must: Inequalities for Stochastic Processes*, New York, McGraw-Hill Book Co.   *35*

Edwards, A. W. F.
1969 "Statistical methods in scientific inference," *Nature*, 222, 1233-1237. Champions a notion of prior likelihood as opposed to prior probability. Contains interesting critical and historical sidelights.

Edwards, Ward
1969 *A Bibliography of Research on Behavior Decision Processes to 1968*, Human Performance Center Memorandum Report No. 7, Ann Arbor, University of Michigan Press.   *283*

Edwards, Ward, Harold Lindman, and Leonard J. Savage
1963 "Bayesian statistical inference for psychological research," *Psychological Review*, 70, 193-242.

Elementary but serious discussion of the personalistic position in statistics. Not particularly confined to psychology.

Edwards, Ward, L. D. Phillips, W. L. Hays, and B. C. Goodman
1968 "Probability information processing systems: Design and evaluation," *IEEE Transactions on Systems Science and Cybernetics*, SSC-4, 248-265.

Concerned with the practical use of personal probabilities involving the opinions of different people, each about his own area of competence.

Edwards, Ward, and A. Tversky
1967 *Decision Making*, Baltimore, Penguin Books.

A small anthology centering around empirical aspects of decision making, including computer-aided improvement of decision.

Ellsberg, Daniel, William Fellner, and Howard Raiffa
1961 "Symposium: Decisions under uncertainty," *Quarterly Journal of Economics*, 75, 643-694.

A key reference for a certain type of departure from the theory of personal probability and utility in this book.

Ericson, William A.
1969 "Subjective Bayesian models in sampling finite populations," *Journal of the Royal Statistical Society, Series B*, 31, 195-224.

A key reference for a new line of thinking about the theory of sampling.

Fabius, J.
1964 "Asymptotic behavior of Bayes' estimates," *Annals of Mathematical Statistics*, 35, 846-856.   214

Fishburn, Peter C.
1964 *Decision and Value Theory*, New York, John Wiley and Sons.

Treats decision and preference extensively in a fashion harmonious with this book, and with a view to applications more to management than to statistics.

Fishburn, Peter C.
1970 *Utility Theory for Decision Making*, New York, John Wiley and Sons.   40, 78, 80

Lucidly treats a great variety of axiomatic approaches to preference, with and without uncertainty, including the axiomatic aspects of this book and later developments.

Fisher, Ronald A.
1934 "Two new properties of mathematical likelihood," *Proceedings of the Royal Society, Series A*, 144, 205-221. (Paper 24 of [F6].)   68

Fisher, Ronald A.
1955 "Statistical methods and scientific induction," *Journal of the Royal Statistical Society, Series B*, 17, 69-78.   iv

Fisher, Ronald A.
1956 *Statistical Methods and Scientific Inference*, New York, Hafner.   262

Fraser, Donald A. S.
1968 *The Structure of Inference*, New York, John Wiley and Sons.

A theory of statistical inference that seems to derive from ideas of fiducial probability and of necessarian reliance on symmetry and ignorance.

Fréchet, Maurice
1955 "Sur l'importance en économétrie de la distinction entre les probabilités rationelles et irrationelles," *Econometrica*, 23, 303-306.

Freedman, David
1962 "Invariants under mixing which generalize de Finetti's theorem," *Annals of Mathematical Statistics*, 33, 916-924.   *53*

Freedman, David
1963 "Invariants under mixing which generalized de Finetti's theorem: Continuous time parameter," *Annals of Mathematical Statistics*, 34, 1194-1216.   *53*

Freedman, David
1965 "On the asymptotic behavior of Bayes estimates in the discrete case II," *Annals of Mathematical Statistics*, 36, 454-456.   *214*

Gauss, Carl Friedrich
1821 "Theoria combinationis observationum erroribus minimis obnoxiae," *Commentationes Societatis Regiae Scientarum Gottingensis Recentiores*, 5, 33-90. In German translation, A. Börsch and P. Simon, *Abhandlungen zur Methode der kleinsten Quadrate*, Berlin, Westdruckerei Joachim Frickert, 1887 (reprinted Wurzburg: Physica-Verlag, 1964). In French translation, J. Bertrand, *Methode des Moindres Carrés*, Paris, Mallet-Bachelier, 1855.   *230*

George, Stephen L.
1968 *An Annotated Bibliography on the Foundations of Statistical Inference*, Institute of Statistics Mimeograph Series No. 572, Raleigh, North Carolina State College Press.   *283*

Giere, Ronald N.
1969 "Bayesian statistics and biased procedures," *Synthese*, 20, 371-387.
A philosophic account of differences between Bayesian and Neyman-Pearsonian statistics, concluding in favor of the latter.

Gnedenko, B. V.
1962 *The Theory of Probability*, New York, Chelsea Publishing Co.
This edition, more than later ones, contains some philosophical passages, a few of them referring to dialectical materialism.

Good, Irving John
1959 "Kinds of probability," *Science*, 129, 443-447.

Good, Irving John
1962 "Subjective probability as the measure of a non-measurable set," pp. 319-329 in (Nagel, Suppes, and Tarski 1962).   *58*

Good, Irving John
1965 *The Estimation of Probabilities: An Essay on Modern Bayesian Methods*, Cambridge, Massachusetts Institute of Technology Press.

Goodman, Leo A.
1953 "A simple method for improving some estimators," *Annals of Mathematical Statistics*, 24, 114-117.   *224*

Grayson, C. Jackson, Jr.
1960 *Decisions Under Uncertainty: Drilling Decisions by Oil and Gas Operators*, Cambridge, Harvard University Press.
Interesting discussion of possible applications of personal probability and utility to a kind of business decision greatly affected by uncertainty.

Greenwood, J. Arthur, Ingram Olkin, and I. Richard Savage (eds.)
1962 *Annals of Mathematical Statistics: Index to Vols. 1-31:
        1930-1960*, St. Paul, Minnesota, North Central Publishing Co.   *283, 284*
        A sophisticated bibliographic apparatus for the period covered,
        including not only author and detailed subject indexes for the
        Annals of Mathematical Statistics, but also indexing citations to
        the *Annals* and citations made in the *Annals* to other journals.
Haag, Jules
1928 "Sur un problème général de probabilités et ses diverses appli-
        cations," pp. 659-674 in *Proceedings of the International Congress
        of Mathematicians, Toronto 1924*, Toronto, University of Toronto
        Press.   *52*
Hacking, Ian
1965 *Logic of Statistical Inference*, Cambridge, Basic Books.
Hacking, Ian
1967 "Slightly more realistic personal probability," *Philosophy of
        Science*, 34, 311-325.   *7*
        Interesting for its own theses but also for its sensitive inter-
        pretation of statistical literature.
Hájek, Jaroslav
1965 "On basic concepts of statistics," *Proceedings of the Fifth
        [1965/66] Berkeley Symposium on Mathematical Statistics and
        Probability*, 1, 139-162.
Hakansson, Nils H.
1970 "Friedman-Savage utility functions consistent with risk aver-
        sion," *Quarterly Journal of Economics*, 84, 472-487.   *104*
Halphen, Etienne
1955 "La notion de vraisemblance," *Publication de l'Institut de Sta-
        tistique de l'Université de Paris*, 4, 41-92.
        Thoughts of an original, and too little known, figure on the
        foundations of probability.
Harrod, Roy
1956 *Foundations of Inductive Logic*, New York, Harcourt, Brace
        and Co.
Hewitt, Edwin, and Leonard J. Savage
1955 "Symmetric measures on Cartesian products," *Transactions of
        the American Mathematical Society*, 80, 470-501.   *53*
Hildreth, C.
1963 "Bayesian statisticians and remote clients," *Econometrica*, 31,
        422-439.
        Brings out, and reflects upon, an important point never made
        explicit in this book : the person in personalistic statistics is hardly
        ever a statistician or even the scientific investigator but is often
        one among the public who appraises a scientific publication.
Hill, Bruce M.
1963 "The three-parameter lognormal distribution and Bayesian
        analysis of a point-source epidemic," *Journal of the American
        Statistical Association*, 58, 72-85.
Hill, Bruce M.
1969 "Foundations for the theory of least squares," *Journal of the*

Royal Statistical Society, Series B, 31, 89-97.
Personalistic treatment of a topic central to modern statistics.

Hilpinen, Risto
1968 *Rules of Acceptance and Inductive Logic* (Acta Philosophica Fennica, Issue 2), Amsterdam, North-Holland Publishing Company.

Hirshleifer, Jack
1961 "The Bayesian approach to statistical decision: An exposition," *The Journal of Business*, 34, 471-489.

Hoadley, Bruce
1970 "A Bayesian look at inverse linear regression," *Journal of the American Statistical Association*, 65, 356-369.
An example of personalistic ideas in the study of a specific problem in statistics.

Huber, Peter J.
1964 "Robust estimation of a location parameter," *Annals of Mathematical Statistics*, 35, 73-101.
An important nonpersonalistic advance in the central problem of statistical robustness.

Jeffrey, Richard C.
1965 *The Logic of Decision*, New York, McGraw-Hill Book Co.
An interesting departure from the theory of personal probability and utility as represented by this book.

Jeffreys, Harold
1955 "The present position of the theory of probability," *British Journal for the Philosophy of Science*, 5, 275-289.

Joiner, Brian L., N. F. Laubscher, Eleanor S. Brown, and Bert Levy
1970 *An Author and Permuted Title Index to Selected Statistical Journals*, National Bureau of Standards Special Publication 321, Washington, United States Department of Commerce. *283, 284*
Brings the cumulative indexes of seven journals up through a little later than 1968 in a form that is convenient and powerful.

Kendall, M. G., and A. Stuart
1958 *The Advanced Theory of Statistics: Vol. 1, Distribution Theory*, London, Charles Griffen and Co.

Kendall, M. G., and A. Stuart
1961 *The Advanced Theory of Statistics: Vol. 2, Inference and Relationship*, London, Charles Griffen and Co.
This item and the one before, together with a projected third volume, *Planning and Analysis, and Time-Series*, will constitute a radically new edition of [K2].

Kraft, C. H., John W. Pratt, and A. Seidenberg
1959 "Intuitive probability on finite sets," *Annals of Mathematical Statistics*, 30, 408-419. *40*

Kullback, Solomon
1961 *Information Theory and Statistics*, New York, John Wiley and Sons. *50*

Kyburg, Henry E., Jr.
1961 *Probability and the Logic of Rational Belief*, Middleton, Conn., Wesleyan University Press.

Kyburg, Henry E., Jr., and Ernest Nagel (eds.)
1963 *Induction: Some Current Issues*, Middletown, Wesleyan University Press.
Kyburg, Henry E., Jr., and Howard E. Smokler
1964 *Studies in Subjective Probability*, New York, John Wiley and Sons.
An anthology consisting of an extract from John Venn, translations of (Borel 1924) and [D2], and reproductions of [R1], [K10], and (Savage 1961). Has a large bibliography.   *274*
Lancaster, Henry O.
1968 *Bibliography of Statistical Bibliographies*, Edinburgh, Oliver and Boyd.   *283*
Lancaster, Henry O.
1969 "A bibliography of statistical bibliographies: A second list," *Review of the International Statistical Institute*, 37, 57-67.   *283*
Lancaster, Henry O.
1970 "A bibliography of statistical bibliographies: A third list," *Review of the International Statistical Institute*, 38, 258-267.   *283*
LeCam, Lucien
1964 "Sufficiency and approximate sufficiency," *Annals of Mathematical Statistics*, 35, 1419-1455.   *134*
Lehmann, E. L.
1958 "Significance level and power," *Annals of Mathematical Statistics*, 29, 1167-1176.   *iv*
Lehmann, E. L.
1959 *Testing Statistical Hypotheses*, New York, John Wiley and Sons.
Excellent illustration of Neyman-Pearson theory, showing older and newer aspects in tension with each other.
Lindley, Dennis V.
1958 "Professor Hogben's 'Crisis'—A survey of the foundations of statistics," *Applied Statistics*, 7, 186-198.
Lindley, Dennis V.
1965 *Introduction to Probability and Statistics: From a Bayesian Viewpoint: Part 1, Probability; Part 2, Inference*, Cambridge, Cambridge University Press.
Linnik, Yu. V.
1968 *Statistical Problems with Nuisance Parameters*, Translations of Mathematical Monographs, 20, Providence, American Mathematical Society.   *262*
Luce, R. Duncan, Robert R. Bush, and Eugene Galanter (eds.)
1965 *Readings in Mathematical Psychology, Volume II*, New York, John Wiley and Sons.
Parts V and VI contain relevant readings.
Luce, R. Duncan, and Howard Raiffa
1957 *Games and Decision*, New York, John Wiley and Sons.
Good account of the theory of games and its contacts with the normative theory of decision.
Luce, R. Duncan, and Patrick Suppes
1965 "Preference, utility, and subjective probability," pp. 249-410

in Vol. 3 of *Handbook of Mathematical Psychology*, eds. R. Duncan Luce, Robert R. Bush, and Eugene Galanter, New York, London, and Sydney, John Wiley and Sons.
> Valuable for itself and as a key reference.

Lukasiewicz, Jan
1970 "Logical foundations of probability theory," in *Jan Lukasiewicz: Selected Works*, ed. L. Borkowski, Amsterdam and London, North-Holland Publishing Company, and Warsaw, PWN, Polish Scientific Publishers.
> English translations of the original German version of 1913. Eloquently presents an early formal-logic approach to probability. Valuable for insights and history.

Lusted, Lee B.
1968 *Introduction to Medical Decision Making*, Springfield, Illinois, C. C. Thomas.
> Hopeful but realistic discussion of a promising field of application for personalistic ideas.

Maritz, J. S.
1970 *Empirical Bayes Methods*, London, Methuen.
> Key reference for a branch of frequentistic statistics with Bayesian elements.

Markowitz, Harry M.
1959 *Portfolio Selection*, Cowles Commission Monograph No. 16, New York, John Wiley and Sons.
> Application of the theory of utility to the theory of investment.

Marschak, Jacob
1968 "Decision making: Economic aspects," pp. 42-45 in the *International Encyclopedia of the Social Sciences*, New York, Macmillan.

Milier-Gruzewska, Halina
1949 "On the law of probability and the characteristic function of the standardized sum of equivalent variables," *Towarzystwo Naukowe Warzawskie*, 42, 99-142.   *53*

Milier-Gruzewska, Halina
1950 "Sulla legge limite delle variabili casuali equivalenti," *Atti Memorie della Accademia Nazionale dei Lincei, Series 8*, 2 (First section), 25-33.   *53*

Miller, Robert B.
1969 *A Selected Bayesian Statistics Bibliography*, Department of Statistics Technical Paper No. 214, Madison, University of Wisconsin Press.   *283*

Morse, Norman, and Richard Sacksteder
1966 "Statistical isomorphism," *Annals of Mathematical Statistics*, 37, 203-214.   *152*

Mosteller, Frederick, and David L. Wallace
1964 *Inference and Disputed Authorship: The Federalist*, Reading, Massachusetts, Addison-Wesley.
> A thorough and illuminating application of statistics illustrating both personalistic and nonpersonalistic approaches.

Nagel, Ernest, Patrick Suppes, and Alfred Tarski (eds.)
1962 *Logic, Methodology and Philosophy of Science*, Stanford, Stanford University Press.
Neyman, Jerzy
1952 *Lectures and Conferences on Mathematical Statistics and Probability*, Washington, Graduate School of the United States Department of Agriculture. (First edition 1938, mimeographed.)
Neyman, Jerzy
1957 "Current problems of mathematical statistics," pp. 349-370 in Vol. 1 of *Proceedings of the International Congress of Mathematicians* [Amsterdam, 1954], Groningen, E. P. Nordhoff.
Neyman, Jerzy
1967 *A Selection of Early Statistical Papers of J. Neyman*, Cambridge, Cambridge University Press.
Neyman, Jerzy, and Egon S. Pearson
1967 *Joint Statistical Papers by J. Neyman and E. S. Pearson*, Berkeley, University of California Press.
Patil, Venkutai H.
1965 "Approximation to the Behrens-Fisher distributions," *Biometrika*, 52, 267-271.
Pearson, Egon S.
1966 *The Selected Papers of E. S. Pearson*, Berkeley and Los Angeles, University of California Press.
Pfanzagl, Johann
1968 *Theory of Measurement*, New York, John Wiley and Sons.
Measurement of utility and personal probability is typical of the concerns of this highly mathematical book.
Plackett, Robert L.
1966 "Current trends in statistical inference," *Journal of the Royal Statistical Society, Series A*, 129, 249-267.
Polya, G.
1954 *Mathematics and Plausible Reasoning. Vol. I, Induction and Analogy in Mathematics. Vol. II, Patterns of Plausible Inference*, Princeton, Princeton University Press, and London, Oxford University Press.
Popper, Karl
1959 *The Logic of Scientific Discovery*, London, Hutchinson. (Reprinted: Science Editions, New York, 1961.)
Pratt, John W.
1961 "Review of E. H. Lehmann's 'Testing Statistical Hypotheses,'" *Journal of the American Statistical Association*, 56, 163-167.
A personalistic review of the highly respected frequentistic work (Lehmann 1959.)
Pratt, John W.
1964 "Risk aversion in the small and in the large," *Econometrica*, 32, 1-2, 122-136.
A valuable contribution to the theory of utility.
Pratt, John W.
1965 "Bayesian interpretation of standard inference statements," *Journal of the Royal Statistical Society, Series B*, 27, 169-203.

262

7

iv

Quine, W. V.
1951 "Two dogmas of empiricism," *Philosophical Review*, 60, 20-43. (Reprinted in *From a Logical Point of View*, 1953, Cambridge, Harvard University Press.) *25*
Raiffa, Howard
1968 *Decision Analysis: Introductory Lectures on Choices under Uncertainty*, Reading, Massachusetts, Addison-Wesley.
Raiffa, Howard, and Robert Schlaifer
1961 *Applied Statistical Decision Theory*, Cambridge, Harvard University Press.
A rather advanced textbook in personalistic statistics.
Rényi, Alfréd
1970 *Foundations of Probability*, San Francisco, Holden-Day.
More mathematical than philosophical but important for the serious philosophical students of probability.
Rényi, Alfréd, and P. Révész
1963 "A study of sequences of equivalent events as special stable sequences," *Publicationes Mathematecae, Debrecen*, 10, 319-325. *53*
Roberts, Harry V.
1963 "Risk, ambiguity, and the Savage axioms: Comment," *Quarterly Journal of Economics*, 77, 327-342.
Ryll-Nardzewski, C.
1957 "On stationary sequences of random variables and the de Finetti's equivalence," *Colloquium Mathematicum*, 4, 149-156. *53*
Salmon, Wesley C.
1966 *The Foundations of Scientific Inference*, Pittsburgh, University of Pittsburgh Press.
Lucid review and study of the problem of induction bequeathed to us by Hume.
Savage, I. Richard
1968 *Statistics: Uncertainty and Behavior*, Boston, Houghton Mifflin Company.
An elementary statistical textbook in the personalistic Bayesian spirit.
Savage, Leonard J.
1960 "Recent tendencies in the foundations of statistics," pp. 540-544 in *Proceedings of the International Congress of Mathematicians* [Edinburgh, 1958], Cambridge, Cambridge University Press.
Savage, Leonard J.
1961 "The foundations of statistics reconsidered," pp. 575-586 in Vol. I of *Proceedings of the Fourth* [1960] *Berkeley Symposium on Mathematical Statistics and Probability*, ed. Jerzy Neyman, Berkeley, University of California Press. *iv*
Savage, Leonard J.
1962 "Bayesian statistics," pp. 161-194 in *Recent Developments in Decision and Information Processes*, eds. Robert E. Machol and Paul Gray, New York, Macmillan Co. *iv*
Savage, Leonard J.
1962 "Subjective probability and statistical practice," pp. 9-35 in (Savage, et al, 1962). *217*

Savage, Leonard J.
1967 "Difficulties in the theory of personal probability," *Philosophy of Science*, 34, 305-310.
Savage, Leonard J.
1967 "Implications of personal probability for induction," *Journal of Philosophy*, 64, 593-607.
Savage, Leonard J.
1970 "Reading suggestions for the foundations of statistics," *The American Statistician*, 24, No. 4, 23-27.
 Somewhat overlaps, but is much shorter than, the present Bibliographic Supplement.   *283*
Savage, Leonard J.
1971 "Elicitation of personal probabilities and expectations," *Journal of the American Statistical Association*, 66, 783-801.
 See (Staël von Holstein 1970).
Savage, Leonard J., et al.
1962 *The Foundations of Statistical Inference: A Symposium*, New York, John Wiley and Sons.
 Valuable for the interchange of ideas among statisticians of diverse experience and outlook.
Scheffé, Henry
1970 "Practical solutions of the Behrens-Fisher problem," *Journal of the American Statistical Association*, 65, 1501-1598.   *262*
Schelling, Thomas C.
1960 *The Strategy of Conflict*, Cambridge, Harvard University Press.
 Pertinent because conflict and group decision are aspects of the same thing. Extramathematical and particularly stimulating.
Schlaifer, Robert
1959 *Probability and Statistics for Business Decisions*, New York, McGraw-Hill Book Co.   *v*
Schmitt, Samuel A.
1969 *Measuring Uncertainty: An Elementary Introduction to Bayesian Statistics*, Reading, Massachusetts, Addison-Wesley.
Shelly, Maynard W., II, and Glenn L. Bryan (eds.)
1964 *Human Judgments and Optimality*, New York, John Wiley and Sons.
 An organized collection of essays by many authors. Interesting in itself and useful as an extensive key reference.
Smith, Cedric A. B.
1961 "Consistency in statistical inference and decision," *Journal of the Royal Statistical Society, Series B*, 23, 1-25.   *58*
Smith, Cedric A. B.
1965 "Personal probability and statistical analysis," *Journal of the Royal Statistical Society, Series A*, 128, 469-489.
Staël von Holstein, Carl-Axel S.
1970 *Assessment and Evaluation of Subjective Probability Distributions*, Stockholm, The Economic Research Institute at the Stockholm School of Economics.   *177*
 Excellent monograph on how to elicit personal probabilities and what to do with them. Reviews and enriches a considerable literature. Related references are (Savage 1971; Winkler 1966).

Stone, Mervyn
1970 "The role of experimental randomization in Bayesian statistics:
Finite sampling and two Bayesians," *Biometrika*, 56, 681-683.
Strawderman, William E.
1971 "Proper Bayes minimax estimators of the multivariate normal
mean," *Annals of Mathematical Statistics*, 42, 385-388.
Key reference for a challenging theoretical development initiated by Charles Stein.
Suppes, Patrick
1960 "Some open problems in the foundations of subjective probability," pp. 162-169 in *Information and Decision Processes*, ed.
Robert Machol, New York, McGraw-Hill Book Co.
Tavanec, P. V., ed.
1970 *Problems of the Logic of Scientific Knowledge*, translated by
T. J. Blakely, New York, Humanities Press.
A rare opportunity to read in English the ideas of some modern
Soviet philosophers about probability and induction.
Tribe, Laurence H.
1971 "Trial by mathematics: Precision and ritual in the legal process," *Harvard Law Review*, 84, 1329-1393.
A key reference on the possible applicability of probabilistic ideas
in the courts, which the author does not find promising.
Tribus, Myron
1969 *Rational Descriptions, Decisions and Designs*, New York, Pergamon Press.
A necessarian approach.
Tukey, John W.
1957 "Some examples with fiducial relevance," *Annals of Mathematical Statistics*, 28, 687-695.   262
Tukey, John W.
1962 "The future of data analysis," *Annals of Mathematical Statistics*, 33, 1-67.
Ulam, Stanislaw
1930 "Zur Masstheorie in der allgemeinen Mengenlehre," *Fundamenta Mathematicae*, 16, 140-150.   41
van Dantzig, David
1950-1 "Review of Carnap's *Logical Foundations of Probability*,"
Synthese, 8, 459-470.
van Dantzig, David
1957 "Statistical priesthood: Savage on personal probabilities," *Statistica Neerlandica*, 11, 1-16.
van Dantzig, David
1957 "Statistical priesthood II: Sir Ronald on scientific inference,"
Statistica Neerlandica, 11, 185-200.
The three preceding references review three different views of
the foundations of probability and statistics from the standpoint
of a fourth.
Vetter, Hermann
1967 *Wahrschein'ichkeit und logischer Spielraum*, Tübingen, J. C. B.
Mohr (Paul Siebeck).

Villegas, C.
  1964 "On qualitative probability σ-algebras," *Annals of Mathematical Statistics*, 35, 1787-1796.   *43*
von Mises, Richard
  1942 "On the correct use of Bayes' formula," *Annals of Mathematical Statistics*, 13, 156-165.
    Illustrates an approach unusual for a frequentist.
von Wright, Georg Henrik
  1962 "Remarks on epistemology of subjective probability," pp. 330-339 in (Nagel, Suppes, and Tarski 1962).
Wald, Abraham (ed.)
  1955 *Selected Papers in Statistics and Probability*, New York, Mc-Graw-Hill Book Co.
Wasserman, Paul, and Fred S. Silander
  1958 *Decision Making: An Annotated Bibliography*, Ithaca, Cornell University Press.   *283*
Wasserman, Paul, and Fred S. Silander
  1964 *Decision-Making: An Annotated Bibliography: Supplement, 1958-63*, Ithaca, Cornell University Press.   *283*
Watts, Donald G. (ed.)
  1967 *The Future of Statistics*, Proceedings of a Conference on the Future of Statistics held at the University of Wisconsin, June 1967, New York and London, Academic Press.
Wetherill, G. B.
  1961 "Bayesian sequential analysis," *Biometrika*, 48, 281-292.
Whittle, Peter
  1957 "Curve and periodogram smoothing," *Journal of the Royal Statistical Society, Series B*, 19, 38-47.
Whittle, Peter
  1958 "On the smoothing of probability density functions," *Journal of the Royal Statistical Society, Series B*, 20, 334-343.
    These two references are suggestive for personalistic technique.
Williams, J. S.
  1966 "The role of probability in fiducial inference," *Sankhyā, Series A*, 28, 271-296.   *262*
Winkler, Robert L.
  1968 "The consensus of subjective probability distributions," *Management Science*, 15, 2, B61-B75.   *177*
Wolfowitz, J.
  1962 "Bayesian inference and axioms of consistent decision," *Econometrica*, 30, 471-479.   *iv*
Wolfowitz, J.
  1970 "Reflections on the future of mathematical statistics," pp. 739-750 in *Essays in Probability and Statistics*, eds. R. C. Bose et al., Chapel Hill, University of North Carolina Press.

Technical Symbols

This index is intended to lead to the definitions of all technical symbols that are defined in the text and used extensively. Some symbols have more than one page reference, corresponding to their use in more than one sense, depending on context.

299

Author Index

This index is intended to lead to every reference made in the text to an author's works or opinions. Only a few of the authors referred to do not have works listed in the bibliography (p. 271).

A few examples illustrate the use of this index: F. J. Anscombe is not referred to in the text proper, but there is a reference to him, beyond the mere listing of his name, in the bibliography under [A4]. A paper of which David Blackwell is a co-author, but whose first listed author is Kenneth J. Arrow, is somewhere referred to without mention of Blackwell's name, but only a bibliographic symbol of the form [An]. A work of S. R. Searle is listed in the bibliography, but not otherwise mentioned.

General Index

See also *Technical Symbols*, p. 299, and *Index of Authors*, pp. 301–3.

Factorability criterion for sufficiency, 130ff
Fair coin, 33
Fiducial probability, 262
Fine, 37, 40
Foundations of sciences, role of, 1
Foundations of statistics, deep, 5
    history of, 1ff
    shallow, 5

Gamble, 70, 71
Gambling, 63, 64, 91, 94
Gambling apparatus, 66
Game, abstract, 184ff
    bilinear, 186ff
    standard, 178ff
    two-person, 178ff
Games, in relation to minimax theories
    of decision, 180ff
    mathematics of, 184ff
    theory of, 156, 178ff
Given, 22, 44
Grand world, 84
Greek fonts, 11
Group, mathematical, 193
Group action, 105
Group decision problem, 172ff
    and observation, 210
Group minimax rule, 207

Hausdorff moment problem, 53, 55, 152
Homogeneous coordinates, 136
Hyper-utility, 75
Hypothesis, alternative, 247
    extreme null, 254
    null, 247

Income, 163
    negative, 164, 169, 170
    and loss, 182, 200
    personal, 173
Inconsistency, 20, 21, 57
Indecision, 21
Independence in qualitative probability, 44, 91
Independent events, 44
Independent random variables, 46
Indifference, 17, 59
    difficulty of testing, 17
Inductive behavior, 159

Inductive inference, 2
Inexact science, 59
Infimum, 80
Infinite sets in applied mathematics, 39, 77
Infinite utility, 81
Information, 50, 153, 235ff
    differential, 236ff
Information inequality, 238
Insufficient reason, principle of, 64, 65, 193
Integral, 263
Interrogation, behavioral, 28
    intermediate mode of, 28
    strictly empirical, 28, 29
Intersection of events, 11
Interval, 266
Interval estimation, 257
    definition of, 259, 260
Interval of gambles, 75
Interval of linearity, 267
Invariance of a game, 194ff
Invariant minimax, 197, 198
Irrelevant, 126
    utterly, 126
Irrelevant event, 44

Journal of American Statistical Association, 270
Judgment, 156

Large numbers, strong law of, 54
    weak law of, 49, 54, 91
Learning, 44, 55
    see also Experience
Lebesgue measure, 41
Likelihood ratio, 48, 135ff, 225
Likelihood-ratio test, 139, 213
Linear function, 267
Logic, 3
    decision and, 6
    empirical interpretation of, 20
        criticism of, 20
    incompleteness of, 59
    normative interpretation of, 20
Logical behavior, implications of, 7, 8, 20
"Look before you leap principle," 16
    criticism of, 16, 17
Loss, 163, 164, 169, 170
    personal, 174

A CATALOG OF SELECTED
DOVER BOOKS
IN SCIENCE AND MATHEMATICS

RELATIVITY, THERMODYNAMICS AND COSMOLOGY, Richard C. Tolman. Landmark study extends thermodynamics to special, general relativity; also applications of relativistic mechanics, thermodynamics to cosmological models. 501pp. 5⅜ × 8½.   65383-8 Pa. $12.95

APPLIED ANALYSIS, Cornelius Lanczos. Classic work on analysis and design of finite processes for approximating solution of analytical problems. Algebraic equations, matrices, harmonic analysis, quadrature methods, much more. 559pp. 5⅜ × 8½.   65656-X Pa. $13.95

SPECIAL RELATIVITY FOR PHYSICISTS, G. Stephenson and C.W. Kilmister. Concise elegant account for nonspecialists. Lorentz transformation, optical and dynamical applications, more. Bibliography. 108pp. 5⅜ × 8½.   65519-9 Pa. $4.95

INTRODUCTION TO ANALYSIS, Maxwell Rosenlicht. Unusually clear, accessible coverage of set theory, real number system, metric spaces, continuous functions, Riemann integration, multiple integrals, more. Wide range of problems. Undergraduate level. Bibliography. 254pp. 5⅜ × 8½.   65038-3 Pa. $7.95

INTRODUCTION TO QUANTUM MECHANICS With Applications to Chemistry, Linus Pauling & E. Bright Wilson, Jr. Classic undergraduate text by Nobel Prize winner applies quantum mechanics to chemical and physical problems. Numerous tables and figures enhance the text. Chapter bibliographies. Appendices. Index. 468pp. 5⅜ × 8½.   64871-0 Pa. $11.95

ASYMPTOTIC EXPANSIONS OF INTEGRALS, Norman Bleistein & Richard A. Handelsman. Best introduction to important field with applications in a variety of scientific disciplines. New preface. Problems. Diagrams. Tables. Bibliography. Index. 448pp. 5⅜ × 8½.   65082-0 Pa. $12.95

MATHEMATICS APPLIED TO CONTINUUM MECHANICS, Lee A. Segel. Analyzes models of fluid flow and solid deformation. For upper-level math, science and engineering students. 608pp. 5⅜ × 8½.   65369-2 Pa. $13.95

ELEMENTS OF REAL ANALYSIS, David A. Sprecher. Classic text covers fundamental concepts, real number system, point sets, functions of a real variable, Fourier series, much more. Over 500 exercises. 352pp. 5⅜ × 8½.   65385-4 Pa. $10.95

PHYSICAL PRINCIPLES OF THE QUANTUM THEORY, Werner Heisenberg. Nobel Laureate discusses quantum theory, uncertainty, wave mechanics, work of Dirac, Schroedinger, Compton, Wilson, Einstein, etc. 184pp. 5⅜ × 8½.
                                                       60113-7 Pa. $5.95

INTRODUCTORY REAL ANALYSIS, A.N. Kolmogorov, S.V. Fomin. Translated by Richard A. Silverman. Self-contained, evenly paced introduction to real and functional analysis. Some 350 problems. 403pp. 5⅜ × 8½.   61226-0 Pa. $9.95

PROBLEMS AND SOLUTIONS IN QUANTUM CHEMISTRY AND PHYSICS, Charles S. Johnson, Jr. and Lee G. Pedersen. Unusually varied problems, detailed solutions in coverage of quantum mechanics, wave mechanics, angular momentum, molecular spectroscopy, scattering theory, more. 280 problems plus 139 supplementary exercises. 430pp. 6½ × 9¼.   65236-X Pa. $12.95

ASYMPTOTIC METHODS IN ANALYSIS, N.G. de Bruijn. An inexpensive, comprehensive guide to asymptotic methods—the pioneering work that teaches by explaining worked examples in detail. Index. 224pp. 5⅜ × 8½.   64221-6 Pa. $6.95

OPTICAL RESONANCE AND TWO-LEVEL ATOMS, L. Allen and J.H. Eberly. Clear, comprehensive introduction to basic principles behind all quantum optical resonance phenomena. 53 illustrations. Preface. Index. 256pp. 5⅜ × 8½.
65533-4 Pa. $7.95

COMPLEX VARIABLES, Francis J. Flanigan. Unusual approach, delaying complex algebra till harmonic functions have been analyzed from real variable viewpoint. Includes problems with answers. 364pp. 5⅜ × 8½.   61388-7 Pa. $8.95

ATOMIC SPECTRA AND ATOMIC STRUCTURE, Gerhard Herzberg. One of best introductions; especially for specialist in other fields. Treatment is physical rather than mathematical. 80 illustrations. 257pp. 5⅜ × 8½.   60115-3 Pa. $6.95

APPLIED COMPLEX VARIABLES, John W. Dettman. Step-by-step coverage of fundamentals of analytic function theory—plus lucid exposition of five important applications: Potential Theory; Ordinary Differential Equations; Fourier Transforms; Laplace Transforms; Asymptotic Expansions. 66 figures. Exercises at chapter ends. 512pp. 5⅜ × 8½.   64670-X Pa. $11.95

ULTRASONIC ABSORPTION: An Introduction to the Theory of Sound Absorption and Dispersion in Gases, Liquids and Solids, A.B. Bhatia. Standard reference in the field provides a clear, systematically organized introductory review of fundamental concepts for advanced graduate students, research workers. Numerous diagrams. Bibliography. 440pp. 5⅜ × 8½.   64917-2 Pa. $11.95

UNBOUNDED LINEAR OPERATORS: Theory and Applications, Seymour Goldberg. Classic presents systematic treatment of the theory of unbounded linear operators in normed linear spaces with applications to differential equations. Bibliography. 199pp. 5⅜ × 8½.   64830-3 Pa. $7.95

LIGHT SCATTERING BY SMALL PARTICLES, H.C. van de Hulst. Comprehensive treatment including full range of useful approximation methods for researchers in chemistry, meteorology and astronomy. 44 illustrations. 470pp. 5⅜ × 8½.   64228-3 Pa. $11.95

CONFORMAL MAPPING ON RIEMANN SURFACES, Harvey Cohn. Lucid, insightful book presents ideal coverage of subject. 334 exercises make book perfect for self-study. 55 figures. 352pp. 5⅜ × 8¼.   64025-6 Pa. $9.95

OPTICKS, Sir Isaac Newton. Newton's own experiments with spectroscopy, colors, lenses, reflection, refraction, etc., in language the layman can follow. Foreword by Albert Einstein. 532pp. 5⅜ × 8½.   60205-2 Pa. $9.95

GENERALIZED INTEGRAL TRANSFORMATIONS, A.H. Zemanian. Graduate-level study of recent generalizations of the Laplace, Mellin, Hankel, K. Weierstrass, convolution and other simple transformations. Bibliography. 320pp. 5⅜ × 8½.   65375-7 Pa. $8.95

THE ELECTROMAGNETIC FIELD, Albert Shadowitz. Comprehensive undergraduate text covers basics of electric and magnetic fields, builds up to electromagnetic theory. Also related topics, including relativity. Over 900 problems. 768pp. 5⅜ × 8¼. 65660-8 Pa. $18.95

FOURIER SERIES, Georgi P. Tolstov. Translated by Richard A. Silverman. A valuable addition to the literature on the subject, moving clearly from subject to subject and theorem to theorem. 107 problems, answers. 336pp. 5⅜ × 8½. 63317-9 Pa. $8.95

THEORY OF ELECTROMAGNETIC WAVE PROPAGATION, Charles Herach Papas. Graduate-level study discusses the Maxwell field equations, radiation from wire antennas, the Doppler effect and more. xiii + 244pp. 5⅜ × 8½. 65678-0 Pa. $6.95

DISTRIBUTION THEORY AND TRANSFORM ANALYSIS: An Introduction to Generalized Functions, with Applications, A.H. Zemanian. Provides basics of distribution theory, describes generalized Fourier and Laplace transformations. Numerous problems. 384pp. 5⅜ × 8½. 65479-6 Pa. $9.95

THE PHYSICS OF WAVES, William C. Elmore and Mark A. Heald. Unique overview of classical wave theory. Acoustics, optics, electromagnetic radiation, more. Ideal as classroom text or for self-study. Problems. 477pp. 5⅜ × 8½. 64926-1 Pa. $12.95

CALCULUS OF VARIATIONS WITH APPLICATIONS, George M. Ewing. Applications-oriented introduction to variational theory develops insight and promotes understanding of specialized books, research papers. Suitable for advanced undergraduate/graduate students as primary, supplementary text. 352pp. 5⅜ × 8½. 64856-7 Pa. $8.95

A TREATISE ON ELECTRICITY AND MAGNETISM, James Clerk Maxwell. Important foundation work of modern physics. Brings to final form Maxwell's theory of electromagnetism and rigorously derives his general equations of field theory. 1,084pp. 5⅜ × 8½. 60636-8, 60637-6 Pa., Two-vol. set $21.90

AN INTRODUCTION TO THE CALCULUS OF VARIATIONS, Charles Fox. Graduate-level text covers variations of an integral, isoperimetrical problems, least action, special relativity, approximations, more. References. 279pp. 5⅜ × 8½. 65499-0 Pa. $7.95

HYDRODYNAMIC AND HYDROMAGNETIC STABILITY, S. Chandrasekhar. Lucid examination of the Rayleigh-Benard problem; clear coverage of the theory of instabilities causing convection. 704pp. 5⅜ × 8¼. 64071-X Pa. $14.95

CALCULUS OF VARIATIONS, Robert Weinstock. Basic introduction covering isoperimetric problems, theory of elasticity, quantum mechanics, electrostatics, etc. Exercises throughout. 326pp. 5⅜ × 8½. 63069-2 Pa. $8.95

DYNAMICS OF FLUIDS IN POROUS MEDIA, Jacob Bear. For advanced students of ground water hydrology, soil mechanics and physics, drainage and irrigation engineering and more. 335 illustrations. Exercises, with answers. 784pp. 6⅛ × 9¼. 65675-6 Pa. $19.95

NUMERICAL METHODS FOR SCIENTISTS AND ENGINEERS, Richard Hamming. Classic text stresses frequency approach in coverage of algorithms, polynomial approximation, Fourier approximation, exponential approximation, other topics. Revised and enlarged 2nd edition. 721pp. 5⅜ × 8½.
65241-6 Pa. $14.95

THEORETICAL SOLID STATE PHYSICS, Vol. I: Perfect Lattices in Equilibrium; Vol. II: Non-Equilibrium and Disorder, William Jones and Norman H. March. Monumental reference work covers fundamental theory of equilibrium properties of perfect crystalline solids, non-equilibrium properties, defects and disordered systems. Appendices. Problems. Preface. Diagrams. Index. Bibliography. Total of 1,301pp. 5⅜ × 8½. Two volumes. Vol. I 65015-4 Pa. $14.95
Vol. II 65016-2 Pa. $14.95

OPTIMIZATION THEORY WITH APPLICATIONS, Donald A. Pierre. Broad-spectrum approach to important topic. Classical theory of minima and maxima, calculus of variations, simplex technique and linear programming, more. Many problems, examples. 640pp. 5⅜ × 8½. 65205-X Pa. $14.95

THE CONTINUUM: A Critical Examination of the Foundation of Analysis, Hermann Weyl. Classic of 20th-century foundational research deals with the conceptual problem posed by the continuum. 156pp. 5⅜ × 8½. 67982-9 Pa. $5.95

ESSAYS ON THE THEORY OF NUMBERS, Richard Dedekind. Two classic essays by great German mathematician: on the theory of irrational numbers; and on transfinite numbers and properties of natural numbers. 115pp. 5⅜ × 8½.
21010-3 Pa. $4.95

THE FUNCTIONS OF MATHEMATICAL PHYSICS, Harry Hochstadt. Comprehensive treatment of orthogonal polynomials, hypergeometric functions, Hill's equation, much more. Bibliography. Index. 322pp. 5⅜ × 8½. 65214-9 Pa. $9.95

NUMBER THEORY AND ITS HISTORY, Oystein Ore. Unusually clear, accessible introduction covers counting, properties of numbers, prime numbers, much more. Bibliography. 380pp. 5⅜ × 8½. 65620-9 Pa. $9.95

THE VARIATIONAL PRINCIPLES OF MECHANICS, Cornelius Lanczos. Graduate level coverage of calculus of variations, equations of motion, relativistic mechanics, more. First inexpensive paperbound edition of classic treatise. Index. Bibliography. 418pp. 5⅜ × 8½. 65067-7 Pa. $11.95

MATHEMATICAL TABLES AND FORMULAS, Robert D. Carmichael and Edwin R. Smith. Logarithms, sines, tangents, trig functions, powers, roots, reciprocals, exponential and hyperbolic functions, formulas and theorems. 269pp. 5⅜ × 8½. 60111-0 Pa. $6.95

THEORETICAL PHYSICS, Georg Joos, with Ira M. Freeman. Classic overview covers essential math, mechanics, electromagnetic theory, thermodynamics, quantum mechanics, nuclear physics, other topics. First paperback edition. xxiii + 885pp. 5⅜ × 8½. 65227-0 Pa. $19.95

HANDBOOK OF MATHEMATICAL FUNCTIONS WITH FORMULAS, GRAPHS, AND MATHEMATICAL TABLES, edited by Milton Abramowitz and Irene A. Stegun. Vast compendium: 29 sets of tables, some to as high as 20 places. 1,046pp. 8 × 10½. 61272-4 Pa. $24.95

MATHEMATICAL METHODS IN PHYSICS AND ENGINEERING, John W. Dettman. Algebraically based approach to vectors, mapping, diffraction, other topics in applied math. Also generalized functions, analytic function theory, more. Exercises. 448pp. 5⅜ × 8¼. 65649-7 Pa. $9.95

A SURVEY OF NUMERICAL MATHEMATICS, David M. Young and Robert Todd Gregory. Broad self-contained coverage of computer-oriented numerical algorithms for solving various types of mathematical problems in linear algebra, ordinary and partial, differential equations, much more. Exercises. Total of 1,248pp. 5⅜ × 8½. Two volumes. Vol. I 65691-8 Pa. $14.95
Vol. II 65692-6 Pa. $14.95

TENSOR ANALYSIS FOR PHYSICISTS, J.A. Schouten. Concise exposition of the mathematical basis of tensor analysis, integrated with well-chosen physical examples of the theory. Exercises. Index. Bibliography. 289pp. 5⅜ × 8½.
65582-2 Pa. $8.95

INTRODUCTION TO NUMERICAL ANALYSIS (2nd Edition), F.B. Hildebrand. Classic, fundamental treatment covers computation, approximation, interpolation, numerical differentiation and integration, other topics. 150 new problems. 669pp. 5⅜ × 8½. 65363-3 Pa. $15.95

INVESTIGATIONS ON THE THEORY OF THE BROWNIAN MOVEMENT, Albert Einstein. Five papers (1905–8) investigating dynamics of Brownian motion and evolving elementary theory. Notes by R. Fürth. 122pp. 5⅜ × 8½.
60304-0 Pa. $4.95

CATASTROPHE THEORY FOR SCIENTISTS AND ENGINEERS, Robert Gilmore. Advanced-level treatment describes mathematics of theory grounded in the work of Poincaré, R. Thom, other mathematicians. Also important applications to problems in mathematics, physics, chemistry and engineering. 1981 edition. References. 28 tables. 397 black-and-white illustrations. xvii + 666pp. 6⅛ × 9¼.
67539-4 Pa. $16.95

AN INTRODUCTION TO STATISTICAL THERMODYNAMICS, Terrell L. Hill. Excellent basic text offers wide-ranging coverage of quantum statistical mechanics, systems of interacting molecules, quantum statistics, more. 523pp. 5⅜ × 8½. 65242-4 Pa. $12.95

ELEMENTARY DIFFERENTIAL EQUATIONS, William Ted Martin and Eric Reissner. Exceptionally clear, comprehensive introduction at undergraduate level. Nature and origin of differential equations, differential equations of first, second and higher orders. Picard's Theorem, much more. Problems with solutions. 331pp. 5⅜ × 8½. 65024-3 Pa. $8.95

STATISTICAL PHYSICS, Gregory H. Wannier. Classic text combines thermodynamics, statistical mechanics and kinetic theory in one unified presentation of thermal physics. Problems with solutions. Bibliography. 532pp. 5⅜ × 8½.
65401-X Pa. $12.95

CATALOG OF DOVER BOOKS

ORDINARY DIFFERENTIAL EQUATIONS, Morris Tenenbaum and Harry Pollard. Exhaustive survey of ordinary differential equations for undergraduates in mathematics, engineering, science. Thorough analysis of theorems. Diagrams. Bibliography. Index. 818pp. 5⅜ × 8½. 64940-7 Pa. $16.95

STATISTICAL MECHANICS: Principles and Applications, Terrell L. Hill. Standard text covers fundamentals of statistical mechanics, applications to fluctuation theory, imperfect gases, distribution functions, more. 448pp. 5⅜ × 8½. 65390-0 Pa. $11.95

ORDINARY DIFFERENTIAL EQUATIONS AND STABILITY THEORY: An Introduction, David A. Sánchez. Brief, modern treatment. Linear equation, stability theory for autonomous and nonautonomous systems, etc. 164pp. 5⅜ × 8¼. 63828-6 Pa. $5.95

THIRTY YEARS THAT SHOOK PHYSICS: The Story of Quantum Theory, George Gamow. Lucid, accessible introduction to influential theory of energy and matter. Careful explanations of Dirac's anti-particles, Bohr's model of the atom, much more. 12 plates. Numerous drawings. 240pp. 5⅜ × 8½. 24895-X Pa. $6.95

THEORY OF MATRICES, Sam Perlis. Outstanding text covering rank, non-singularity and inverses in connection with the development of canonical matrices under the relation of equivalence, and without the intervention of determinants. Includes exercises. 237pp. 5⅜ × 8½. 66810-X Pa. $7.95

GREAT EXPERIMENTS IN PHYSICS: Firsthand Accounts from Galileo to Einstein, edited by Morris H. Shamos. 25 crucial discoveries: Newton's laws of motion, Chadwick's study of the neutron, Hertz on electromagnetic waves, more. Original accounts clearly annotated. 370pp. 5⅜ × 8½. 25346-5 Pa. $10.95

INTRODUCTION TO PARTIAL DIFFERENTIAL EQUATIONS WITH AP-PLICATIONS, E.C. Zachmanoglou and Dale W. Thoe. Essentials of partial differential equations applied to common problems in engineering and the physical sciences. Problems and answers. 416pp. 5⅜ × 8½. 65251-3 Pa. $10.95

BURNHAM'S CELESTIAL HANDBOOK, Robert Burnham, Jr. Thorough guide to the stars beyond our solar system. Exhaustive treatment. Alphabetical by constellation: Andromeda to Cetus in Vol. 1; Chamaeleon to Orion in Vol. 2; and Pavo to Vulpecula in Vol. 3. Hundreds of illustrations. Index in Vol. 3. 2,000pp. 6⅛ × 9¼. 23567-X, 23568-8, 23673-0 Pa., Three-vol. set $41.85

CHEMICAL MAGIC, Leonard A. Ford. Second Edition, Revised by E. Winston Grundmeier. Over 100 unusual stunts demonstrating cold fire, dust explosions, much more. Text explains scientific principles and stresses safety precautions. 128pp. 5⅜ × 8½. 67628-5 Pa. $5.95

AMATEUR ASTRONOMER'S HANDBOOK, J.B. Sidgwick. Timeless, compre-hensive coverage of telescopes, mirrors, lenses, mountings, telescope drives, micrometers, spectroscopes, more. 189 illustrations. 576pp. 5⅜ × 8¼. (Available in U.S. only) 24034-7 Pa. $9.95

SPECIAL FUNCTIONS, N.N. Lebedev. Translated by Richard Silverman. Famous Russian work treating more important special functions, with applications to specific problems of physics and engineering. 38 figures. 308pp. 5⅜ × 8½.
60624-4 Pa. $8.95

OBSERVATIONAL ASTRONOMY FOR AMATEURS, J.B. Sidgwick. Mine of useful data for observation of sun, moon, planets, asteroids, aurorae, meteors, comets, variables, binaries, etc. 39 illustrations. 384pp. 5⅜ × 8¼. (Available in U.S. only)
24033-9 Pa. $8.95

INTEGRAL EQUATIONS, F.G. Tricomi. Authoritative, well-written treatment of extremely useful mathematical tool with wide applications. Volterra Equations, Fredholm Equations, much more. Advanced undergraduate to graduate level. Exercises. Bibliography. 238pp. 5⅜ × 8½.
64828-1 Pa. $7.95

POPULAR LECTURES ON MATHEMATICAL LOGIC, Hao Wang. Noted logician's lucid treatment of historical developments, set theory, model theory, recursion theory and constructivism, proof theory, more. 3 appendixes. Bibliography. 1981 edition. ix + 283pp. 5⅜ × 8½.
67632-3 Pa. $8.95

MODERN NONLINEAR EQUATIONS, Thomas L. Saaty. Emphasizes practical solution of problems; covers seven types of equations. ". . . a welcome contribution to the existing literature. . . ."—*Math Reviews.* 490pp. 5⅜ × 8½. 64232-1 Pa. $11.95

FUNDAMENTALS OF ASTRODYNAMICS, Roger Bate et al. Modern approach developed by U.S. Air Force Academy. Designed as a first course. Problems, exercises. Numerous illustrations. 455pp. 5⅜ × 8½.
60061-0 Pa. $9.95

INTRODUCTION TO LINEAR ALGEBRA AND DIFFERENTIAL EQUATIONS, John W. Dettman. Excellent text covers complex numbers, determinants, orthonormal bases, Laplace transforms, much more. Exercises with solutions. Undergraduate level. 416pp. 5⅜ × 8½.
65191-6 Pa. $10.95

INCOMPRESSIBLE AERODYNAMICS, edited by Bryan Thwaites. Covers theoretical and experimental treatment of the uniform flow of air and viscous fluids past two-dimensional aerofoils and three-dimensional wings; many other topics. 654pp. 5⅜ × 8½.
65465-6 Pa. $16.95

INTRODUCTION TO DIFFERENCE EQUATIONS, Samuel Goldberg. Exceptionally clear exposition of important discipline with applications to sociology, psychology, economics. Many illustrative examples; over 250 problems. 260pp. 5⅜ × 8½.
65084-7 Pa. $7.95

LAMINAR BOUNDARY LAYERS, edited by L. Rosenhead. Engineering classic covers steady boundary layers in two- and three-dimensional flow, unsteady boundary layers, stability, observational techniques, much more. 708pp. 5⅜ × 8½.
65646-2 Pa. $18.95

LECTURES ON CLASSICAL DIFFERENTIAL GEOMETRY, Second Edition, Dirk J. Struik. Excellent brief introduction covers curves, theory of surfaces, fundamental equations, geometry on a surface, conformal mapping, other topics. Problems. 240pp. 5⅜ × 8½.
65609-8 Pa. $8.95

ROTARY-WING AERODYNAMICS, W.Z. Stepniewski. Clear, concise text covers aerodynamic phenomena of the rotor and offers guidelines for helicopter performance evaluation. Originally prepared for NASA. 537 figures. 640pp. 6⅛ × 9¼.
64647-5 Pa. $15.95

DIFFERENTIAL GEOMETRY, Heinrich W. Guggenheimer. Local differential geometry as an application of advanced calculus and linear algebra. Curvature, transformation groups, surfaces, more. Exercises. 62 figures. 378pp. 5⅜ × 8½.
63433-7 Pa. $8.95

INTRODUCTION TO SPACE DYNAMICS, William Tyrrell Thomson. Comprehensive, classic introduction to space-flight engineering for advanced undergraduate and graduate students. Includes vector algebra, kinematics, transformation of coordinates. Bibliography. Index. 352pp. 5⅜ × 8½. 65113-4 Pa. $8.95

A SURVEY OF MINIMAL SURFACES, Robert Osserman. Up-to-date, in-depth discussion of the field for advanced students. Corrected and enlarged edition covers new developments. Includes numerous problems. 192pp. 5⅜ × 8½.
64998-9 Pa. $8.95

ANALYTICAL MECHANICS OF GEARS, Earle Buckingham. Indispensable reference for modern gear manufacture covers conjugate gear-tooth action, gear-tooth profiles of various gears, many other topics. 263 figures. 102 tables. 546pp. 5⅜ × 8½. 65712-4 Pa. $14.95

SET THEORY AND LOGIC, Robert R. Stoll. Lucid introduction to unified theory of mathematical concepts. Set theory and logic seen as tools for conceptual understanding of real number system. 496pp. 5⅜ × 8¼. 63829-4 Pa. $12.95

A HISTORY OF MECHANICS, René Dugas. Monumental study of mechanical principles from antiquity to quantum mechanics. Contributions of ancient Greeks, Galileo, Leonardo, Kepler, Lagrange, many others. 671pp. 5⅜ × 8½.
65632-2 Pa. $14.95

FAMOUS PROBLEMS OF GEOMETRY AND HOW TO SOLVE THEM, Benjamin Bold. Squaring the circle, trisecting the angle, duplicating the cube: learn their history, why they are impossible to solve, then solve them yourself. 128pp. 5⅜ × 8½. 24297-8 Pa. $4.95

MECHANICAL VIBRATIONS, J.P. Den Hartog. Classic textbook offers lucid explanations and illustrative models, applying theories of vibrations to a variety of practical industrial engineering problems. Numerous figures. 233 problems, solutions. Appendix. Index. Preface. 436pp. 5⅜ × 8½. 64785-4 Pa. $10.95

CURVATURE AND HOMOLOGY, Samuel I. Goldberg. Thorough treatment of specialized branch of differential geometry. Covers Riemannian manifolds, topology of differentiable manifolds, compact Lie groups, other topics. Exercises. 315pp. 5⅜ × 8½. 64314-X Pa. $9.95

HISTORY OF STRENGTH OF MATERIALS, Stephen P. Timoshenko. Excellent historical survey of the strength of materials with many references to the theories of elasticity and structure. 245 figures. 452pp. 5⅜ × 8½. 61187-6 Pa. $11.95

GEOMETRY OF COMPLEX NUMBERS, Hans Schwerdtfeger. Illuminating, widely praised book on analytic geometry of circles, the Moebius transformation, and two-dimensional non-Euclidean geometries. 200pp. 5⅜ × 8¼.
63830-8 Pa. $8.95

MECHANICS, J.P. Den Hartog. A classic introductory text or refresher. Hundreds of applications and design problems illuminate fundamentals of trusses, loaded beams and cables, etc. 334 answered problems. 462pp. 5⅜ × 8½. 60754-2 Pa. $9.95

TOPOLOGY, John G. Hocking and Gail S. Young. Superb one-year course in classical topology. Topological spaces and functions, point-set topology, much more. Examples and problems. Bibliography. Index. 384pp. 5⅜ × 8¼.
65676-4 Pa. $9.95

STRENGTH OF MATERIALS, J.P. Den Hartog. Full, clear treatment of basic material (tension, torsion, bending, etc.) plus advanced material on engineering methods, applications. 350 answered problems. 323pp. 5⅜ × 8½. 60755-0 Pa. $8.95

ELEMENTARY CONCEPTS OF TOPOLOGY, Paul Alexandroff. Elegant, intuitive approach to topology from set-theoretic topology to Betti groups; how concepts of topology are useful in math and physics. 25 figures. 57pp. 5⅜ × 8½.
60747-X Pa. $3.50

ADVANCED STRENGTH OF MATERIALS, J.P. Den Hartog. Superbly written advanced text covers torsion, rotating disks, membrane stresses in shells, much more. Many problems and answers. 388pp. 5⅜ × 8½. 65407-9 Pa. $9.95

COMPUTABILITY AND UNSOLVABILITY, Martin Davis. Classic graduate-level introduction to theory of computability, usually referred to as theory of recurrent functions. New preface and appendix. 288pp. 5⅜ × 8½. 61471-9 Pa. $7.95

GENERAL CHEMISTRY, Linus Pauling. Revised 3rd edition of classic first-year text by Nobel laureate. Atomic and molecular structure, quantum mechanics, statistical mechanics, thermodynamics correlated with descriptive chemistry. Problems. 992pp. 5⅜ × 8½. 65622-5 Pa. $19.95

AN INTRODUCTION TO MATRICES, SETS AND GROUPS FOR SCIENCE STUDENTS, G. Stephenson. Concise, readable text introduces sets, groups, and most importantly, matrices to undergraduate students of physics, chemistry, and engineering. Problems. 164pp. 5⅜ × 8½. 65077-4 Pa. $6.95

THE HISTORICAL BACKGROUND OF CHEMISTRY, Henry M. Leicester. Evolution of ideas, not individual biography. Concentrates on formulation of a coherent set of chemical laws. 260pp. 5⅜ × 8½. 61053-5 Pa. $6.95

THE PHILOSOPHY OF MATHEMATICS: An Introductory Essay, Stephan Körner. Surveys the views of Plato, Aristotle, Leibniz & Kant concerning propositions and theories of applied and pure mathematics. Introduction. Two appendices. Index. 198pp. 5⅜ × 8½. 25048-2 Pa. $7.95

THE DEVELOPMENT OF MODERN CHEMISTRY, Aaron J. Ihde. Authoritative history of chemistry from ancient Greek theory to 20th-century innovation. Covers major chemists and their discoveries. 209 illustrations. 14 tables. Bibliographies. Indices. Appendices. 851pp. 5⅜ × 8½. 64235-6 Pa. $18.95

DE RE METALLICA, Georgius Agricola. The famous Hoover translation of greatest treatise on technological chemistry, engineering, geology, mining of early modern times (1556). All 289 original woodcuts. 638pp. 6¾ × 11.
60006-8 Pa. $18.95

SOME THEORY OF SAMPLING, William Edwards Deming. Analysis of the problems, theory and design of sampling techniques for social scientists, industrial managers and others who find statistics increasingly important in their work. 61 tables. 90 figures. xvii + 602pp. 5⅜ × 8½.
64684-X Pa. $15.95

THE VARIOUS AND INGENIOUS MACHINES OF AGOSTINO RAMELLI: A Classic Sixteenth-Century Illustrated Treatise on Technology, Agostino Ramelli. One of the most widely known and copied works on machinery in the 16th century. 194 detailed plates of water pumps, grain mills, cranes, more. 608pp. 9 × 12.
28180-9 Pa. $24.95

LINEAR PROGRAMMING AND ECONOMIC ANALYSIS, Robert Dorfman, Paul A. Samuelson and Robert M. Solow. First comprehensive treatment of linear programming in standard economic analysis. Game theory, modern welfare economics, Leontief input-output, more. 525pp. 5⅜ × 8½.
65491-5 Pa. $14.95

ELEMENTARY DECISION THEORY, Herman Chernoff and Lincoln E. Moses. Clear introduction to statistics and statistical theory covers data processing, probability and random variables, testing hypotheses, much more. Exercises. 364pp. 5⅜ × 8½.
65218-1 Pa. $9.95

THE COMPLEAT STRATEGYST: Being a Primer on the Theory of Games of Strategy, J.D. Williams. Highly entertaining classic describes, with many illustrated examples, how to select best strategies in conflict situations. Prefaces. Appendices. 268pp. 5⅜ × 8½.
25101-2 Pa. $7.95

MATHEMATICAL METHODS OF OPERATIONS RESEARCH, Thomas L. Saaty. Classic graduate-level text covers historical background, classical methods of forming models, optimization, game theory, probability, queueing theory, much more. Exercises. Bibliography. 448pp. 5⅜ × 8¼.
65703-5 Pa. $12.95

CONSTRUCTIONS AND COMBINATORIAL PROBLEMS IN DESIGN OF EXPERIMENTS, Damaraju Raghavarao. In-depth reference work examines orthogonal Latin squares, incomplete block designs, tactical configuration, partial geometry, much more. Abundant explanations, examples. 416pp. 5⅜ × 8¼.
65685-3 Pa. $10.95

THE ABSOLUTE DIFFERENTIAL CALCULUS (CALCULUS OF TENSORS), Tullio Levi-Civita. Great 20th-century mathematician's classic work on material necessary for mathematical grasp of theory of relativity. 452pp. 5⅜ × 8½.
63401-9 Pa. $9.95

VECTOR AND TENSOR ANALYSIS WITH APPLICATIONS, A.I. Borisenko and I.E. Tarapov. Concise introduction. Worked-out problems, solutions, exercises. 257pp. 5⅜ × 8¼.
63833-2 Pa. $7.95

THE FOUR-COLOR PROBLEM: Assaults and Conquest, Thomas L. Saaty and Paul G. Kainen. Engrossing, comprehensive account of the century-old combinatorial topological problem, its history and solution. Bibliographies. Index. 110 figures. 228pp. 5⅜ × 8½. 65092-8 Pa. $6.95

CATALYSIS IN CHEMISTRY AND ENZYMOLOGY, William P. Jencks. Exceptionally clear coverage of mechanisms for catalysis, forces in aqueous solution, carbonyl- and acyl-group reactions, practical kinetics, more. 864pp. 5⅜ × 8½. 65460-5 Pa. $19.95

PROBABILITY: An Introduction, Samuel Goldberg. Excellent basic text covers set theory, probability theory for finite sample spaces, binomial theorem, much more. 360 problems. Bibliographies. 322pp. 5⅜ × 8½. 65252-1 Pa. $8.95

LIGHTNING, Martin A. Uman. Revised, updated edition of classic work on the physics of lightning. Phenomena, terminology, measurement, photography, spectroscopy, thunder, more. Reviews recent research. Bibliography. Indices. 320pp. 5⅜ × 8¼. 64575-4 Pa. $8.95

PROBABILITY THEORY: A Concise Course, Y.A. Rozanov. Highly readable, self-contained introduction covers combination of events, dependent events, Bernoulli trials, etc. Translation by Richard Silverman. 148pp. 5⅜ × 8¼. 63544-9 Pa. $5.95

AN INTRODUCTION TO HAMILTONIAN OPTICS, H. A. Buchdahl. Detailed account of the Hamiltonian treatment of aberration theory in geometrical optics. Many classes of optical systems defined in terms of the symmetries they possess. Problems with detailed solutions. 1970 edition. xv + 360pp. 5⅜ × 8½. 67597-1 Pa. $10.95

STATISTICS MANUAL, Edwin L. Crow, et al. Comprehensive, practical collection of classical and modern methods prepared by U.S. Naval Ordnance Test Station. Stress on use. Basics of statistics assumed. 288pp. 5⅜ × 8½. 60599-X Pa. $6.95

DICTIONARY/OUTLINE OF BASIC STATISTICS, John E. Freund and Frank J. Williams. A clear concise dictionary of over 1,000 statistical terms and an outline of statistical formulas covering probability, nonparametric tests, much more. 208pp. 5⅜ × 8½. 66796-0 Pa. $6.95

STATISTICAL METHOD FROM THE VIEWPOINT OF QUALITY CONTROL, Walter A. Shewhart. Important text explains regulation of variables, uses of statistical control to achieve quality control in industry, agriculture, other areas. 192pp. 5⅜ × 8½. 65232-7 Pa. $7.95

THE INTERPRETATION OF GEOLOGICAL PHASE DIAGRAMS, Ernest G. Ehlers. Clear, concise text emphasizes diagrams of systems under fluid or containing pressure; also coverage of complex binary systems, hydrothermal melting, more. 288pp. 6½ × 9¼. 65389-7 Pa. $10.95

STATISTICAL ADJUSTMENT OF DATA, W. Edwards Deming. Introduction to basic concepts of statistics, curve fitting, least squares solution, conditions without parameter, conditions containing parameters. 26 exercises worked out. 271pp. 5⅜ × 8½. 64685-8 Pa. $8.95

CATALOG OF DOVER BOOKS

TENSOR CALCULUS, J.L. Synge and A. Schild. Widely used introductory text covers spaces and tensors, basic operations in Riemannian space, non-Riemannian spaces, etc. 324pp. 5⅜ × 8¼. 63612-7 Pa. $8.95

A CONCISE HISTORY OF MATHEMATICS, Dirk J. Struik. The best brief history of mathematics. Stresses origins and covers every major figure from ancient Near East to 19th century. 41 illustrations. 195pp. 5⅜ × 8½. 60255-9 Pa. $7.95

A SHORT ACCOUNT OF THE HISTORY OF MATHEMATICS, W.W. Rouse Ball. One of clearest, most authoritative surveys from the Egyptians and Phoenicians through 19th-century figures such as Grassman, Galois, Riemann. Fourth edition. 522pp. 5⅜ × 8½. 20630-0 Pa. $10.95

HISTORY OF MATHEMATICS, David E. Smith. Nontechnical survey from ancient Greece and Orient to late 19th century; evolution of arithmetic, geometry, trigonometry, calculating devices, algebra, the calculus. 362 illustrations. 1,355pp. 5⅜ × 8½. 20429-4, 20430-8 Pa., Two-vol. set $23.90

THE GEOMETRY OF RENÉ DESCARTES, René Descartes. The great work founded analytical geometry. Original French text, Descartes' own diagrams, together with definitive Smith-Latham translation. 244pp. 5⅜ × 8½. 60068-8 Pa. $7.95

THE ORIGINS OF THE INFINITESIMAL CALCULUS, Margaret E. Baron. Only fully detailed and documented account of crucial discipline: origins; development by Galileo, Kepler, Cavalieri; contributions of Newton, Leibniz, more. 304pp. 5⅜ × 8½. (Available in U.S. and Canada only) 65371-4 Pa. $9.95

THE HISTORY OF THE CALCULUS AND ITS CONCEPTUAL DEVELOPMENT, Carl B. Boyer. Origins in antiquity, medieval contributions, work of Newton, Leibniz, rigorous formulation. Treatment is verbal. 346pp. 5⅜ × 8½. 60509-4 Pa. $8.95

THE THIRTEEN BOOKS OF EUCLID'S ELEMENTS, translated with introduction and commentary by Sir Thomas L. Heath. Definitive edition. Textual and linguistic notes, mathematical analysis. 2,500 years of critical commentary. Not abridged. 1,414pp. 5⅜ × 8½. 60088-2, 60089-0, 60090-4 Pa., Three-vol. set $29.85

GAMES AND DECISIONS: Introduction and Critical Survey, R. Duncan Luce and Howard Raiffa. Superb nontechnical introduction to game theory, primarily applied to social sciences. Utility theory, zero-sum games, n-person games, decision-making, much more. Bibliography. 509pp. 5⅜ × 8½. 65943-7 Pa. $12.95

THE HISTORICAL ROOTS OF ELEMENTARY MATHEMATICS, Lucas N.H. Bunt, Phillip S. Jones, and Jack D. Bedient. Fundamental underpinnings of modern arithmetic, algebra, geometry and number systems derived from ancient civilizations. 320pp. 5⅜ × 8½. 25563-8 Pa. $8.95

CALCULUS REFRESHER FOR TECHNICAL PEOPLE, A. Albert Klaf. Covers important aspects of integral and differential calculus via 756 questions. 566 problems, most answered. 431pp. 5⅜ × 8½. 20370-0 Pa. $8.95

CATALOG OF DOVER BOOKS

CHALLENGING MATHEMATICAL PROBLEMS WITH ELEMENTARY SOLUTIONS, A.M. Yaglom and I.M. Yaglom. Over 170 challenging problems on probability theory, combinatorial analysis, points and lines, topology, convex polygons, many other topics. Solutions. Total of 445pp. 5⅜ × 8½. Two-vol. set.
Vol. I 65536-9 Pa. $7.95
Vol. II 65537-7 Pa. $6.95

FIFTY CHALLENGING PROBLEMS IN PROBABILITY WITH SOLUTIONS, Frederick Mosteller. Remarkable puzzlers, graded in difficulty, illustrate elementary and advanced aspects of probability. Detailed solutions. 88pp. 5⅜ × 8½.
65355-2 Pa. $4.95

EXPERIMENTS IN TOPOLOGY, Stephen Barr. Classic, lively explanation of one of the byways of mathematics. Klein bottles, Moebius strips, projective planes, map coloring, problem of the Koenigsberg bridges, much more, described with clarity and wit. 43 figures. 210pp. 5⅜ × 8½. 25933-1 Pa. $5.95

RELATIVITY IN ILLUSTRATIONS, Jacob T. Schwartz. Clear nontechnical treatment makes relativity more accessible than ever before. Over 60 drawings illustrate concepts more clearly than text alone. Only high school geometry needed. Bibliography. 128pp. 6⅛ × 9¼. 25965-X Pa. $6.95

AN INTRODUCTION TO ORDINARY DIFFERENTIAL EQUATIONS, Earl A. Coddington. A thorough and systematic first course in elementary differential equations for undergraduates in mathematics and science, with many exercises and problems (with answers). Index. 304pp. 5⅜ × 8½. 65942-9 Pa. $8.95

FOURIER SERIES AND ORTHOGONAL FUNCTIONS, Harry F. Davis. An incisive text combining theory and practical example to introduce Fourier series, orthogonal functions and applications of the Fourier method to boundary-value problems. 570 exercises. Answers and notes. 416pp. 5⅜ × 8½. 65973-9 Pa. $9.95

THE THEORY OF BRANCHING PROCESSES, Theodore E. Harris. First systematic, comprehensive treatment of branching (i.e. multiplicative) processes and their applications. Galton-Watson model, Markov branching processes, electron-photon cascade, many other topics. Rigorous proofs. Bibliography. 240pp. 5⅜ × 8½. 65952-6 Pa. $6.95

AN INTRODUCTION TO ALGEBRAIC STRUCTURES, Joseph Landin. Superb self-contained text covers "abstract algebra": sets and numbers, theory of groups, theory of rings, much more. Numerous well-chosen examples, exercises. 247pp. 5⅜ × 8½. 65940-2 Pa. $7.95

Postulates of a Personalistic

The seven postulates (P1 through P7) scattered through the first five chapters of this book are reproduced here for ready reference along with a minimum of explanatory material. The language of the postulates is here changed somewhat for conciseness and to show an alternative mode of expression, but the logical content of each postulate is left unaltered.

The formal subject matter of the theory

The states, a set S of elements $s, s', \cdots$ with subsets $A, B, C, \cdots$ (page 11).

The consequences, a set F of elements $f, g, h, \cdots$ (page 14).

Acts, arbitrary functions **f, g, h,** $\cdots$ from S to F (page 14).

The relation "is not preferred to" between acts, $\leq$ (page 18).

The postulates, and definitions on which they depend

Definitions of terms not in general mathematical use are given here as D1 through D5; for others consult the General Index (page 289) and the Technical Symbols (page 283).

P1   The relation $\leq$ is a simple ordering (page 18).

D1 $\mathbf{f} \leq \mathbf{g}$ given B, if and only if $\mathbf{f}' \leq \mathbf{g}'$ for every $\mathbf{f}'$ and $\mathbf{g}'$ that agree with $\mathbf{f}$ and $\mathbf{g}$, respectively, on B and with each other on $\sim B$ and $\mathbf{g}' \leq \mathbf{f}'$ either for all such pairs or for none (page 22).